Algorithms and Combinatorics 18

Springer
Berlin
Heidelberg
New York
Barcelona
Hong Kong
London
Milan
Paris
Singapore
Tokyo

Jiří Matoušek

Geometric Discrepancy

An Illustrated Guide

Springer

Jiří Matoušek
Department of Applied Mathematics
Charles University
Malostranské nám. 25
11800 Praha
Czech Republic
e-mail: matousek@kam.mff.cuni.cz

Cataloging-in-Publication Data applied for.

Die Deutsche Bibliothek – CIP-Einheitsaufnahme

Matoušek, Jiří:
Geometric discrepancy: an illustrated guide / Jiří Matoušek. – Berlin; Heidelberg; New York;
Barcelona; Hong Kong; London; Milan; Paris; Singapore; Tokyo: Springer, 1999
(Algorithms and combinatorics; 18)
ISBN 3-540-65528-X

Mathematics Subject Classification (1991):
11K38, 05D99, 65C99

ISSN 0937-5511
ISBN 3-540-65528-X Springer-Verlag Berlin Heidelberg New York

Printed in Germany

Typesetting: Typeset in LaTEX by the author using a Springer TEX macro package
SPIN 10699021 46/3143 - 5 4 3 2 1 0 - Printed on acid-free paper

Preface

Discrepancy theory is also called the theory of *irregularities of distribution*. Here are some typical questions: What is the "most uniform" way of distributing n points in the unit square? How big is the "irregularity" necessarily present in any such distribution? For a precise formulation of these questions, we must quantify the irregularity of a given distribution, and discrepancy is a numerical parameter of a point set serving this purpose.

Such questions were first tackled in the thirties, with a motivation coming from number theory. A more or less satisfactory solution of the basic discrepancy problem in the plane was completed in the late sixties, and the analogous higher-dimensional problem is far from solved even today. In the meantime, discrepancy theory blossomed into a field of remarkable breadth and diversity. There are subfields closely connected to the original number-theoretic roots of discrepancy theory, areas related to Ramsey theory and to hypergraphs, and also results supporting eminently practical methods and algorithms for numerical integration and similar tasks. The applications include financial calculations, computer graphics, and computational physics, just to name a few.

This book is an introductory textbook on discrepancy theory. It should be accessible to early graduate students of mathematics or theoretical computer science. At the same time, about half of the book consists of material that up until now was only available in original research papers or in various surveys.

Some number of people may be interested in discrepancy theory with some specific application in mind, or because they want to do research in it. But, in my opinion, discrepancy theory can also serve as an example of "live mathematics" for students completing the basic math courses. The problems in discrepancy are natural, easy to state, and relatively narrowly focused. The solutions, although some of them are quite deep and clever, can often be explained in several pages with all the details. Hence, the beginner need not feel overwhelmed by the volume of material or by a technical machinery towering above him like a Gothic cathedral. At the same time, many notions and theorems the student has to learn in the basic curriculum can be seen in action here (such as calculus, geometry of the Euclidean space, harmonic analysis, elementary number theory, probability theory and the probabilistic method in combinatorics, hypergraphs, counting and asymptotic estimates, linear algebra, finite fields, polynomial algebra, and algorithm design). The Fourier series is encountered not because the next item in the course outline is called the Fourier series, but because one needs it to answer a seemingly unrelated question about points in the unit square. In my opinion, such examples "from the outside" are very important and refreshing in learning a mathematical discipline, but the basic courses can seldom include them.

Based on the book, it is possible to teach a one-semester or two-semester "special topic" course (experiments in this direction have been kindly performed by Joram Lindenstrauss and by Nati Linial). For a general course on discrepancy, I suggest covering Section 1.1 (perhaps omitting Weyl's criterion), the Van der Corput and Halton–Hammersley constructions (Sec. 2.1), maybe Beck's upper bound for discs (Sec. 3.1), definitely Roth's lower bound (Sec. 6.1), the notion of combinatorial discrepancy (Sec. 1.3), basic combinatorial upper bounds (Sec. 4.1), the lower bound using eigenvalues (Sec. 4.2), and the partial coloring method (Sec. 4.5). If time permits, the next recommendations are Halász' lower bound proof (Sec. 6.2) and Alexander's lower bound (Sec. 6.4 or 6.5). I leave further extension to the instructor's judgment. For those wishing to pursue the subject of quasi-Monte Carlo methods, the main recommended parts are Section 1.4 and the whole of Chapter 2. Convinced combinatorialists are invited to read mainly Chapters 4 and 5. The latter discusses the Vapnik–Chervonenkis dimension, which is of considerable interest in statistics, computational learning theory, computational geometry, etc.

Sections usually consist of three parts: the main text (what I would talk about in a course), bibliographic references and remarks intended mainly for specialists, and exercises. The exercises are classified by difficulty as no-star, one-star, and two-star (but this classification is quite subjective). No-star exercises should be more or less routine, and two-star ones often contain a clever idea that had once been enough for a publication, although the difficulty may now be greatly reduced by a suggestive formulation. More difficult exercises are usually accompanied by hints given at the end of the book. Rather than seriously expecting anyone to solve a large part of the exercises, I used the exercise-hint combination as a way of packing lots of results into a much smaller space than would be required for writing them out according to the customary way of mathematical presentation. This, of course, greatly enlarges the danger of introducing errors and making false claims, so the reader who wants to use such information should check carefully if the hint really works.

The book contains two tables summarizing some important asymptotic bounds in discrepancy theory, an index, and a list of references with cross-references to the pages where they are cited. I consider this last provision convenient for the reader, but it has the unfortunate aspect that the authors mentioned in the references can immediately find where their work is cited and conclude that their results were misquoted and insufficiently appreciated. I apologize to them; my only excuse is that such shortcomings are not intentional and that I simply did not have as much time to devote to each of the referenced papers and books as it would have deserved.

If you find errors in the book, especially serious ones, please let me know (Email: `matousek@kam.mff.cuni.cz`). I plan to post the list of errors on the world-wide web (`http://www.ms.mff.cuni.cz/~matousek`).

Acknowledgment. For invaluable advice and/or very helpful comments on preliminary versions of this book, I would like to thank József Beck, Johannes Blömer, William L. Chen, Vsevolod Lev, Joram Lindenstrauss, János Pach, Maxim Skriganov, Vera T. Sós, Joel Spencer, Shu Tezuka, and Henryk Woźniakowski. I am grateful to Anand Srivastav and his associates for great work in the organization of a discrepancy theory workshop in Kiel, where I learned many things now stored in this book. I also wish to thank many other people for their friendly support; this group is too precious and too fuzzy to be defined by enumeration.

Prague, January 1999 *Jiří Matoušek*

Table of Contents

Notation

For a real number x, $\lfloor x \rfloor$ denotes the largest integer $\leq x$, $\lceil x \rceil$ means the smallest integer $\geq x$, and $\{x\} = x - \lfloor x \rfloor$ is the fractional part of x.

The letters $\mathbf{N}$, $\mathbf{Z}$, $\mathbf{Q}$, and $\mathbf{R}$ are reserved for the set of all natural numbers, integers, rationals, and reals, respectively. The symbol $\mathbf{R}^d$ denotes the d-dimensional Euclidean space. For a point $x = (x_1, x_2, \ldots, x_d) \in \mathbf{R}^d$, $\|x\| = \sqrt{x_1^2 + x_2^2 + \cdots + x_d^2}$ is the Euclidean norm of x, and for $x, y \in \mathbf{R}^d$, $\langle x, y \rangle = x_1y_1 + x_2y_2 + \cdots + x_dy_d$ is the scalar product. The symbol $B(x, r)$ denotes the ball of radius r centered at x in some metric space (usually in $\mathbf{R}^d$ with the Euclidean distance), i.e. the set of all points with distance at most r from x.

If X is a set, the symbol $|X|$ denotes the number of elements (or cardinality) of X. For a measurable set $A \subseteq \mathbf{R}^d$, $\mathrm{vol}(A)$ is the d-dimensional Lebesgue measure of A. (We use this notation to indicate that in all specific instances in geometric discrepancy, we deal with very simple sets for which the Lebesgue measure is just volume or area in the usual intuitive sense.) Since we will often consider the intersection of some sets with the unit cube $[0,1]^d$, we introduce the notation

$$\mathrm{vol}_\sqcup(A) = \mathrm{vol}(A \cap [0,1]^d).$$

Let f and g be real functions (of one or several variables). The notation $f = O(g)$ means that there exists a number C such that $|f| \leq C|g|$ for all values of the variables. Normally C should be an absolute constant, but if f and g depend on some parameter(s) which we explicitly declare to be fixed (such as the space dimension d), then C may depend on these parameters as well. The notation $f = \Omega(g)$ is equivalent to $g = O(f)$, and $f = \Theta(g)$ means that both $f = O(g)$ and $f = \Omega(g)$. Finally $f(n) = o(g(n))$ means $\lim_{n\to\infty} \frac{f(n)}{g(n)} = 0$.

For a random variable X, the symbol $\mathbf{E}[X]$ denotes the expectation of X, and $\Pr[A]$ stands for the probability of an event A.

1. Introduction

In this chapter, we introduce the concept of discrepancy. We formulate a basic problem concerning discrepancy for rectangles, we show its connections to the discrepancy of infinite sequences in the unit interval, and we briefly comment on the historical roots of discrepancy in the theory of uniform distribution (Section 1.1). In Section 1.2, we introduce discrepancy in a general geometric setting, as well as some variations of the basic definition. Section 1.3 defines discrepancy in a seemingly different situation, namely for set systems on finite sets, and shows a close relationship to the previously discussed "Lebesgue-measure" discrepancy. Finally, Section 1.4 is a mosaic of notions, results, and comments illustrating the numerous and diverse connections and applications of discrepancy theory. Most of the space in that section is devoted to applications in numerical integration and similar problems, which by now constitute an extensive branch of applied mathematics, with conventions and methods quite different from "pure" discrepancy theory.

1.1 Discrepancy for Rectangles and Uniform Distribution

The word *discrepancy* means "disagreement" (from Latin *discrepare*—to sound discordantly). In our case it is a "disagreement between the ideal situation and the actual one," namely a "deviation from a perfectly uniform distribution."

We will investigate how uniformly an n-point set can be distributed in the d-dimensional unit cube $[0,1]^d$. For $d=1$, the set of n equidistant points as in the following picture

hardly finds serious competitors as a candidate for the most uniformly distributed n-point set in the unit interval. But already in dimension 2, one can come up with several reasonable criteria of uniform distribution, and sets that are very good for some may be quite bad for others.

Here is one such criterion: "uniformly" means, for the moment, "uniformly with respect to axis-parallel rectangles." Let P be an n-point set in the unit square $[0,1]^2$. Let us consider an axis-parallel rectangle[1] $R = [a_1, b_1) \times [a_2, b_2) \subseteq [0,1]^2$:

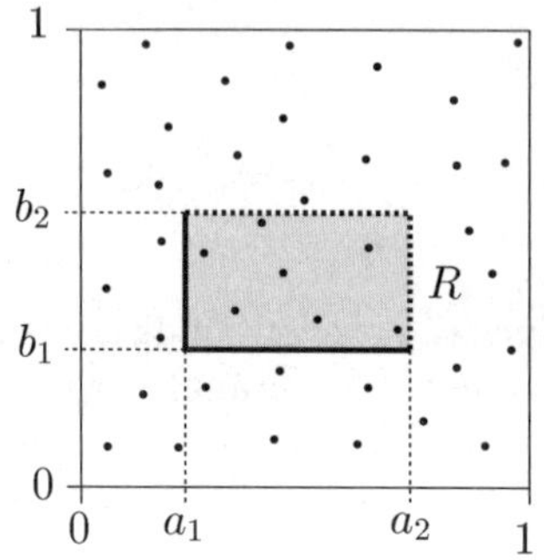

For a uniformly distributed set P, we expect that the number of points of P that fall in the rectangle R is approximately $n \cdot \mathrm{vol}(R)$, where $\mathrm{vol}(R)$ denotes the area of R. (Note that $n \cdot \mathrm{vol}(R)$ is the expected number of points hitting R if we pick n points in the unit square uniformly and independently at random.) Let us call P *justly distributed* if the deviation

$$\left| n \cdot \mathrm{vol}(R) - |P \cap R| \right|$$

is at most 100 for all axis-parallel rectangles R. Do arbitrarily large justly distributed set exist? (Or, should the constant 100 be too small, we can ask if the deviation can be bounded by some other constant, possibly large but independent of n, P, and R.) This is one of the fundamental questions that gave birth to discrepancy theory. Since we do not hope to keep the reader in suspense until the end of the book by postponing the answer, we can just as well state it right away: no, just distribution is impossible for sufficiently large sets. Any distribution of n points in the unit square has to display a significant irregularity for some rectangle R, and the magnitude of the irregularity must grow to infinity as $n \to \infty$. For this particular two-dimensional problem, it is even known fairly precisely how large this irregularity must be, and we will see the corresponding lower and upper bound proofs later in this book. The proofs may perhaps seem simple, but one should not forget that the presentation is based on the work of outstanding mathematicians and that originally the problem looked formidably difficult. To put these results into a better perspective, we remark that already the obvious generalization of the problem in dimension 3 has so far defied all attempts at obtaining a quantitatively precise answer.

[1] For technical reasons, we take semi-open rectangles—the left side and the bottom side are included, the right and top sides are not. For the discrepancy this doesn't matter much; we only accept this convention for simplifying some formulas in the sequel.

Here is some notation for expressing these questions and answers. First we introduce the symbol $D(P, R)$ for the deviation of P from uniform distribution on a particular rectangle R, namely

$$D(P, R) = n \cdot \mathrm{vol}(R) - |P \cap R|.$$

Let $\mathcal{R}_2$ denote the set of all axis-parallel rectangles in the unit square. The quantity

$$D(P, \mathcal{R}_2) = \sup_{R \in \mathcal{R}_2} |D(P, R)|$$

is called the *discrepancy of* P for axis-parallel rectangles, and the function

$$D(n, \mathcal{R}_2) = \inf_{\substack{P \subset [0,1]^2 \\ |P| = n}} D(P, \mathcal{R}_2)$$

quantifies the smallest possible discrepancy of an n-point set. The above question about a just distribution can thus be re-formulated as follows:

1.1 Problem. *Is* $D(n, \mathcal{R}_2)$ *bounded above by a constant for all* n*, or does* $\limsup_{n \to \infty} D(n, \mathcal{R}_2) = \infty$ *hold?*

In this book, we will judge the uniformity of distribution exclusively in terms of discrepancy, but we should remark that there are also other sensible criteria of uniform distribution. For example, one such criterion might be the minimum distance of two points in the considered set. This concept is also studied quite extensively (in the theory of ball packings, in coding theory, and so on), but it is quite distinct from the uniform distribution measured by discrepancy. For example, the set in the unit square maximizing the minimum interpoint distance is (essentially) a triangular lattice:

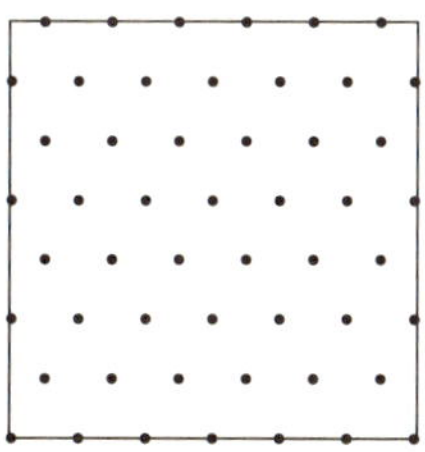

As it turns out, this set is quite bad from the discrepancy point of view: it has discrepancy about $\sqrt{n}$, while in Chapter 2 we will learn how to produce sets with only $O(\log n)$ discrepancy. On the other hand, a set with a very good discrepancy may contain two very close points.

Uniform Distribution of Infinite Sequences. The question about the "most uniform" distribution in the one-dimensional interval $[0, 1]$ is trivial for an n-point set, but it becomes quite interesting for an infinite sequence $u = (u_1, u_2, \ldots)$ of points in $[0, 1]$. Here we want that if the points of u are added one by one in the natural order, they "sweep out" all subintervals of $[0,1]$ as evenly as possible. This is actually the setting where discrepancy theory began. So let us outline the definitions concerning uniform distribution of sequences.

The sequence $u = (u_1, u_2, \ldots)$ is called *uniformly distributed* in $[0,1]$ if we have, for each subinterval $[a,b) \subset [0,1]$,

$$\lim_{n\to\infty} \left(\frac{1}{n} \bigl| \{u_1, \ldots, u_n\} \cap [a,b) \bigr| \right) = b - a. \tag{1.1}$$

Uniformly distributed sequences have the following seemingly stronger property (which is actually not difficult to prove from the just given definition of uniform distribution). For any Riemann-integrable function $f\colon [0,1] \to \mathbf{R}$, we have

$$\lim_{n\to\infty} \left(\frac{1}{n} \sum_{i=1}^{n} f(u_i) \right) = \int_0^1 f(x)\, dx. \tag{1.2}$$

Note that (1.1) is a particular case of the last equation, with the characteristic function of the interval $[a,b)$ in the role of f. Thus, in order to test the validity of (1.2) for all Riemann-integrable functions f, it suffices to consider all characteristic functions of intervals in the role of f.

Another interesting class of functions which are sufficient for testing (1.1) are the trigonometric polynomials, i.e. functions of the form

$$f(x) = \sum_{k=0}^{n} \Bigl(a_k \sin(2\pi k x) + b_k \cos(2\pi k x) \Bigr)$$

with real or complex coefficients $a_0, a_1, \ldots, a_n$ and $b_0, b_1, \ldots, b_n$. More conveniently, a trigonometric polynomial can be written using the complex exponential: $f(x) = \sum_{k=-n}^{n} c_k e^{2\pi \mathrm{i} k x}$, with i standing for the imaginary unit. From a basic approximation theorem involving trigonometric polynomials (a suitable version of Weierstrass' approximation theorem), it can be shown that if (1.2) holds for all trigonometric polynomials f, then the sequence u is uniformly distributed. Since any trigonometric polynomial is a linear combination of the functions $x \mapsto e^{2\pi \mathrm{i} k x}$ for various integers k, and since for $k = 0$, the condition (1.2) with the function $f(x) = e^{2\pi \mathrm{i} 0 x} = 1$ is trivially satisfied by any sequence u, the following criterion is obtained: a sequence $u = (u_1, u_2, \ldots)$ is uniformly distributed in $[0,1]$ if and only if we have, for all integers $k \neq 0$,

$$\lim_{n\to\infty} \left(\frac{1}{n} \sum_{j=1}^{n} e^{2\pi \mathrm{i} k u_j} \right) = \int_0^1 e^{2\pi \mathrm{i} k x}\, dx = 0.$$

This result is called *Weyl's criterion*. Here is a simple but lovely application:

1.2 Theorem. *For each irrational number α, the sequence $u = (u_1, u_2, \ldots)$ given by $u_n = \{\alpha n\}$ is uniformly distributed in $[0,1]$. (Here $\{x\}$ denotes the fractional part of x.)*

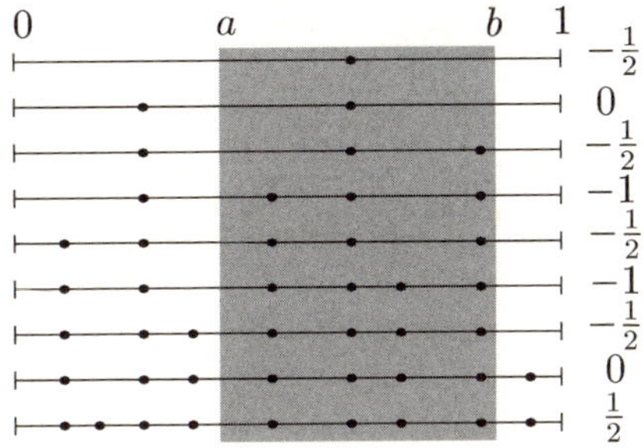

Fig. 1.1. Adding the terms of a sequence one by one; the numbers on the right are the deviations $n(b-a) - |\{u_1, \ldots, u_n\} \cap [a,b)|$ for the marked interval $[a,b)$ of length $\frac{1}{2}$.

Proof. We use Weyl's criterion. This is particularly advantageous here since we have $e^{2\pi i k u_n} = e^{2\pi i k \alpha n}$, with the unpleasant "fractional part" operation disappearing. Putting $A_k = e^{2\pi i k \alpha}$, we calculate

$$\sum_{j=1}^{n} e^{2\pi i k u_j} = \sum_{j=1}^{n} A_k^j = \frac{A_k^{n+1} - A_k}{A_k - 1}.$$

We have $|A_k| = 1$, and since α is irrational, $k\alpha$ is never an integer for a nonzero k, and so $A_k \neq 1$. Therefore, $\left|\frac{A_k^{n+1}-A_k}{A_k-1}\right| \leq \frac{2}{|A_k-1|}$ is bounded by a number independent of n, and we have

$$\lim_{n\to\infty} \left(\frac{1}{n} \sum_{j=1}^{n} e^{2\pi i k u_j} \right) = 0$$

as required. □

Discrepancy of Sequences: a "Dynamic" Setting. We now know that all the sequences $(\{n\alpha\})$ with α irrational are uniformly distributed, but if one looks into the matter more closely, one finds that some are more uniformly distributed than the others. Discrepancy was first introduced as a quantitative measure of non-uniformity of distribution for infinite sequences. We define the *discrepancy of an infinite sequence* u in $[0,1]$ as the function

$$\Delta(u,n) = \sup_{0\leq a\leq b\leq 1} \Big| n(b-a) - \big|\{u_1, \ldots, u_n\} \cap [a,b)\big| \Big|$$

(see Fig. 1.1). The original formulation of Problem 1.1 actually was: does there exist a sequence u with $\Delta(u,n)$ bounded by a constant for all n?

Let us sketch the connection of this formulation concerning infinite sequences to the formulation with axis-parallel rectangles. First, suppose that u is some given sequence in $[0,1]$. We claim that for every natural number n, there exists an n-point set $P \subset [0,1]^2$ with

$$D(P,\mathcal{R}_2) \leq 2\max\{\Delta(u,k)\colon k = 1,2,\ldots,n\} + 2. \tag{1.3}$$

A suitable set P can be defined as the "graph" of the first n terms of u. Namely, we put

$$P = \left\{(\tfrac{1}{n}, u_1), (\tfrac{2}{n}, u_2), (\tfrac{3}{n}, u_3), \ldots, (\tfrac{n}{n}, u_n)\right\}.$$

We leave it as Exercise 1(a) to verify that (1.3) indeed holds for this P. Conversely, suppose that we have an n-point set P in $[0,1]^2$. Let us list the points of P in the order of increasing x-coordinates; that is, write $P = \{(x_1,y_1),(x_2,y_2),\ldots,(x_n,y_n)\}$, where $x_1 \leq x_2 \leq \cdots \leq x_n$. Then it is not difficult to verify that if u is a sequence with $y_1, y_2, \ldots, y_n$ as the first n terms, we have

$$\Delta(u,k) \leq 2D(P,\mathcal{R}_2) \quad \text{for all } k = 1,2,\ldots,n \tag{1.4}$$

(Exercise 1(b)). Finally, with a little more work one can show that if we have $D(n,\mathcal{R}_2) \leq f(n)$ for some nondecreasing function f and for all n, then there exists a sequence u with $\Delta(u,n) = O(f(n))$. Therefore, the question about an infinite sequence with bounded discrepancy and Problem 1.1 are equivalent in a strong sense—even the quantitative bounds are the same up to a small multiplicative constant.

The difference between the discrepancy $D(P,\mathcal{R}_2)$ of a finite point set and the discrepancy $\Delta(u,n)$ of an infinite sequence is not so much in the finite/infinite distinction (note that $\Delta(u,n)$ is well-defined even for a finite sequence with at least n terms), but rather, it distinguishes a "static" and a "dynamic" setting. In the definition of the discrepancy for rectangles, we deal with the behavior of the whole set P, whereas in the definition of $\Delta(u,n)$, we look at all the initial segments $\{u_1\}$, $\{u_1,u_2\}$,..., $\{u_1,u_2,\ldots,u_n\}$ simultaneously. If we start with the empty interval $[0,1]$ and add the points of the sequence one by one in the natural order, the current set should be uniformly distributed all the time. Note that the discrepancy of a sequence can change drastically by rearranging the terms into a different order (while the discrepancy of a set does not depend on any ordering of the points). As the above reductions show, the dynamic problem in dimension 1 is more or less equivalent to the static problem in dimension 2, and similar reductions are possible between dynamic settings in dimension d and static settings in dimension $d+1$. In this book, we will mostly treat the static case.

Bibliography and Remarks. Discrepancy theory grew out of the theory of uniform distribution. A nice and accessible book where this development can be followed is Hlawka [Hla84]. The fact that the discrepancy for axis-parallel rectangles grows to infinity, in the equivalent formulation dealing with one-dimensional infinite sequences, was conjectured by Van der Corput [Cor35a], [Cor35b] and first proved by Van Aardenne-Ehrenfest [AE45], [AE49]. Her lower bound for the discrepancy was improved by Roth [Rot54], who invented the two-dimensional formulation of Problem 1.1 and used it to establish a much stronger lower bound for the discrepancy in question[2] (see Section 6.1).

A foundational paper in the theory of uniform distribution is due to Weyl [Wey16]. Earlier, uniform distribution of one-dimensional sequences $(\{n\alpha\})$ with irrational α was proved by several authors by elementary means, but the criterion involving exponential sums enabled Weyl to establish a multidimensional analogue—uniform distribution of Kronecker sequences; see Section 2.5.

Weyl's criterion uses trigonometric polynomials for testing uniform distribution of a sequence. There are also sophisticated results in this spirit bounding the discrepancy of a sequence in terms of certain trigonometric sums. The most famous of such theorems is perhaps the Erdős–Turán inequality: for any sequence $u = (u_1, u_2, \ldots)$ of points in $[0, 1]$ and any integer $H \geq 1$, we have

$$\Delta(u, n) \leq \frac{10n}{H+1} + \frac{4}{\pi} \sum_{h=1}^{H} \frac{1}{h} \left| \sum_{k=1}^{n} e^{2\pi \mathrm{i} h u_k} \right|$$

(Hlawka [Hla84] has a masterly exposition). A multidimensional version of this inequality is due to Koksma, and various other estimates of this type are known (see e.g. [DT97]). Such inequalities are useful but in general they need not give tight bounds and sometimes the trigonometric sums may be too difficult to estimate.

There is an extensive literature and many beautiful results concerning the uniform distribution and various kinds of discrepancy of specific sequences, such as the sequences $(\{n\alpha\})$ for irrational α and their higher-dimensional analogues. A minor sample of theorems will be mentioned in Section 2.5; much more material and citations can be found in the books Drmota and Tichy [DT97] or Kuipers and Niederreiter [KN74], or also in the lively surveys Sós [Sós83a] and Beck and Sós [BS95].

Some of these results are closely connected to ergodic theory and similar branches of mathematics. Some well-known low-discrepancy sequences can be obtained from the initial point by iterating a suitable ergodic transform, and the ergodicity of the transform is directly related to the uniform distribution of the sequence. For example, for α irrational and for any $x_0 \in \{0, 1\}$, the sequence $(\{n\alpha + x_0\})_{n=0}^{\infty}$ is uniformly distributed in $[0, 1)$. Consequently we have, for any Riemann-integrable function f and all $x_0 \in [0, 1)$,

$$\lim_{n\to\infty} \left(\frac{1}{n} \sum_{i=1}^{n} f(T^i x_0) \right) = \int_0^1 f(x)\,\mathrm{d}x, \tag{1.5}$$

[2] The significance of this paper of Roth is also documented by the subsequent popularity of its title in discrepancy theory—look, for instance, at the list of references in [BC87].

where $T:[0,1) \to [0,1)$ is given by $Tx = \{x + \alpha\}$. This T is obviously a measure-preserving transform of $[0,1)$, and (1.5) is the conclusion of Birkhoff's ergodic theorem for this T (more precisely, the ergodic theorem would only imply (1.5) for *almost all* x_0). And indeed, T is and important example of an ergodic transform. The connection of other low-discrepancy sequences to ergodic transforms has been investigated by Lambert [Lam85]. On the other hand, some results first discovered for the $(\{n\alpha\})$ sequences were later generalized to flows (see [DT97] or [Sós83a] for some references).

The notion of uniform distribution of a sequence can be generalized considerably, for instance to sequences in a compact topological group X, by requiring that $\lim_{n\to\infty}\left(\frac{1}{n}\sum_{i=1}^{n} f(u_i)\right) = \int_X f(x)\,\mathrm{d}x$ for all continuous functions f. Or, instead of a discrete sequence u, one can look at the uniform distribution of a function $u:[0,\infty) \to \mathbf{R}^d$, where uniform distribution can be defined by the condition $\lim_{t\to\infty}\left(\frac{1}{t}\int_0^t f(u(t))\,\mathrm{d}t\right) = \int_{\mathbf{R}^d} f(x)\,\mathrm{d}x$, and so on.

Books and Surveys. A basic source on older results in geometric discrepancy theory, more comprehensive in this respect than the present text, is a book by Beck and Chen [BC87]. A newer excellent overview, with many references but only a few proofs, is a handbook chapter by Beck and Sós [BS95]. Alexander et al. [ABC97] also give a brief but delightful survey. An extensive recent monograph with an incredible count of 2000 bibliography references is Drmota and Tichy [DT97]. It covers results from many directions, but its main focus is the classical uniform distribution theory (investigation of the discrepancy of particular sequences etc.). The books Spencer [Spe87], Alon and Spencer [AS92], Montgomery [Mon94], and Pach and Agarwal [PA95] have nice but more narrowly focused chapters on discrepancy. A monograph on discrepancy and its relations to theoretical computer science by Chazelle [Cha98b] should appear not much later than the present book. Discrepancy theory is now an extensive subject with many facets, reaching to a number of mathematical disciplines. The amount of available material and literature makes any account of the size of a normal book necessarily incomplete. It is no wonder that more narrowly focused subfields tend to single out and the communication and flow of results and ideas between these areas are often nontrivial.

Exercises

1. Let $u = (u_1, u_2, \ldots)$ be an infinite sequence of real numbers in the interval $[0,1]$.
   (a) Verify that if an n-point set P is constructed from u as in the text above then (1.3) holds. (Consider the rectangles $[0, \frac{i}{n}) \times [a,b)$ first.)

(b)* Show that if P is a given n-point set in $[0,1]^2$ and u is a sequence with the first n terms defined as in the text above then (1.4) holds.
(c)* Show that if $D(n,\mathcal{R}_2) \leq f(n)$ for some nondecreasing function f and for all n, then there exists a sequence u with $\Delta(u,n) = O(f(n))$, where the constant of proportionality depends on f.

2. (Three-distance theorem)
(a)** Let α be a real number, let n be a natural number, and let $0 \leq z_1 \leq z_2 \leq \cdots \leq z_n < 1$ be the first n terms of the sequence $(\{i\alpha\})_{i=1}^{\infty}$ listed in a nondecreasing order. Prove that the differences $|z_{j+1} - z_j|$, $j = 1, 2, \ldots, n-1$, attain at most three distinct values. Moreover, if there are three values $\delta_1 < \delta_2 < \delta_3$, then $\delta_3 = \delta_1 + \delta_2$. It may be instructive to imagine that the real axis with the numbers 0, α, 2α,..., $n\alpha$ on it is wound around a circle of unit length, which produces a picture similar to the following one (here $\alpha = 1/\sqrt{2}$):

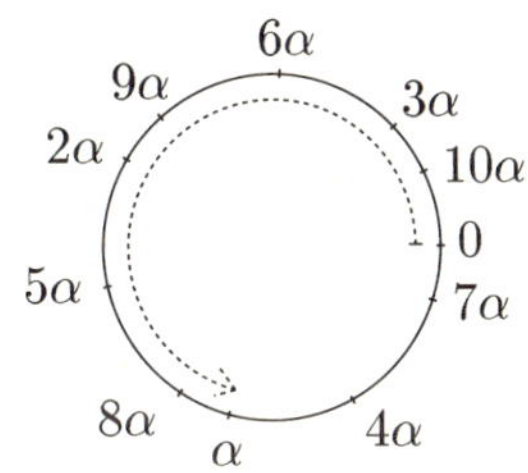

(b)** Let α be irrational, and p be the permutation of the set $\{1, 2, \ldots, n\}$ such that $0 < \{p(1)\alpha\} < \{p(2)\alpha\} < \cdots < \{p(n)\alpha\} < 1$. Show that the whole of p can be determined by the knowledge of $p(1)$ and $p(n)$ (without knowing α). (This illustrates that the sequence $(\{i\alpha\})_{i=1}^{\infty}$ is highly non-random in many respects, although it might perhaps look random-like at first sight.)
These results are due to Sós [Sós58], and we refer to that paper for a solution of this exercise.

1.2 Geometric Discrepancy in a More General Setting

Discrepancy is also studied for classes of geometric figures other than the axis-parallel rectangles, such as the set of all balls, or the set of all boxes, and so on. For discrepancy, only the part of a set $A \in \mathcal{A}$ lying in the unit cube $[0,1]^d$ is important. We are interested in finding an n-point set $P \subset [0,1]^d$ such that the fraction of points of P lying in A is a good approximation of the volume of $A \cap [0,1]^d$, and the discrepancy measures the accuracy of such an approximation. For more convenient notation, let us write $\mathrm{vol}_\Box(A)$ for $\mathrm{vol}(A \cap [0,1]^d)$.

For an n-point set $P \subset [0,1]^d$ and $A \in \mathcal{A}$, we put

$$D(P, A) = n \cdot \mathrm{vol}_\square(A) - |P \cap A|$$
$$D(P, \mathcal{A}) = \sup_{A \in \mathcal{A}} |D(P, A)|.$$

The quantity $D(P, \mathcal{A})$ is called the *discrepancy of P for $\mathcal{A}$*. Further we define the *discrepancy function* of $\mathcal{A}$:

$$D(n, \mathcal{A}) = \inf_{\substack{P \subset [0,1]^d \\ |P|=n}} D(P, \mathcal{A}).$$

Hence, in order to show that $D(n, \mathcal{A})$ is small (an upper bound), we must exhibit one n-point set and verify that it is good for all $A \in \mathcal{A}$. To prove that $D(n, \mathcal{A})$ is large (lower bound), we have to demonstrate that for any n-point set P, given by the enemy, there exists a bad $A \in \mathcal{A}$.

A Warning Concerning Notational Conventions. Let us stress that in this book, the discrepancy is measured in units of *points of P*. Often it is more natural to work with the *relative* error, i.e. with the quantity $\frac{1}{n}D(P, A)$, and this is also what one finds in a significant part of the literature. Indeed, $\frac{1}{n}D(P, A)$ is the relative error made by approximating $\mathrm{vol}_\square(A)$ by the fraction of points of P falling into A, and in many applications, the relative error is prescribed and we are looking for the smallest point set providing the desired accuracy. This interpretation becomes somewhat obscured by the definition of discrepancy we use, with the unit of one point. Nevertheless, we stick to the more traditional way, which usually leads to nicer formulas.

Two Basic Types of Behavior of the Discrepancy Function. One can try to classify the various classes $\mathcal{A}$ of geometric shapes according to the behavior of their discrepancy function. Here is perhaps the most significant (and a bit vague) division. On the one hand, we have classes consisting of scaled and translated copies of a fixed polygon or polytope, such as the class $\mathcal{R}_d$ of all axis-parallel boxes (no rotation is allowed). Two such families in the plane are indicated below:

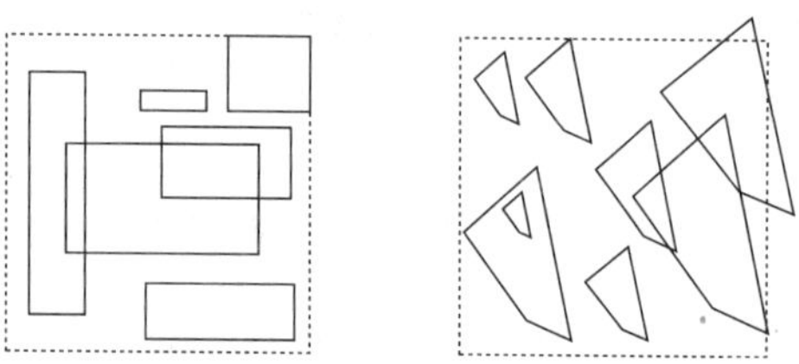

discrepancy
about $\log n$

For such classes, as a rule, the discrepancy function is bounded from above and from below by some constant powers of $\log n$. On the other hand, for rotationally invariant classes, such as halfspaces or rectangular boxes in arbitrarily rotated positions, the discrepancy function behaves like a fractional power of n, and in higher dimensions it is quite close to $\sqrt{n}$. Similar behavior occurs for translated, or translated and scaled, copies of a set with a smooth curved boundary, such as a disc or an ellipsoid. Three examples are schematically depicted below:

discrepancy about $n^{1/4}$

The middle example, the family of all triangles with two sides parallel to the axes, is particularly striking when compared with the case of axis-parallel rectangles. There are point sets giving small discrepancy, of the order $\log n$, for all axis-parallel rectangles, but if we slice each rectangle by its diagonal, some of the resulting triangles have *much* larger discrepancy.

Surprisingly, no natural classes of geometric objects are known with an intermediate behavior of discrepancy (larger than a power of $\log n$ but smaller than any fixed power of n).

The just indicated basic classification of shapes also strongly influences the subdivision of this book into chapters and sections: the two cases, classes with polylogarithmic discrepancy and classes with much larger discrepancy, usually involve distinct techniques and are mostly treated separately. Another general wisdom to remember for the study of discrepancy is this: *look at the boundary*. The irregularity of distribution always "happens" close to the boundary of the considered set, and the boundary length and shape influence the magnitude of the irregularity. The area of the considered sets, for example, is much less significant. Again, I know of no suitable exact formulation of this principle, but we will see some examples throughout the book.

More Generalizations and Variations. Clearly, discrepancy can be defined in yet more general situations. One obvious generalization is to replace the unit cube $[0,1]^d$ by other domains (a frequently investigated case is the d-dimensional unit sphere S^d), or even by complicated sets like fractals. In this book, we mostly keep working with the unit cube, since this setting seems appropriate for the first encounter with most of the ideas, and also most of the known results are formulated for the unit cube situation.

Later we will meet interesting generalizations of discrepancy in other directions, such as average discrepancy, combinatorial discrepancy, discrepancy of weighted point sets, discrepancies with respect to classes of functions, toroidal discrepancy, etc.

Decomposing Geometric Shapes for Bounding Discrepancy. We now mention a simple observation, which often allows us to simplify the class of sets for which the discrepancy is studied.

1.3 Observation. *If A, B are disjoint (and measurable) sets, then $|D(P, A\cup B)| = |D(P,A) + D(P,B)| \leq |D(P,A)| + |D(P,B)|$, for an arbitrary finite set P. Similarly for $A \subseteq B$, we have $|D(P, B\setminus A)| \leq |D(P,A)| + |D(P,B)|$.* □

As an example, we express axis-parallel rectangles (and boxes in higher dimensions) using simpler sets. For a point $x = (x_1, \ldots, x_d) \in [0,1]^d$, we define the *corner*[3] *with vertex at x* as the set

$$C_x = [0, x_1) \times [0, x_2) \times \cdots \times [0, x_d).$$

This corner is also written $C_{(x_1,x_2,\ldots,x_d)}$. Let $\mathcal{C}_d = \{C_x \colon x \in [0,1]^d\}$ be the set of all d-dimensional corners.

1.4 Observation. *For any finite set $P \subseteq [0,1]^d$ we have*

$$D(P, \mathcal{C}_d) \leq D(P, \mathcal{R}_d) \leq 2^d D(P, \mathcal{C}_d)$$

($\mathcal{R}_d$ stands for the set of all (semi-open) axis-parallel boxes in dimension d).

Sketch of Proof. The first inequality is obvious (each corner is an axis-parallel box). To see the second inequality, we express any axis-parallel box R using 2^d corners. For instance, in the plane we have

$$[a_1, b_1) \times [a_2, b_2) = \left(C_{(b_1,b_2)} \setminus C_{(a_1,b_2)}\right) \setminus \left(C_{(b_1,a_2)} \setminus C_{(a_1,a_2)}\right),$$

pictorially

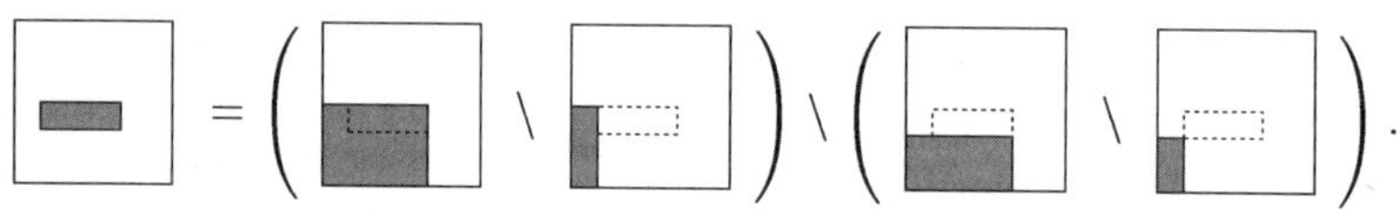

Finding the expression for a d-dimensional box using 2^d corners is left as an exercise. □

Thus, if we are not interested in the exact constant of proportionality, we can estimate the discrepancy for corners instead of that for axis-parallel rectangles.

Let us remark that the discrepancy for corners is frequently treated in the literature, and it is often denoted by D^* and called, for historical reasons, the *star-discrepancy*.

Average Discrepancy. In our definition above, the discrepancy $D(P, \mathcal{A})$ is taken as a supremum over all sets $A \in \mathcal{A}$, so it is a discrepancy *in the worst case*. In order to show a lower bound for discrepancy of some point set, it suffices to exhibit a single bad set A from the class $\mathcal{A}$ of allowed shapes. In most of the known proofs, one actually shows that a "random" or "average"

[3] In the literature, corners are sometimes called *anchored boxes*.

set from $\mathcal{A}$ must be bad. For this, and also for various applications, we need to define an average discrepancy. Since the set $\mathcal{A}$ is, in general, infinite, in order to speak of an average over $\mathcal{A}$, we have to fix a measure ν on $\mathcal{A}$. For convenience we assume that it is a probability measure, i.e. that $\nu(\mathcal{A}) = 1$. For the time being, we give only one example of such a measure ν, namely a measure on the set $\mathcal{C}_d$ of all d-dimensional corners. Since each corner C_x is determined by its vertex $x \in [0,1]^d$, we can define the measure of a set of corners $\mathcal{K} \subseteq \mathcal{C}_d$ as $\mathrm{vol}(\{x \in [0,1]^d\colon C_x \in \mathcal{K}\})$.

For a given number p, $1 \le p < \infty$, and for a probability measure ν on $\mathcal{A}$ we define the *p-th degree average discrepancy* (also called the *L_p-discrepancy*) as follows:

$$\begin{aligned} D_{p,\nu}(P,\mathcal{A}) &= \left(\int_{\mathcal{A}} |D(P,A)|^p \mathrm{d}\nu(A)\right)^{1/p} \\ D_{p,\nu}(n,\mathcal{A}) &= \inf_{\substack{P\subset[0,1]^d \\ |P|=n}} D_{p,\nu}(P,\mathcal{A}). \end{aligned}$$

If the measure ν is clear from the context, we only write D_p instead of $D_{p,\nu}$. For example, the concrete formula for the L_p-discrepancy for corners is

$$D_p(P,\mathcal{C}_d) = \left(\int_{[0,1]^d} |D(P,C_x)|^p \,\mathrm{d}x\right)^{1/p}.$$

It is easy to see that for any p and any ν, we have $D_{p,\nu}(P,\mathcal{A}) \le D(P,\mathcal{A})$ (the integral over a region of unit measure is upper-bounded by the maximum of the integrated function). By a well-known inequality for L_p-norms, we also have $D_{p,\nu}(P,\mathcal{A}) \le D_{p',\nu}(P,\mathcal{A})$ whenever $p \le p'$.

Some people may find it convenient to think about the L_p-discrepancy using a probabilistic interpretation. If the set P is fixed and $A \in \mathcal{A}$ is chosen at random according to the probability measure ν, then the discrepancy $D(P,A)$ is a random variable, and $D_{p,\nu}(P,\mathcal{A})^p$ is its pth (absolute) moment. (Note that the expectation of $D(P,A)$ need not be 0 in general, and so the L_2-discrepancy is not the same as the variance.) The L_2-discrepancy is the most important one among the various average discrepancies. It is usually the easiest to handle analytically, mainly because we need not take any absolute values in the definition.

In many papers, mainly in more practically oriented ones, the L_2-discrepancy for corners is used as the main measure of non-uniformity of distribution of a point set. (Part of its popularity can be attributed to its efficient computability; see Section 2.4 and, in particular, Exercise 2.4.11.) However, it can be argued that the L_2-discrepancy for corners does not capture the intuitive notion of uniform distribution too well, especially in higher dimensions. Roughly speaking, it exaggerates the importance of points lying close to the vertex $(0,0,\ldots,0)$ of the unit cube, and, in high dimension, a "typical" corner has a very small volume. Moreover, the directions of the coordinate axes play a significant role in the definition, and the L_2-discrepancy for corners says very little concerning the uniform distribution with respect to halfplanes, for instance. Modifications have been proposed that address some of these shortcomings; more details are given in the remarks below.

Bibliography and Remarks. The question of discrepancy for classes of shapes other than axis-parallel boxes was first raised by Erdős [Erd64]. For the special case where the point set is (a part of) the lattice $\mathbf{Z}^2$ or other lattices, discrepancy for right-angled triangles was considered much earlier (Hardy and Littlewood [HL22a], [HL22b]), and an extensive theory concerning the number of lattice points in various convex bodies has been developed ([Skr98] provides a list of references).

The rough classification of shapes according to the behavior of their discrepancy function emerged from fine works of several researchers, most notably of Roth, Schmidt, and Beck; references will be given in the subsequent chapters.

The L_2-discrepancy for corners was introduced by Roth [Rot54], first as a technical device for a lower-bound proof. Since then, it has been used widely in numerous theoretical and empirical studies. As was remarked above, it has some disadvantages. If the dimension is not very small in terms of the number of points, say if $n \leq 2^d$ (which is often the case in applications), then the L_2-discrepancy for corners gives very little information about uniform distribution, essentially because the average volume of a corner is very small; see Exercise 2.4.5 or [Mat98b]. A notion of L_2-discrepancy favoring larger corners can be found in Hickernell [Hic98], [Hic96]. We will consider a particular instance of Hickernell's notion in the discussion of Zaremba's inequality (1.8) in Section 1.4 and in Exercise 2.4.6.

Another counterintuitive feature of the L_2-discrepancy for corners is the lack of translation invariance: $D_2(P, \mathcal{C}_d)$ may be very different from $D_2(\{P+x\}, \mathcal{C}_d)$, where $\{P+x\}$ arises from P by translation by the vector x and then reducing all coordinates of each point modulo 1 (Lev [Lev95] makes this observation and notes some other undesirable properties). In fact, a surprising result of [Lev96] shows that for any n-point set $P \subset [0,1]^d$, there exists a translation vector x such that

$$D_2(\{P+x\}, \mathcal{C}_d) = \Omega(D(P, \mathcal{C}_d)),$$

with the constant of proportionality depending on d. That is, for any point set there is a translated copy whose L_2-discrepancy is nearly as bad as the worst-case discrepancy!

An alternative notion, advocated in [Lev95], is the L_2-discrepancy with respect to the class

$$\tilde{\mathcal{R}}_d = \{\{B_{x,a}\} : x, a \in [0,1]^d\},$$

where $B_{x,a}$ stands for the box $[x_1, x_1 + a_1) \times [x_2, x_2 + a_2) \times \cdots \times [x_d, x_d + a_d)$ and $\{B_{x,a}\}$ is $B_{x,a}$ reduced modulo 1 in each coordinate:

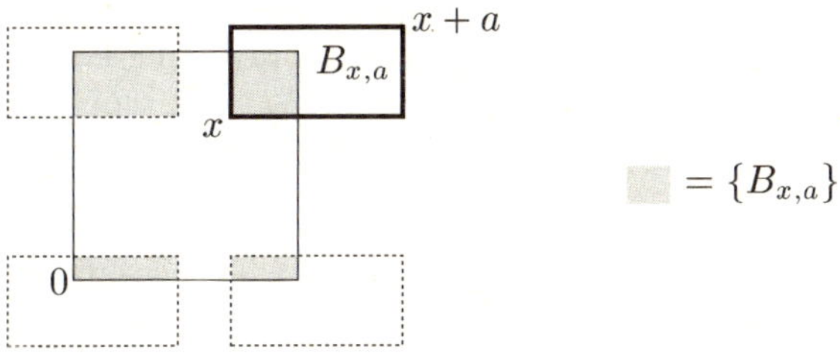

The measure of a set $R \subseteq \tilde{\mathcal{R}}_d$ is the $2d$-dimensional Lebesgue measure of the corresponding set of pairs $(x, a) \in [0,1]^d \times [0,1]^d$. The discrepancy $D_2(P, \tilde{\mathcal{R}}_d)$ is translation-invariant. Moreover, it always lies between suitable constant multiples of another discrepancy-like quantity, the *diaphony* of P (where the constants depend on the dimension; see Exercise 7.1.6). This rather technical-looking notion, introduced by Zinterhof [Zin76], is motivated by many proofs where estimates on the (usual) discrepancy are obtained via Fourier analysis. The diaphony of P is

$$\left(\sum_{m \in \mathbf{Z}^d \setminus \{0\}} \frac{|\hat{P}(m)|^2}{\prod_{k=1}^d \max(|m_k|^2, 1)} \right)^{1/2},$$

where $\hat{P}(m)$ is the exponential sum $\sum_{p \in P} e^{-2\pi \mathrm{i} \langle m, p \rangle}$, with i denoting the imaginary unit and $\langle .,. \rangle$ the usual scalar product in $\mathbf{R}^d$. Thus, the L_2-discrepancy $D_2(P, \tilde{\mathcal{R}}_d)$ provides a convenient geometric interpretation of diaphony (up to a constant factor, that is). Lev [Lev95] suggests to call the discrepancy for $\tilde{\mathcal{R}}_d$ the *Weyl discrepancy*, because Weyl's foundational paper [Wey16] also considers the (worst-case) discrepancy for intervals taken modulo 1, i.e. on the unit circle. A formula for an efficient computation of this kind of discrepancy can be found in Exercise 7.1.8.

To conclude this discussion of alternatives to the L_2-discrepancy for corners, let us remark that the latter has its advantages too: it is well-established in the literature, and it is perhaps more intuitive and sometimes technically simpler than the alternative notions mentioned above (Hickernell's generalized discrepancy or the discrepancy for $\tilde{\mathcal{R}}_d$). For most of the questions studied in this book, the differences between these notions are not very important. In any case, for measuring the irregularity of distribution, the choice of the "right" discrepancy should be guided by the particular application, and there is probably no single optimal definition.

The discrepancy for the class $\tilde{\mathcal{R}}_d$ of boxes reduced modulo 1 is a special case of the so-called *toroidal discrepancy*. For an arbitrary class

$\mathcal{A}$ of shapes, we can define the corresponding class $\tilde{\mathcal{A}} = \{\{A\} : A \in \mathcal{A}\}$. That is, instead of cutting off the parts of a set A protruding out from the unit cube, we wrap them around. In other words, the unit cube is replaced by (or interpreted as) the torus $\mathbf{R}^d/\mathbf{Z}^d$, which has some technical advantages for methods involving Fourier analysis. Toroidal discrepancy has been used for a long time, especially in proofs of lower bounds (e.g., in Schmidt [Sch69c]). We will return to this in the remarks to Section 7.1.

The second most important domain in which discrepancy is studied, besides the unit cube, is probably the unit sphere S^d. Various notions of discrepancy for this situation, and their applications to numerical integration, are surveyed in Grabner et al. [GKT97]. A very natural and much investigated notion is the discrepancy for *spherical caps* (i.e. intersections of S^d with halfspaces). A little more about this will be said in the remarks to Section 3.1.

Exercises

1. Prove that any axis-parallel box $R = [a_1, b_1) \times \ldots \times [a_d, b_d)$ can be expressed by 2^d corners, using the operations of disjoint union and "encapsulated difference" (meaning the difference of two sets A, B with $B \subseteq A$).
2. Let $\mathcal{A}$ be some class of measurable sets in $\mathbf{R}^d$.
   (a) Prove that for each n, $D(n, \mathcal{A}) - 1 \leq D(n+1, \mathcal{A}) \leq D(n, \mathcal{A}) + 1$.
   (b) Is the function $D(n, \mathcal{A})$ necessarily nondecreasing in n?
3. Check that $D_p(P, \tilde{\mathcal{R}}_2) = D_p(\{P + x\}, \tilde{\mathcal{R}}_d)$ for any finite $P \subset [0,1]^d$, any $x \in \mathbf{R}^d$, and any $p \in [1, \infty)$, where $\tilde{\mathcal{R}}_d$ is as in the remarks above.

1.3 Combinatorial Discrepancy

In this section, we start considering a seemingly different problem. Let X be an n-element set, and let $\mathcal{S}$ be a system of subsets[4] of X. We want to color each point of X either red or blue, in such a way that any of the sets of $\mathcal{S}$ has roughly the same number of red points and blue points, as in the following (schematic and possibly misleading) picture:

[4] Sometimes we will write "a set system $(X, \mathcal{S})$," meaning that X is a set and $\mathcal{S}$ is a system of subsets of X. This notation is analogous to the standard notation (V, E) for graphs, where V is the vertex set and E is the edge set. In fact, our notion of "set system" is fully synonymous to the notion of "hypergraph," and for hypergraphs, the notation $(X, \mathcal{S})$ is quite standard. On the other hand, when the underlying set is understood (usually it is the union of all sets in $\mathcal{S}$), we will say "the set system $\mathcal{S}$" only.

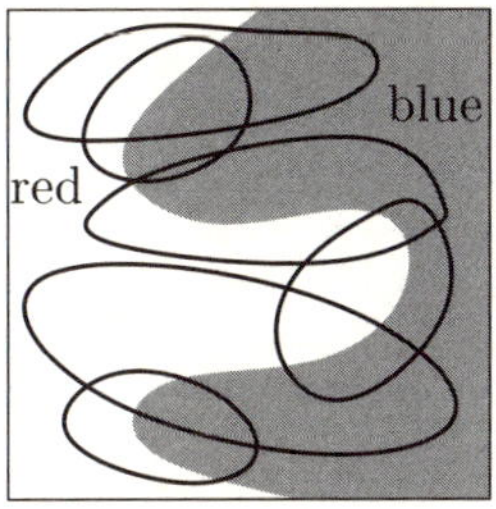

Easy considerations reveal that it is not always possible to achieve an exact splitting of each set. The error for some sets may even be arbitrarily large—in fact, if we take all subsets of X for $\mathcal{S}$, then there will always be a completely monochromatic set of size at least $\frac{n}{2}$. The maximum deviation from an even splitting, over all sets of $\mathcal{S}$, is the *discrepancy* of the set system $\mathcal{S}$. We now express this formally. A *coloring* of X is any mapping $\chi: X \to \{-1, +1\}$. The *discrepancy* of $\mathcal{S}$, denoted by $\mathrm{disc}(\mathcal{S})$, is the minimum, over all colorings χ, of

$$\mathrm{disc}(\chi, \mathcal{S}) = \max_{S \in \mathcal{S}} |\chi(S)|,$$

where we use the shorthand $\chi(S)$ for $\sum_{x \in S} \chi(x)$. (If $+1$'s are red and -1's are blue then $\chi(S)$ is the number of red points in S minus the number of blue points in S.)

To distinguish this notion of discrepancy from the one introduced previously, we sometimes speak of *combinatorial discrepancy*.[5] Our earlier notion of discrepancy, where we approximate the continuous Lebesgue measure by a discrete point set, may be referred to as *Lebesgue-measure discrepancy* (also "measure-theoretic discrepancy" or "continuous discrepancy" in the literature). Here we mention just a few facts and definitions concerning combinatorial discrepancy; Chapter 4 is devoted to a more systematic treatment.

Combinatorial discrepancy can be transferred to a geometric setting as well. The following is a typical example of a geometrically defined problem in combinatorial discrepancy: given an n-point set P in the plane, we want to color each point of P red or blue in such a way that the maximum difference, over all halfplanes h, in the number of red points and blue points in h is as small as possible. Such a problem can be re-phrased using the combinatorial discrepancy of the set system induced by halfplanes.

If $(X, \mathcal{A})$ is a set system, with X possibly infinite, and $Y \subseteq X$ is a set, we define the *set system induced by $\mathcal{A}$ on Y* as the set system

$$\mathcal{A}|_Y = \{A \cap Y \colon A \in \mathcal{A}\}.$$

(We remark that $\mathcal{A}|_Y$ is sometimes called the *trace* of $\mathcal{A}$ on Y.) In a geometric setting, $\mathcal{A}$ is a system of subsets of $\mathbf{R}^d$, such as the system of all halfspaces, or the system of all balls, and so on, and we will investigate the combinatorial

[5] Also the name "red-blue discrepancy" is used in the literature.

discrepancy of set systems $\mathcal{A}|_P$, where $P \subseteq \mathbf{R}^d$ is a finite set. For a more convenient notation, we will also write $\mathrm{disc}(P, \mathcal{A})$ for $\mathrm{disc}(\mathcal{A}|_P)$. Explicitly, $\mathrm{disc}(P, \mathcal{A})$ is the minimum, over all colorings $\chi: P \to \{-1, 1\}$, of

$$\mathrm{disc}(\chi, P, \mathcal{A}) = \max_{A \in \mathcal{A}} \bigl|\chi(P \cap A)\bigr|.$$

Further we define the *discrepancy function* of $\mathcal{A}$ by

$$\mathrm{disc}(n, \mathcal{A}) = \max_{|P|=n} \mathrm{disc}(P, \mathcal{A}).$$

Combinatorial discrepancy in a geometric setting is worth investigating for its own sake, but moreover, there is a close connection with the Lebesgue-measure discrepancy. Roughly speaking, upper bounds for the combinatorial discrepancy for some class $\mathcal{A}$ imply upper bounds for the Lebesgue-measure discrepancy for $\mathcal{A}$. (The reverse direction does not work in general.) This relation is used in many proofs; currently it appears convenient to prove many upper bounds in the combinatorial setting, and some lower bounds, even for combinatorial discrepancy, are proved via the Lebesgue-measure setting. Before giving a precise formulation of this relationship, we introduce another useful notion.

A Common Generalization. The ε-approximation, a notion with origins in probability theory, can be regarded as a generalization of both the Lebesgue-measure discrepancy and the combinatorial discrepancy. It is defined in the following setting: X is some finite or infinite ground set, μ is a measure on X with $\mu(X) < \infty$, and $\mathcal{S}$ is a system of μ-measurable subsets of X. Let $\varepsilon \in [0, 1]$ be a real number. We say that a finite subset $Y \subseteq X$ is an *ε-approximation* for the set system $(X, \mathcal{S})$ with respect to the measure μ if we have, for all $S \in \mathcal{S}$,

$$\left| \frac{|Y \cap S|}{|Y|} - \frac{\mu(S)}{\mu(X)} \right| \le \varepsilon.$$

This means that the fraction of the points of Y lying in S should approximate the relative measure of S with accuracy no worse than ε. If the phrase "with respect to μ" is omitted, we always mean the counting measure on X given by $\mu(S) = |S|$ (we thus also assume that X is a finite set). For example, if X are the inhabitants of some country, the sets in $\mathcal{S}$ are various interest groups, and Y are the members of the parliament, then Y being a $\frac{1}{100}$-approximation means that all interest groups are represented proportionally, with deviation at most 1% of the total population.

The connection to the Lebesgue-measure discrepancy is fairly obvious:

1.5 Observation. *If $\mathcal{A}$ is a class of Lebesgue-measurable sets in $\mathbf{R}^d$ and $P \subset [0, 1]^d$ is an n-point set, then $D(P, \mathcal{A}) \le \varepsilon n$ if and only if P is an ε-approximation*[6] *for $(\mathbf{R}^d, \mathcal{A})$ with respect to the measure* $\mathrm{vol}_\square$. $\square$

[6] If we measured discrepancy as a relative error, rather than in the units of points, and if the term ε-approximation were not well-established, we could naturally

The relationship of ε-approximations to combinatorial discrepancy is a bit more complicated.

1.6 Lemma (Combinatorial discrepancy and ε-approximations). *Let $\mathcal{S}$ be a system of subsets of a $2n$-point set X.*

(i) If $Y \subset X$ is an n-point set that is an ε-approximation for $(X, \mathcal{S})$ then $\operatorname{disc}(\mathcal{S}) \le 2\varepsilon n$. (By the above agreement, we mean ε-approximation with respect to the counting measure on X.)

(ii) If $\mathcal{S}$ is such that $X \in \mathcal{S}$ and $\operatorname{disc}(\mathcal{S}) \le \varepsilon n$ then there exists an n-point set $Y \subset X$ that is an ε-approximation for $(X, \mathcal{S})$.

Proof. In (i), the mapping χ with $\operatorname{disc}(\chi, \mathcal{S}) \le \varepsilon n$ is given by $\chi(x) = 1$ for $x \in Y$ and $\chi(x) = -1$ for $x \notin Y$. Indeed, for any $S \in \mathcal{S}$ we have

$$\chi(S) = |S \cap Y| - (|S| - |S \cap Y|) = 2|S \cap Y| - |S|, \tag{1.6}$$

and since we assume

$$\left| \frac{|Y \cap S|}{|Y|} - \frac{|S|}{|X|} \right| = \frac{1}{2n} \Big| 2|Y \cap S| - |S| \Big| \le \varepsilon,$$

the required bound $|\chi(S)| \le 2\varepsilon n$ follows.

As for (ii), consider a coloring χ with $\operatorname{disc}(\chi, \mathcal{S}) \le \varepsilon n$, and let Y_0 be the larger of the two color classes $\chi^{-1}(1)$ and $\chi^{-1}(-1)$. Since we assume $X \in \mathcal{S}$, we have, using (1.6) with $S = X$, $|\chi(X)| = \big|2|Y_0| - 2n\big| \le \varepsilon n$, and consequently $n \le |Y_0| \le n + \frac{\varepsilon}{2} n$. Let Y be a set of exactly n points arising from Y_0 by removing some arbitrary $|Y_0| - n \le \frac{\varepsilon}{2} n$ points. For $S \in \mathcal{S}$, we calculate

$$\begin{aligned} \left| \frac{|Y \cap S|}{|Y|} - \frac{|S|}{|X|} \right| &= \frac{1}{2n} \Big| 2|Y \cap S| - |S| \Big| \\ &\le \frac{1}{n} \Big| |Y \cap S| - |Y_0 \cap S| \Big| + \frac{1}{2n} \Big| 2|Y_0 \cap S| - |S| \Big| \\ &\le \frac{\varepsilon}{2} + \frac{1}{2n} |\chi(S)| \le \varepsilon. \end{aligned}$$

□

Somewhat imprecisely, this proof can be summarized by saying "if χ is a coloring with small discrepancy, then each of the color classes $\chi^{-1}(1)$ and $\chi^{-1}(-1)$ makes a good ε-approximation." But the two color classes of a coloring need not be exactly of the same size, and this is a technical nuisance in the proof.

Another fairly trivial but useful observation about ε-approximations is

---

call an ε-approximation for a set system $(X, \mathcal{S})$ with respect to a measure μ a *set with discrepancy at most ε for $(X, \mathcal{S})$ with respect to μ.* This gives a fairly general definition of discrepancy, although certainly not the most general reasonable definition.

1.7 Observation (Iterated approximation). *Let Y_0 be an ε-approximation for $(X, \mathcal{S})$ with respect to some measure μ, and let Y_1 be a δ-approximation for the set system $(Y_0, \mathcal{S}|_{Y_0})$. Then Y_1 is an $(\varepsilon + \delta)$-approximation for $(X, \mathcal{S})$ with respect to μ.* □

After this digression concerning ε-approximations, we return to discrepancy.

Upper Bounds for Combinatorial Discrepancy Imply Upper Bounds for Lebesgue-Measure Discrepancy. Here is one possible precise formulation of the relationship of the combinatorial and Lebesgue-measure discrepancies.

1.8 Proposition (Transference lemma). *Let $\mathcal{A}$ be a class of Lebesgue-measurable sets in $\mathbf{R}^d$ containing a set A_0 with $[0,1]^d \subseteq A_0$. Suppose that $D(n, \mathcal{A}) = o(n)$ as $n \to \infty$, and that* $\operatorname{disc}(n, \mathcal{A}) \leq f(n)$ *for all n, where $f(n)$ is a function satisfying $f(2n) \leq (2 - \delta)f(n)$ for all n and some fixed $\delta > 0$. Then we have*

$$D(n, \mathcal{A}) = O(f(n)).$$

On the other hand, if we know that $D(n, \mathcal{A}) = o(n)$ and $D(n, \mathcal{A}) \geq f(n)$ for all n, with a class $\mathcal{A}$ and a function $f(n)$ as above, then $\operatorname{disc}(n, \mathcal{A}) \geq cf(n)$ *holds for infinitely many n with a suitable constant $c = c(\delta) > 0$.*

All sublinear bounds $f(n)$ for discrepancy we are likely to encounter, such as $n^{1/2}$, $\log n$, etc., satisfy the condition in the proposition. Also the requirement that $D(n, \mathcal{A}) = o(n)$ is usually quite weak: it only requires that the Lebesgue measure on the sets of $\mathcal{A}$ can be approximated with an arbitrarily good relative accuracy by the uniform measure concentrated on a finite point set, but there is no condition on the size of the finite set. Except for quite wild sets, a fine enough regular grid of points suffices for such an approximation. So essentially the proposition says that $D(n, \mathcal{A}) = O(\operatorname{disc}(n, \mathcal{A}))$, except possibly for some pathological situations.

Proof of Proposition 1.8. Let $f(n)$ be a function as in the proposition and let n be a given number. We set $\varepsilon = \frac{f(n)}{n}$ and we choose a sufficiently large natural number k so that

$$\frac{D(2^k n, \mathcal{A})}{2^k n} \leq \varepsilon.$$

In other words, there exists a set P_0 of $2^k n$ points that is an ε-approximation for $(\mathbf{R}^d, \mathcal{A})$ with respect to the measure $\mathrm{vol}_\square$. We have thus approximated the continuous measure $\mathrm{vol}_\square$ by the possibly very large but finite set P_0.

Next, we are going to reduce the size of this approximating set to n by a repeated halving, using Lemma 1.6(ii). Namely, we consider the set system $(P_0, \mathcal{A}|_{P_0})$ and we take a coloring χ_0 for it with discrepancy at most $f(|P_0|) = f(2^k n)$. By Lemma 1.6(ii), such a coloring yields a subset $P_1 \subset$

P_0 of $2^{k-1}n$ points that is an ε_0-approximation for $(P_0, \mathcal{A}|_{P_0})$, where $\varepsilon_0 = f(2^k n)/2^k n$. We repeat this step with the set system $(P_1, \mathcal{A}|_{P_1})$, obtaining a set $P_2 \subset P_1$ of $2^{k-2}n$ points that is an ε_1-approximation for $(P_1, \mathcal{A}|_{P_1})$ with $\varepsilon_1 = f(2^{k-1}n)/2^{k-1}n$, and so on. Schematically, this procedure is indicated in the following picture:

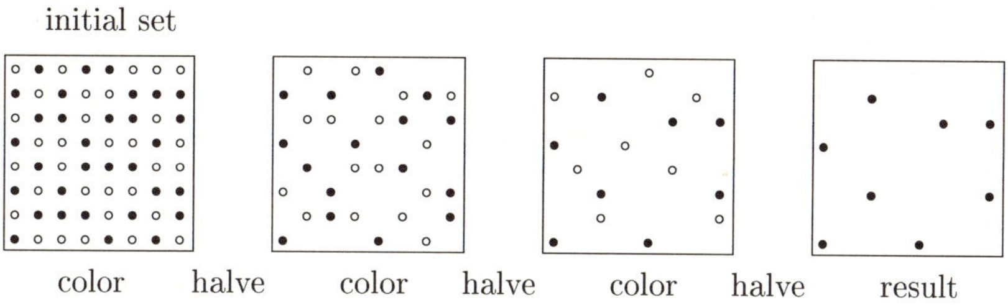

We make k such halving steps. The resulting set P_k has n points, and by Observation 1.7, it is an η-approximation for the original set system $(X, \mathcal{S})$ with respect to the measure $\mathrm{vol}_\square$, where

$$\eta = \varepsilon + \sum_{i=0}^{k-1} \varepsilon_i = \frac{f(n)}{n} + \sum_{i=0}^{k-1} \frac{f(2^{k-i}n)}{2^{k-i}n}$$
$$\leq \frac{f(n)}{n}\left(1 + \sum_{j=0}^{\infty}\left(\frac{2-\delta}{2}\right)^j\right) = O\left(\frac{f(n)}{n}\right).$$

In view of Observation 1.5, this implies the first part of the proposition. The second part is a contraposition of the first part and we leave it to the reader. □

Of course, the same proof could be phrased without introducing ε-approximations, but without such a notion, it would become somewhat obscure.

Combinatorial L_p-Discrepancy. This is similar to the L_p-discrepancy in the Lebesgue-measure setting. For a set system $\mathcal{S}$ on a finite set X and a coloring χ of X, we put

$$\mathrm{disc}_p(\chi, \mathcal{S}) = \left(\frac{1}{|\mathcal{S}|}\sum_{S\in\mathcal{S}} |\chi(S)|^p\right)^{1/p}.$$

More generally, if $\mathcal{A}$ is a family of subsets of a set X, ν is a probability measure on $\mathcal{A}$, P is a finite subset of X, and χ is a coloring of P, we set

$$\mathrm{disc}_{p,\nu}(\chi, P, \mathcal{A}) = \left(\int_{\mathcal{A}} |\chi(P\cap A)|^p \,\mathrm{d}\nu(A)\right)^{1/p}.$$

Thus, each subset S of P induced by $\mathcal{A}$ is counted with weight equal to $\nu(\{A \in \mathcal{A} \colon A \cap P = S\})$.

Bibliography and Remarks. A close relation of the combinatorial discrepancy to the Lebesgue-measure one has been folklore for some time; M. Simmonovits (private communication) attributes this observation to V. T. Sós. A written version of this idea appears in Beck [Bec81a], where it is used to lower-bound the combinatorial discrepancy for axis-parallel rectangles in the plane (*Tusnády's problem*) using classical lower bounds on the Lebesgue-measure discrepancy. A quite general version of this "transference principle," dealing with classes of convex sets in the plane, was formulated by Lovász et al. [LSV86] (with the proof phrased slightly differently from our proof of Proposition 1.8).

The ε-approximations were defined and used by Vapnik and Chervonenkis [VC71] (the name itself was given by Haussler and Welzl [HW87]). We will hear more about them in Chapter 5.

Exercises

1. Prove the second part of Proposition 1.8 (beginning with "On the other hand,...") from the first part.
2. Let $\mathcal{K}_2$ denote the collection of all closed convex sets in the plane. Show that $D(n, \mathcal{K}_2) = o(n)$ and $\mathrm{disc}(n, \mathcal{K}_2) \geq \frac{n}{2}$.
3. Find a class $\mathcal{A}$ of measurable sets in the plane such that $D(n, \mathcal{A}) = \Omega(n)$.

1.4 On Applications and Connections

Sets with small discrepancy, that is, "very uniformly distributed," have considerable theoretical and practical significance. Moreover, discrepancy theory uses various nice and important mathematical ideas and techniques (some of which we intend to demonstrate in the subsequent chapters), and these ideas have numerous applications in other branches of mathematics. Also, in theoretical computer science, discrepancy theory methods became crucial in many results in recent years. In this section, we mainly discuss relations of discrepancy to numerical integration and to Ramsey theory. A few more applications and connections will be addressed in the remarks.

Numerical Integration. One of the most important applications of low-discrepancy sets is to numerical integration in higher dimensions. In numerical integration, the definite integral of a given function over some region, such as the unit cube, is approximated by the arithmetic mean of the function's values at suitably chosen points. A basic problem is which points are to be chosen for calculating the function's values so that the error of the approximation is as small as possible. The points of a regular grid, or other straightforward generalizations of classical one-dimensional quadrature rules, do not work well in higher dimensions. On the other hand, point sets with small discrepancy are suitable candidates from both theoretical and practical points of view.

A well-known estimate for the integration error via discrepancy is the so-called *Koksma–Hlawka inequality*. Let $f\colon [0,1]^d \to \mathbf{R}$ be the integrated function and let $P \subset [0,1]^d$ be an n-point set used for the approximation. Then the inequality says

$$\left| \int_{[0,1]^d} f(x)\,\mathrm{d}x - \frac{1}{n} \sum_{p \in P} f(p) \right| \le \frac{1}{n} D(P, \mathcal{C}_d) V(f). \tag{1.7}$$

On the right-hand side, the first term (the discrepancy for corners) only depends on P, while the second term $V(f)$ is determined solely by f. For $d = 1$, $V(f)$ is the *variation* of f; for a continuously differentiable function f, we have $V(f) = \int_0^1 |f'(x)|\,\mathrm{d}x$. For higher dimensions, $V(f)$ denotes an appropriate generalization of variation, the so-called *variation in the sense of Hardy and Krause*, which we will not define here. Although the Koksma–Hlawka inequality is tight in the worst case, it is often very far from being tight for functions encountered in practice.

By now, there is a large body of theory concerning error estimates in the Koksma–Hlawka spirit. These inequalities bound the maximum (or average) integration error for functions from some class in terms of various kinds of discrepancy of the point set used to approximate the integral. Some of them even exactly characterize discrepancy as the worst-case integration error, or the average-case integration error, for very natural classes of functions.

Such results can be considered as a part of a general theory of optimal numerical integration. Here, roughly speaking, a function f from some suitable class is given by a black box, which is a hypothetical device computing $f(x)$ for any given input point x. The basic question is, what is the minimum necessary number of calls to the black box that allows one to calculate the integral of f with error at most ε. Here one need not restrict oneself to the particular algorithm approximating the integral of f by the average $\frac{1}{n}\sum_P f(p)$. It is allowed to combine the values of f obtained from the black box in any other, perhaps nonlinear, way. Moreover, the points are input to the black box one by one, with each point possibly depending on the values of f at the previous points (an *adaptive* algorithm). However, it turned out that for "reasonable" classes[7] of functions, neither nonlinearity nor adaptivity helps. However, it may be truly helpful to combine the computed function's values with weights other than $\frac{1}{n}$.

Discrepancy of Weighted Point Sets. As the reader may know from numerical analysis, more sophisticated one-dimensional quadrature rules (Simp-

[7] Here "reasonable" means closed on convex combinations and on the operation $f \mapsto -f$.

son's rule, Gauss quadrature, etc.) use non-uniform weights. They approximate the integral by

$$\sum_{p \in P} w(p) f(p),$$

where P is a suitable n-point set and $w(p) \in \mathbf{R}$ are real weights, generally distinct from $\frac{1}{n}$. Such formulas achieve error bounds that are not attainable with the uniform weights $\frac{1}{n}$. Not surprisingly, in the literature related to numerical integration, discrepancy is often investigated for *weighted* point sets. For a point set P with a weight function $w: P \to \mathbf{R}$, the quantity $|P \cap A|$ is replaced by $w(P \cap A) = \sum_{p \in P \cap A} w(p)$ in the definition of discrepancy in Section 1.2. Thus, we approximate the continuous measure $\mathrm{vol}_\square$ by a (signed) measure concentrated on an n-point set. Actually, there are at least four different notions of discrepancy involving weighted point sets: we may require the weights to be nonnegative and to sum up to $n = |P|$, or we may drop one of these two conditions or both (negative weights are not as absurd as it might seem, since some of the classical quadrature formulas, such as the Newton–Cotes rule, involve negative coefficients). For discrepancy theory, the generalization to weighted point sets is usually not too significant—most of the lower bounds in this book, say, go through for weighted point sets without much difficulty.

Discrepancy for Classes of Functions. The discrepancy of a point set, say the discrepancy for axis-parallel boxes, can obviously be viewed as the maximum integration error for a class of functions, namely for the class of characteristic functions of axis-parallel boxes. But in practice, one often integrates functions with much better smoothness properties, for example continuous functions, functions with continuous derivatives of rth order, or functions with "nice" Fourier series. In such cases, the integration method should ideally take some advantage of the nice behavior of the function. Therefore, it is natural to consider various "smoother analogues" of discrepancy as the maximum integration error for suitable classes of functions, hopefully resembling the functions we are likely to encounter in applications. Specifically, let $\mathcal{F}$ be a class of real Lebesgue-integrable functions on $[0,1]^d$. For a function $f \in \mathcal{F}$, we can set

$$D(P, f) = n \int_{[0,1]^d} f(x)\, \mathrm{d}x - \sum_{p \in P} f(p),$$

and proceed to define $D(P, \mathcal{F}) = \sup_{f \in \mathcal{F}} |D(P, f)|$ and so on. Note that this definition includes the discrepancy for a class $\mathcal{A}$ of sets as a special case: use the characteristic functions of the sets in $\mathcal{A}$ as $\mathcal{F}$. Interestingly, for some natural classes $\mathcal{F}$ of *smooth* functions, the standard notion of discrepancy for axis-parallel boxes is recovered (such an alternative characterization of the L_2-discrepancy for boxes is presented in the remarks below).

Another example of a class $\mathcal{F}$ considered in the literature are the characteristic functions of axis-parallel boxes smoothed out by r-fold integration:

for a parameter $r \geq 0$ and for a point $y \in [0,1]^d$, define a function h_y by setting $h_y(x) = \prod_{k=1}^{d} \max(0, x_k - y_k)^r$, and let $\mathcal{F} = \{h_y \colon y \in [0,1]^d\}$. The resulting notion of discrepancy is called the *r-smooth discrepancy*. In general, the goal is to choose the discrepancy-defining function class small and simple, so that the corresponding discrepancy notion can be handled reasonably, but in such a way that it provides strong "Koksma–Hlawka" type inequalities, i.e. supplies good error bounds for numerical integration of a possibly much wider class of functions. A modern approach to this issue uses the so-called *reproducing kernels* in Hilbert spaces of functions; a little more on this can be found in the remarks below.

Irregularities of Partitions and Ramsey Theory. The preceding part of this section discussed things quite close to practical applications. Now, for a change, let us mention relationship of discrepancy theory to some fast-growing areas of combinatorics. The fact that some set system $(X, \mathcal{S})$ has large combinatorial discrepancy can be rephrased as follows: for any coloring of X by two colors, there is a set where one color prevails significantly. This relates the discrepancy problem to the question of *2-colorability* of a set system. Namely, $(X, \mathcal{S})$ is 2-colorable if there is a coloring of X by two colors with no set of $\mathcal{S}$ completely monochromatic. So, in a sense, the lack of 2-colorability can be regarded as an ultimate case of large discrepancy: for any coloring by two colors, there is a set with one color prevailing completely.

As an example, let us consider the set $X = \{1, 2, \ldots, n\}$ and the set system $\mathcal{A}$ of all arithmetic progressions on X; that is, of all the sets $\{a, a+d, a+2d, \ldots\} \cap X$, $a, d \in \mathbf{N}$. A theorem of Roth states that the discrepancy of $\mathcal{A}$ is at least of the order $n^{1/4}$. On the other hand, if $\mathcal{A}_k$ denotes the subsystem of all $A \in \mathcal{A}$ of size at most k, a famous theorem of Van der Waerden asserts that for any k, there exists an $n = n(k)$ such that $\mathcal{A}_k$ is not 2-colorable. That is, if a sufficiently large initial segment of the natural numbers is colored by two colors, then there is always a long monochromatic arithmetic progression. Van der Waerden's theorem is one of the significant results in the so-called Ramsey theory. In a typical Ramsey-theory question, we consider some sufficiently large combinatorial or algebraic structure X (such as a graph, a finite vector space, etc.) and we color some small substructures of X by two colors (so we may color graph edges, lines in a vector space, etc.). We ask if there always exists a substructure of a given size with all the small substructures having the same color (so we may look for a large subgraph in the given graph with all edges monochromatic, or for a k-dimensional vector subspace with all lines of the same color, and so on). These problems can be formulated as questions about 2-colorability of certain set systems. (Of course, questions involving colorings with more than two colors are studied as well.)

Both discrepancy theory and Ramsey theory can thus be regarded as parts of theory of "irregularities of partition." For each Ramsey-theory question, we automatically get a corresponding discrepancy-theoretic question for the same set system, and vice versa. The case of arithmetic progressions is a model example. Clearly, some questions that are interesting for Ramsey theory have a trivial or not so interesting discrepancy theory counterpart, and similarly for the other way round. Even if both versions are interesting, the methods of solution maybe vastly different (this is what happens for arithmetic progressions). Nevertheless, this connection can be inspiring and useful to keep in mind.

Another area related to combinatorial discrepancy theory but with mathematical life of its own is the theory of totally unimodular set systems and matrices. Here one is interested in set systems whose each subsystem has discrepancy at most 1. This subject will be briefly touched on in Section 4.4.

More Applications. Without going into details, let us mention that discrepancy theory has also been applied in such diverse areas as computer graphics, image processing, statistics, complexity of algorithms (in particular, replacing probabilistic algorithms by deterministic ones), graph theory, number theory, spectral theory of operators, and Tarski's problem of "squaring the circle" (partition the circle of area 1 into finitely many parts and move each part rigidly so that they together fill the unit square without overlap).

Bibliography and Remarks.

Quasi-Monte Carlo. Sets with low discrepancy can be used for numerical integration in higher dimensions, thus competing with (and often beating) random point sets employed in the popular *Monte Carlo method.* The replacement of random point sets by deterministic (or semi-random) constructions, with a presumably greater "uniformity," is usually called *quasi-Monte Carlo methods.* Such methods are not limited to integration; they can help in numerical solution to differential and integral equations, in optimization, and in other problems.

A concise survey of quasi-Monte Carlo methods is Spanier and Maize [SM94]; newer ones are James et al. [JHK97] and Morokoff and Caflisch [MC95], both written from a practical (computational physicist's) point of view. Tezuka [Tez95] has another brief introduction, also more on the practitioner's side. A more theoretically oriented and considerably more comprehensive (and also technically more demanding) is a monograph by Niederreiter [Nie92].

The study of efficient algorithms for approximating the integral of a function given by a black box is a part of the theory of *information-based complexity* . This theory considers the complexity of algorithms for "continuous" problems, such as computing derivatives, integrals, evaluating various linear operators on function spaces, etc. Two books covering this area are Traub et al. [TWW88] and the newer Traub and Werschulz [TW98].

The area of quasi-Monte Carlo methods is certainly related to discrepancy, but it has somewhat distinct flavor and distinct goals. In "pure" discrepancy theory, as it has been developing so far, one is mainly interested in asymptotic results, such as that for any fixed dimension d, one can construct an n-point set in $[0,1]^d$ with discrepancy $O(\log^{d-1} n)$ for axis-parallel boxes, where the constant of proportionality depends on d. Since a random point set would only give about $O(\sqrt{n})$ discrepancy, the just mentioned construction is better—period. But if one wants to use such a construction for numerical integration, asymptotic results do not suffice. One has to ask—for how large n is the discrepancy of the constructed set significantly better than the discrepancy of a random point set? Even for not too large dimension, such as 10, an astronomically large n may be required to show the superiority over the random points for some asymptotically good constructions. Moreover, one cannot simply say "the smaller discrepancy, the better set," since the Koksma–Hlawka inequality and its relatives often grossly overestimate the error. Nevertheless, point sets constructed as examples of low-discrepancy sets for axis-parallel boxes proved quite successful in many practical applications.

In this book, we restrict ourselves to a few occasional remarks concerning relations of discrepancy to quasi-Monte Carlo methods. For studying the quasi-Monte Carlo papers, a warning concerning different conventions might perhaps be helpful. In the discrepancy-theory literature, one usually looks at the discrepancy of *sets* (as defined above), while for quasi-Monte Carlo methods, the authors more often work with low-discrepancy *sequences* (where every initial segment is required to be a low-discrepancy set). There is a simple theoretical relation between these two settings. Essentially, good sets in dimension d correspond to good sequences in dimension $d-1$ (see Section 1.1 for the case $d=2$). But from the practical point of view, the sequences are often preferable.

What Dimension? For various applications in physics, several authors have argued that the advantage of quasi-Monte Carlo methods over the Monte Carlo method (random points) becomes negligible from the practical point of view for dimensions over 20, say (e.g., [JT93]) Also, Sloan and Woźniakowski [SW97] show that numerical integration using fewer than 2^d sample points in dimension d is hopeless in the worst case for certain quite nice classes of functions: no algorithm can do better in the worst case than the trivial algorithm that always outputs 0 as the answer! The threshold 2^d is very sharp, since there exist algorithms with much smaller error using exactly 2^d points.

On the other hand, quasi-Monte Carlo methods have recently been applied successfully for problems of very high dimensions in financial computations (where even small errors may cost big money!); see, for instance, [PT95], [NT96]. A typical dimension appearing in these applications is 360, which is the number of months in 30 years—a typical period for which U.S. banks provide loans. These very high-dimensional integrals can be seen as approximations to infinite-dimensional path-integrals (also some path-integrals in physics have been handled successfully; see [MC95] for references). Here the success of the quasi-Monte Carlo approach should probably be attributed to a special low-dimensional structure of the integrated functions. A partial theoretical explanation of this phenomenon was found by Sloan and Woźniakowski [SW98].

Error of Integration and Discrepancies. The notion of discrepancy with respect to a given class of functions is very natural in the context of quasi-Monte Carlo methods. It appears in numerous papers, often without references to previous literature with similar concepts. The earliest reference I found is Hlawka [Hla75], who considered the one-dimensional case with $\mathcal{F} = \{x \mapsto x^k \colon k = 0, 1, 2, \ldots\}$. This *polynomial discrepancy* and its higher-dimensional analogues have been studied further by Schmidt, Klinger, Tichy, and others; recent results and references can be found in [KT97]. The r-smooth discrepancy mentioned in the text was considered by Paskov [Pas93].

The one-dimensional Koksma–Hlawka inequality is due to Koksma [Kok43], and the multidimensional version was derived by Hlawka [Hla61].

We have not defined the variation in the sense of Hardy and Krause occurring in the Koksma–Hlawka inequality. Now we pick another among the numerous generalizations and modifications of the Koksma–Hlawka inequality and state it precisely. We begin with some notation, which will allow us to formulate the results more compactly and make them look less frightening. Through this and a few subsequent paragraphs, f is a real function on $[0,1]^d$. We recall the notation, for a finite $P \subset [0,1]^d$, $D(P,f) = n \int_{[0,1]^d} f(x)\,\mathrm{d}x - \sum_{p \in P} f(p)$, which is n-times the integration error. We let $[d] = \{1, 2, \ldots, d\}$, and for an index set $I = \{i_1, i_2, \ldots, i_k\} \subseteq [d]$, we put

$$\frac{\partial^{|I|} f(x)}{\partial x_I} = \frac{\partial^k f(x_1, x_2, \ldots, x_d)}{\partial x_{i_1} \partial x_{i_2} \cdots \partial x_{i_k}}.$$

The notation Q_I stands for the $|I|$-dimensional cube

$$Q_I = \{x \in [0,1]^d \colon x_i = 1 \text{ for all } i \notin I\}.$$

And here is the promised inequality, derived by Zaremba [Zar68], involving the L_2-discrepancy for corners:

$$|D(P,f)| \leq D_{2,proj}(P) V_2(f). \tag{1.8}$$

The quantity $V_2(f)$ only depends on f:

$$V_2(f) = \left(\sum_{\emptyset \neq I \subseteq [d]} \int_{Q_I} \left(\frac{\partial^{|I|} f(x)}{\partial x_I} \right)^2 \mathrm{d}x \right)^{1/2}.$$

And $D_{2,proj}$ is a certain L_2-discrepancy of P for corners, taking into account all the coordinate projections of P:

$$D_{2,proj}(P)^2 = \sum_{\emptyset \neq I \subseteq [d]} D_2(\pi_I(P), \mathcal{C}_{|I|})^2,$$

with π_I denoting the projection on the coordinates (x_i: $i \in I$). Some assumptions on f are needed for Zaremba's inequality, of course; for instance, it is enough that the mixed partial derivative $\frac{\partial^d}{\partial x_{[d]}}$ exist and be continuous on $[0,1]^d$, but this requirement can be further relaxed. A proof is indicated in Exercise 1.

Reproducing Kernels. Next, we indicate a fairly general approach to deriving Koksma–Hlawka type inequalities, which subsumes many earlier results and notions of discrepancy. We essentially follow Hickernell [Hic98] and Sloan and Woźniakowski [SW98] (the exposition in [SW98] is somewhat simpler). Hoogland and Kleiss [HK96] and James et al. [JHK97] present interesting and somewhat related ideas (using generating functions and Feynmann diagrams).

Let $(X, \langle.,.\rangle)$ be a Hilbert space of (some) real-valued functions on $[0,1]^d$. A *reproducing kernel* on X is a bivariate function $\eta: X \times X \to \mathbf{R}$ such that the function $\eta_x: y \mapsto \eta(y,x)$ is in X for all $x \in [0,1]^d$, and the scalar product with η_x represents the evaluation at x: for all $f \in X$ and $x \in [0,1]^d$, we have $f(x) = \langle f, \eta_x \rangle$. (To see what is going on here, one can work out simple examples in Exercise 2 below, and, for instance, [Wah90] provides a more comprehensive introduction to reproducing kernels.) For a reproducing kernel to exist, it is necessary and sufficient that the evaluation operators $T_x: f \mapsto f(x)$ be all bounded (by the Riesz representation theorem). For example, the perhaps most usual function space $L_2([0,1])$ with scalar product $\langle f, g \rangle = \int_0^1 f(x)g(x)\,dx$ does not have any reproducing kernel (why?). Spaces with reproducing kernels mostly involve functions with some smoothness requirements (such as various Sobolev spaces), and the formulas for the scalar product usually contain derivatives.

For a fixed point set $P \subset [0,1]^d$, the integration error $D(P,f)$ is a linear functional on X, and it can be represented as $D(P,f) = \langle \xi_P, f \rangle$, where $\xi_P(x) = D(P, \eta_x)$. The Cauchy–Schwarz inequality then gives

$$|D(P,f)| \leq \|\xi_P\|_X \cdot \|f\|_X.$$

Here $\|.\|_X$ is the norm derived from the scalar product in X. The quantity $\|f\|_X$ is an abstract version of $V(f)$ from the Koksma–Hlawka inequality, and $\|\xi_P\|_X$ can be interpreted as a discrepancy of P. Moreover, ξ_P is a worst-case integrand, where the inequality holds with equality, and so we get a characterization of the discrepancy as a worst-case integration error. These ideas mechanize the process of deriving

Koksma–Hlawka type bounds greatly, but one has to find interesting spaces and reproducing kernels and calculate concrete formulas.

Using the Cauchy–Schwarz inequality in the above considerations leads to various notions of L_2-discrepancy; to obtain notions of L_p-discrepancy, one uses Hölder's inequality (see [Hic98], [SW98]).

Characterizations of discrepancy as an integration error found nice applications in discrepancy theory. Examples are Frolov [Fro80] and, in particular, Wasilkowski and Woźniakowski [WW95], [WW97], who upper-bound the L_2-discrepancy for corners using algorithms for approximate numerical integration.

As an example, we state a characterization of the usual L_2-discrepancy for corners as integration error for a natural class of smooth functions:

$$D_2(P, \mathcal{C}_d) = \sup_f |D(P, f)|. \tag{1.9}$$

The supremum is taken over all functions f such that $f(x) = 0$ for any x with at least one component equal to 1, the mixed partial derivative $\frac{\partial^d f}{\partial x_{[d]}}$ exists and is continuous, and

$$\int_{[0,1]^d} \left(\frac{\partial^d f(x)}{\partial x_{[d]}} \right)^2 \mathrm{d}x \leq 1.$$

The scalar product is $\langle f, g \rangle = \int_{[0,1]^d} \frac{\partial^d f(x)}{\partial x_{[d]}} \cdot \frac{\partial^d f(x)}{\partial x_{[d]}} \,\mathrm{d}x$, and the reproducing kernel is very simple: $\eta(x, y) = \prod_{k=1}^d \min(1 - x_k, 1 - y_k)$ (Exercise 2). Zaremba's inequality (1.8), for instance, can be obtained by this approach as well, together with an example showing it to be tight. The appropriate the scalar product has to consider other mixed derivatives as well: $\langle f, g \rangle = \sum_{I \subseteq [d]} \int_{Q_I} \frac{\partial^{|I|} f(x)}{\partial x_I} \cdot \frac{\partial^{|I|} g(x)}{\partial x_I} \,\mathrm{d}x$. The reproducing kernel is then $\prod_{k=1}^d \min(2 - x_k, 2 - y_k)$; see [SW98].

Random Functions and Average-Case Error. There are also characterizations of discrepancy as the expected (average-case) integration error for a *random* function. Let us begin with some motivation of this approach. For the Monte Carlo method of integration, one can estimate the error (with a reasonable confidence) by choosing several random sets and comparing the results. This cannot easily be done for a quasi-Monte Carlo method that produces just one set of a given size, say. For this reason, error estimates have been theoretically investigated in another setting, namely when the point set is fixed and the integrated function is chosen "at random." Since natural classes of functions usually form infinite-dimensional spaces with no "canonical" measure on them, it is not clear what should a random function mean.

Woźniakowski [Woź91] obtained a very nice result for one possible definition of a "random function," the so-called *Wiener sheet measure* or multidimensional *Brownian motion*. Instead of a precise definition of the appropriate class of functions and of the measure on it, we present an informal description of a random function from this class. To approximately plot the graph of a one-dimensional random function, start at the point $(1,0)$ and proceed in N steps. In each step, choose one of the possibilities "up" or "down" at random with equal probability, and go left by $\frac{1}{N}$ and either up or down, according to the random choice, by $\frac{1}{\sqrt{N}}$:

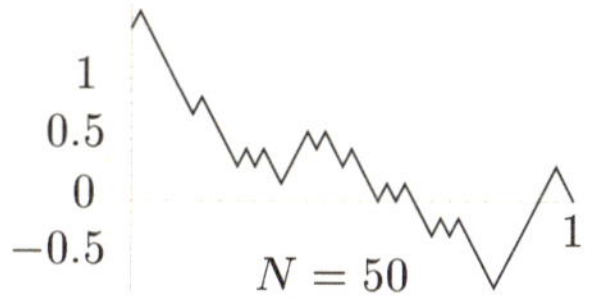

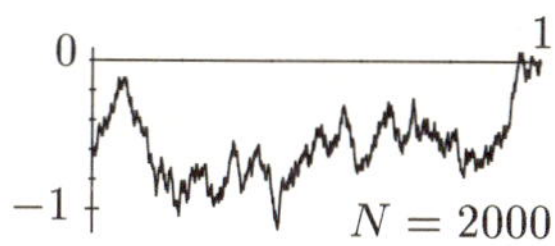

For large N, the resulting plot is approximately the graph of a random f. (The boundary condition $f(1) = 0$ is a consequence of the choice of boxes anchored at 0 in the definition of discrepancy.) For a 2-dimensional random function, subdivide the unit square $[0,1]^2$ into an $N \times N$ square grid, and for each square s of this grid, independently choose a number $\delta_s \in \{-\frac{1}{N}, +\frac{1}{N}\}$ at random, both possibilities having probability $\frac{1}{2}$. Now define the values of f at all the vertices of the little squares: for each grid square, require the condition indicated in the following picture

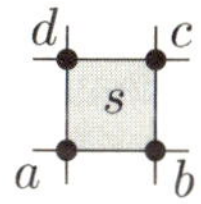

$$f(a) - f(b) + f(c) - f(d) = \delta_s$$

and also use the boundary condition $f(x,1) = f(1,y) = 0$ for all x, y. (It is easy to see that given the δ_s's, these conditions determine the values of f at all the vertices uniquely.) A result for $N = 80$ is shown below:

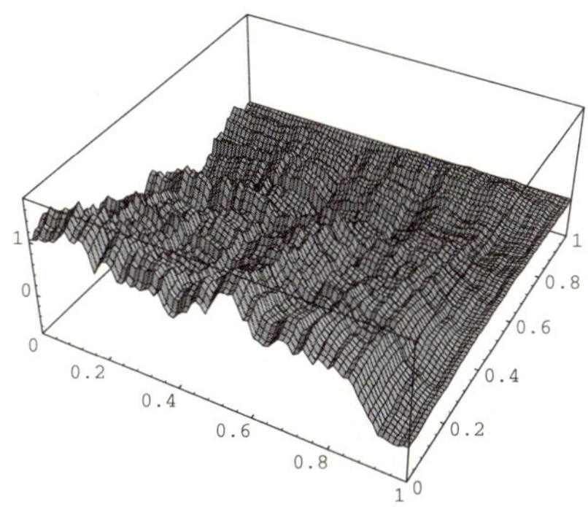

This can be generalized to dimension d in a straightforward manner.

Woźniakowski proved that for any fixed n-point set $P \subset [0,1]^d$ and f random in this sense, the expected integration error satisfies

$$\mathbf{E}\left[D(P,f)\right] = D_2(P,\mathcal{C}_d).$$

Earlier, similar results for various notions of a random function in the one-dimensional case were established by Sacks and Ylvisaker [SY70]. Woźniakowski [Woź91] has the d-dimensional statement and, moreover, directly relates the L_2-discrepancy to the average-case algorithmic complexity of numerical integration. Alternative derivations of this average-case characterization of discrepancy, as well as some generalizations, can be found in [MC94] and [JHK97].

Functions occurring in practical problems seldom resemble random functions in the sense discussed above; for instance, the latter ones are continuous but typically nowhere differentiable. Paskov [Pas93] derived an analogue of Woźniakowski's result for random functions with a given degree r of smoothness.

A general relation of the worst-case and average-case error estimates is considered in Wahba [Wah90] or in Traub et al. [TWW88].

Ramsey Theory. Nice overviews of Ramsey theory are Graham et al. [GRS90] or Nešetřil [Neš95]. An inspiring account of the connections of discrepancy theory to Ramsey theory is Sós [Sós83a] (a shorter version is in [BS95]). The $\Omega(n^{1/4})$ lower bound for discrepancy of arithmetic progressions is from Roth [Rot64]. This bound is asymptotically tight; we will say more about this problem in Sections 4.2, 4.5, and 4.6.

Number Theory. As an example of a result related to number theory, we can quote Beck's solution of a problem of Erdős concerning "flat" polynomials on the unit circle. Using discrepancy theory methods, Beck [Bec91a] proved that there are constants $c, \alpha > 0$ such that whenever $\xi_1, \xi_2, \ldots, \xi_n$ are complex numbers with $|\xi_i| = 1$ for all i and we define polynomials $p_1(z)$, $p_2(z)$,..., $p_n(z)$ by setting $p_i(z) = \prod_{j=1}^{i}(z - \xi_j)$, then

$$\max_{1\leq i\leq n} \max_{|z|=1} |p_i(z)| \geq cn^\alpha.$$

Geometry. The beautiful "squaring the circle" result mentioned in the text is due to Laczkovich [Lac90].

Here are two combinatorial geometry problems related to discrepancy. One of them asks for an n-point set on the unit d-dimensional sphere such that the sum of all the $\binom{n}{2}$ Euclidean distances determined by these points is maximal. An exact solution appears very difficult in most cases. Stolarsky [Sto73] discovered a relation of this problem to a certain kind of discrepancy, and Beck [Bec84] used results in discrepancy theory to give good asymptotic bounds for the maximum sum of distances.

Another problem concerns the approximation of the unit ball in $\mathbf{R}^d$ by a *zonotope*. A zonotope is a special type of a convex polytope that can be defined as a d-dimensional projection of an n-dimensional cube, and for the approximation we want to have n as small as possible. This problem has several equivalent formulations, one of them being a "tomography" question (Betke and McMullen [BM83]): find the minimum number n of directions $y_1, \ldots, y_n \in S^{d-1}$ (where S^{d-1} is the unit sphere in $\mathbf{R}^d$), such that the surface area of any convex body K in $\mathbf{R}^d$ can be determined, up to a relative error of ε, by the knowledge of the volumes of the $(d-1)$-dimensional projections of K on the hyperplanes $\{x \in \mathbf{R}^d\colon \langle y_i, x\rangle = 0\}$.[8] Using harmonic analysis techniques similar to those employed for discrepancy lower bounds, Bourgain et al. [BLM89] established lower bounds for this problem, and these were shown to be asymptotically tight or almost tight in a sequence of papers (also applying various discrepancy theory methods): Bourgain and Lindenstrauss [BL88], Wagner [Wag93], Bourgain and Lindenstrauss [BL93], and Matoušek [Mat96b].

Graph Theory. Discrepancy of a certain kind also appears in graph theory. For instance, Chung [Chu97] defines the discrepancy of a graph G as

$$\max_{S \subseteq V} \left| e(S,S) - \rho |S|^2 \right|,$$

where V is the vertex set of G, $e(S,S)$ is the number of ordered pairs $(u,v) \in S \times S$ such that $\{u,v\}$ is an edge of G, and $\rho = e(V,V)/|V|^2$ is the *density* of G. Thus, in a graph with small discrepancy, the number of edges on each subset S is close to the expected number of edges on a random subset of size $|S|$. The discrepancy of a graph can be bounded in terms of the second largest eigenvalue of its adjacency matrix. For graphs of density about $\frac{1}{2}$, the best possible discrepancy is of the order $n^{3/2}$. If a graph is a good *expander* then it has small discrepancy. Expanders are a very important type of "random-like" graphs, with numerous applications (in communication networks, parallel computing, sorting networks, pseudorandom generators, error-correcting codes, and so on), and the reader can learn about them in [AS92] or in [Chu97], for instance.

Computer Science. As we mentioned at the beginning of this section, discrepancy theory methods gained in importance in computer science

[8] To appreciate this formulation, one should know that if we are given the volumes of all the $(d-1)$-dimensional projections of a convex body K then the surface area is determined exactly by *Cauchy's surface area formula*. For instance, in $\mathbf{R}^3$, the surface area equals 4 times the expected area of the projection in a random direction; see [San76].

in recent years, as is successfully illustrated by the book [Cha98b]. Whenever a small sample is needed that represents well a large collection of objects, which is a very frequent situation in the search for efficient algorithms, connections to discrepancy theory may appear, and they often do.

For instance, geometric discrepancy turned out to be relevant in several results in computational geometry. This field of computer science considers the design of efficient algorithms for computing with geometric objects in the Euclidean space, usually of a low dimension (see also the remarks to Section 5.2). Lower bounds for geometric discrepancy have been used by Chazelle [Cha94] to show lower bounds for the computational complexity of a geometric database problem (the so-called range searching), and for another version of this problem, a set with low discrepancy for axis-parallel rectangles has been employed in a lower-bound proof (see [Cha98b]). The ε-approximations in geometrically defined set systems play a key role in the so-called *derandomization* of computational geometry algorithms; that is, replacing probabilistic algorithms by deterministic ones. For more information see the survey [Mat96a] or the book [Cha98b].

In another subfield of computer science, derandomizing combinatorial algorithms, the questions have discrepancy-theory flavor too but often they concern spaces of very high dimensions. For example, Linial et al. [LLSZ97] construct n-point subsets of the d-dimensional grid $\{1, 2, \ldots, q\}^d$ uniformly distributed with respect to the *combinatorial rectangles*, where a combinatorial rectangle is a set of the form $S_1 \times S_2 \times \cdots \times S_d$, with $S_1, \ldots, S_d \subseteq \{1, 2, \ldots, q\}$ being arbitrary subsets. (In out terminology, they construct an ε-approximation for combinatorial rectangles.) In contrast to the "classical" discrepancy theory setting, where the dimension is considered fixed, they need to investigate the situation where d is large (comparable to n and q, say). The main challenge here is to approach the quality of a random set by a deterministic construction, while in classical discrepancy, one can usually beat random sets. Also various constructions of *approximately k-wise independent* random variables on small probability spaces can be viewed as (explicit) constructions of small ε-approximations for certain set systems. An introduction to k-wise independence in derandomization can be found in Alon and Spencer [AS92] or in Motwani and Raghavan [MR95], and a sample of papers devoted to such constructions are [AGHP92], [EGL$^+$92].

Another interesting example of an explicit construction of an ε-approximation is provided by Razborov et al. [RSW93], who describe a set $A \subset \{1, 2, \ldots, n-1\}$ of size bounded by a polynomial in $\log n$ and in $\frac{1}{\varepsilon}$ that is an ε-approximation for the system of all arithmetic progressions (modulo n). Here it is easy to show that a random A will work with high probability; the point is to avoid randomness. This result was applied by Alon and Mansour [AM95] in a fast deterministic algorithm for interpolating multivariate polynomials.

Exercises

1. (Zaremba's inequality) Let $f\colon [0,1]^d \to \mathbf{R}$ have a continuous mixed partial derivative $\frac{\partial^d f}{\partial x_{[d]}}$ (notation as in the remarks above).
(a)* Derive the following identity for the integration error, by repeated integration by parts:
$$D(P,f) = \sum_{\emptyset \neq I \subseteq [d]} (-1)^{|I|} \int_{Q_I} D(P,C_x) \cdot \frac{\partial^{|I|} f(x)}{\partial x_I} \, \mathrm{d}x.$$
Try to get at least the cases $d=1$ (where nothing too interesting happens) and $d=2$.
(b) Using Cauchy–Schwarz, derive Zaremba's inequality (1.8) from (a).
2. (Reproducing kernels) Consider the Hilbert space X of absolutely continuous functions $f\colon [0,1] \to \mathbf{R}$ such that $f(1) = 0$ and $f' \in L_2(0,1)$ (i.e. $\int_0^1 f'(x)^2 \, \mathrm{d}x < \infty$). The scalar product is $\langle f,g \rangle = \int_0^1 f'(x)g'(x) \, \mathrm{d}x$. Recall that an absolutely continuous function f is differentiable almost everywhere, and we have $\int_a^b f'(x) \, \mathrm{d}x = f(b) - f(a)$. Functions with a continuous first derivative form a dense subspace in X.
(a) Check that $\eta(x,y) = \min(1-x, 1-y)$ is a reproducing kernel in X.
(b) Calculate ξ_P, and check that the corresponding discrepancy is just the L_2-discrepancy of P for corners.
(c)* Generalize (a) and (b) to an arbitrary dimension d (try at least $d=2$), with the reproducing kernel $\eta(x,y) = \prod_{k=1}^d \min(1-x_k, 1-y_k)$, scalar product $\langle f,g \rangle = \int_{[0,1]^d} \frac{\partial^d f(x)}{\partial x_{[d]}} \cdot \frac{\partial^d g(x)}{\partial x_{[d]}} \, \mathrm{d}x$, and functions f satisfying $f(x) = 0$ for all x with at least one component equal to 1. Derive (1.9).
Remark. The functions with $\frac{\partial^d f}{\partial x_{[d]}}$ continuous form a dense set in the appropriate Hilbert space in (c). To describe the functions in the resulting (Sobolev) space, one needs the notion of distributional derivatives, and the definitions are not entirely simple (see a book dealing with Sobolev spaces, such as [Ada75], [Wah90]). But for this exercise, such a description is not really needed, and all the functions actually encountered in the proof are piecewise polynomial.

2. Low-Discrepancy Sets for Axis-Parallel Boxes

What should a planar set with small discrepancy for axis-parallel rectangles look like? Maybe the first thing coming to mind would be the regular $\sqrt{n}\times\sqrt{n}$ grid, placed in the unit square in an appropriate scale, as in Fig. 2.1(a). It is easy to see that this gives discrepancy of the order $\sqrt{n}$. Another attempt might be n independent random points in the unit square as in Fig. 2.1(b), but these typically have discrepancy about $\sqrt{n}$ as well. (In fact, with high probability, the discrepancy is of the order $\sqrt{n \log\log n}$; a result well-known to probabilists under the name *law of the iterated logarithm* comes into play.) It turns out that a far better discrepancy can be achieved, of the order $\log n$. This chapter is devoted to various constructions of such sets and to their higher-dimensional generalizations. In dimension d, for d arbitrary but fixed, the best known sets have discrepancy for axis-parallel boxes of the order $\log^{d-1} n$.

In Section 2.1, we show perhaps the simplest construction of such sets. First we treat the planar case, due to Van der Corput, and then the d-dimensional generalization due to Hammersley and Halton.

Section 2.2 focuses on the L_2-discrepancy. We explain a modification of the Halton–Hammersely construction which provides sets with L_2-discrepan-

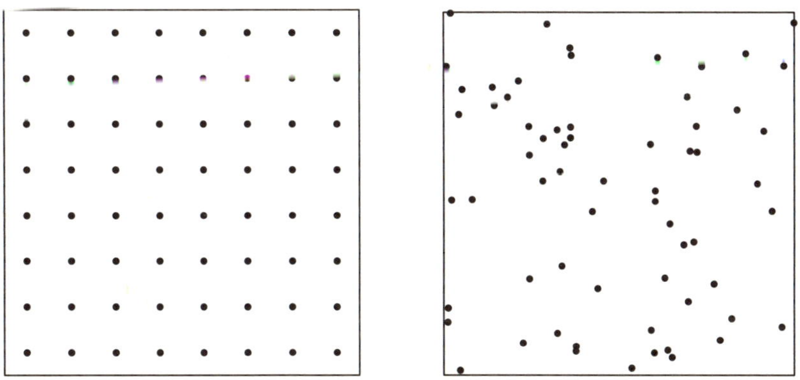

(a) (b)

Fig. 2.1. The grid points (a) and random points (b) for $n = 64$.

cy for corners of the order $O(\log^{(d-1)/2} n)$ only. This is considerably better than the worst-case discrepancy.

In the rest of this chapter, we treat some alternative constructions of low-discrepancy sets. These do not achieve better asymptotic bounds than those from the first two sections, but they involve nice mathematics and they are significant for applications. Using such constructions, the computation for some problems in physics, financial mathematics, and other areas can sometimes be accelerated by a factor of hundreds or thousands compared to the traditional Monte-Carlo approach employing random point sets. The practical behavior of constructions of low-discrepancy sets in such computations is also influenced by other factors besides discrepancy for axis-parallel boxes and the differences among various methods can be vast. None of the known approaches is fully satisfactory in all respects, and the research in this area is still active. Our treatment is just a brief introduction concentrating on the discrepancy bounds.

In Section 2.3, we discuss a class of constructions which can be seen as a generalization of some ideas from the Van der Corput and Halton–Hammersely constructions. The constructions are based on suitable collections of matrices over a finite field. Section 2.4 discusses the so-called scrambling, a way of adding randomness to the constructions from the preceding sections. This approach is quite recent and the sets produced in this way are among the most successful ones in practical applications. We also show that random scrambling leads to sets with asymptotically optimal L_2-discrepancy, giving an alternative proof of the result from Section 2.2.

In Section 2.5, we look at a different class of constructions of low-discrepancy sets, with strong number-theoretic flavor. These are based on *lattices*, i.e. images of the integer grid $\mathbf{Z}^d$ under suitable bijective linear maps.

2.1 Sets with Good Worst-Case Discrepancy

The following is perhaps the simplest example of a set with only logarithmic discrepancy for axis-parallel rectangles in the plane.

2.1 Example (Van der Corput set). This set, $P = \{p_0, p_1, \ldots, p_{n-1}\} \subset [0,1]^2$, is described by the formula

$$p_i = \left(\tfrac{i}{n}, r(i)\right), \quad i = 0, 1, \ldots, n-1, \tag{2.1}$$

where $r(i)$ is a function defined as follows. We write the number i in binary, then we write its digits in the reverse order (e.g., for $i = 13$, binary 1101, we would write 1011), and finally we prefix this by "0" and "." (for $i = 13$ we obtain 0.1011). The result is read as a real number from the interval $[0, 1)$ written in binary, and this number is the value of $r(i)$.

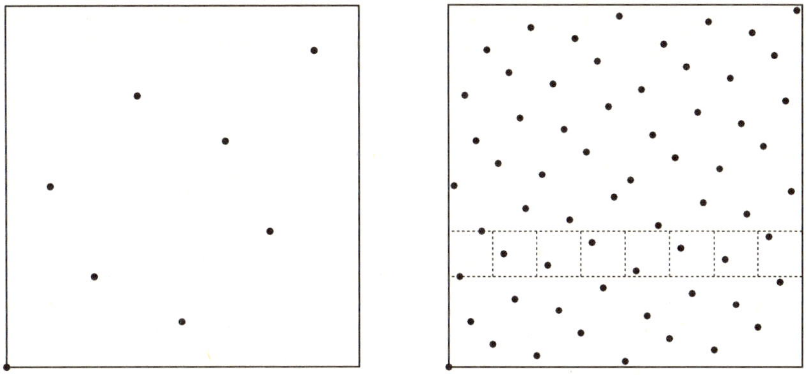

(a) (b)

Fig. 2.2. The Van der Corput set for $n = 8$ (a) and for $n = 64$ (b).

As a reminder of this construction, the sequence $r(0), r(1), r(2), \ldots$ is sometimes called the *bit reversal sequence*. More formally, the definition of $r(i)$ can be written as follows: if $i = a_0 + 2a_1 + 2^2a_2 + 2^3a_3 + \cdots$, where $a_j \in \{0,1\}$, then

$$r(i) = \frac{a_0}{2} + \frac{a_1}{2^2} + \frac{a_2}{2^3} + \frac{a_3}{2^4} + \cdots.$$

2.2 Proposition. *The discrepancy of the n-point Van der Corput set P for axis-parallel rectangles satisfies $D(P, \mathcal{R}_2) = O(\log n)$.*

Proof. By a *canonical interval* we mean one of the intervals $[0,1)$, $[0,\frac{1}{2})$, $[\frac{1}{2},1)$, $[0,\frac{1}{4})$, $[\frac{1}{4},\frac{1}{2})$, $[\frac{1}{2},\frac{3}{4})$, $[\frac{3}{4},1)$, $[0,\frac{1}{8})$, $[\frac{1}{8},\frac{1}{4})$,...; in general an interval of the form

$$\left[\frac{k}{2^q}, \frac{k+1}{2^q}\right) \quad \text{with} \quad 0 \le k < 2^q,$$

as in the following picture:

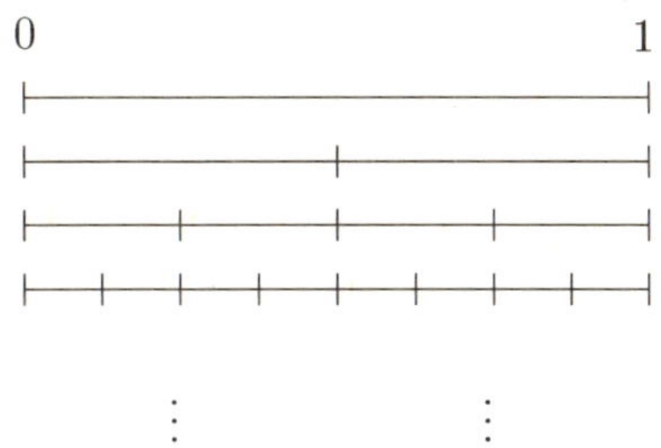

The proof of Proposition 2.2 follows from Claim I and Claim II below.

Claim I. *For any rectangle R of the form $[0,a) \times I$, where I is a canonical interval and $a \in (0,1]$ is arbitrary (as in the following illustration),*

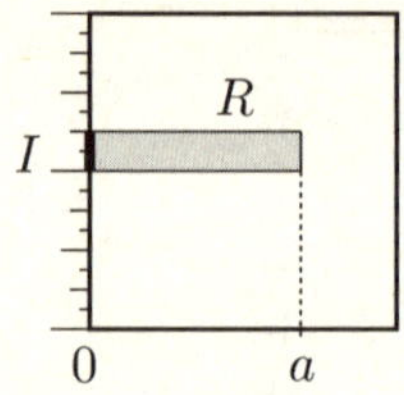

we have $|D(P,R)| \leq 1$.

Proof. Let $I = [k/2^q, (k+1)/2^q)$. The points $p_i \in P$ whose y-coordinates lie in I are those with $r(i) \in I$. This means that the first q binary digits of $r(i)$ are fixed while the remaining ones can be arbitrary. This in turn says that the q least significant binary digits of i are fixed. In other words, we get $i \equiv \bar{k} \pmod{2^q}$, for a certain integer $\bar{k}$ (we have $\bar{k} = 2^q r(k)$ but this is not important here). Therefore, the x-coordinates of the points of P lying in the strip $S = [0,1) \times I$ are regularly spaced with step $\frac{2^q}{n}$. If we subdivide the strip S into rectangles of length $\frac{2^q}{n}$ in the x-direction

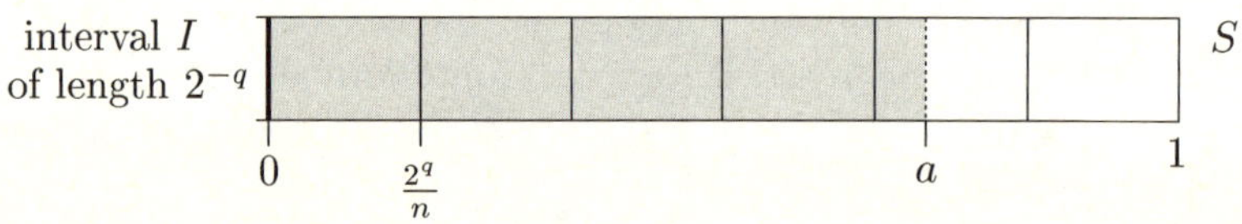

then each such rectangle has area $\frac{1}{n}$ and exactly one point of P inside, so its discrepancy is 0. One such strip S, with $q = 3$, is drawn by dotted line in Fig. 2.2(b).

The strip $[0,a) \times I$ (shaded in the picture above) can be partitioned in several of these zero-discrepancy rectangles plus a remainder with discrepancy at most 1. This proves Claim I.

Claim II. *Any corner* $C_{(x,y)}$ *can be expressed as a disjoint union of at most* $\lceil \log_2 n \rceil$ *rectangles as in Claim I plus a set* M *with* $|D(P,M)| \leq 1$.

Proof. Let m be the smallest integer with $2^m \geq n$ and let y_0 be the largest integer multiple of 2^{-m} not exceeding y. Then M is the rectangle $[0,x) \times [y_0, y)$, as in the following picture (with $m = 4$):

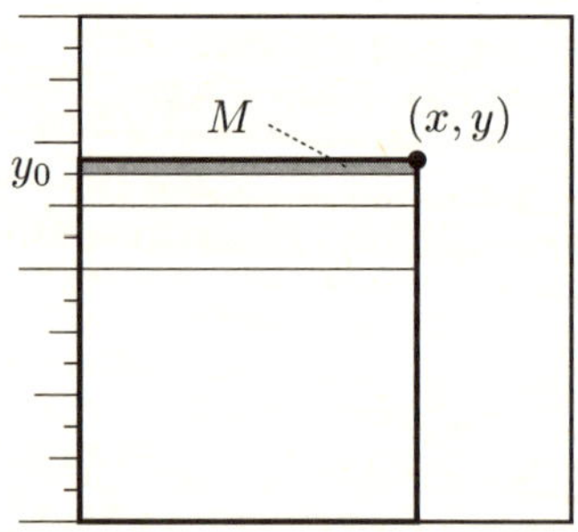

The area of M is at most $y - y_0 < 2^{-m} \le \frac{1}{n}$, and M contains at most one point of P, since any two y-coordinates of points in P differ by at least 2^{-m}. Consequently, $|D(P, M)| \le 1$.

Next, we observe that the interval $[0, y_0)$ can be partitioned into at most m canonical intervals. This can be seen by induction on m: by possibly removing one canonical interval of length 2^{-m} from the end of the interval $[0, y_0)$, we obtain an interval $[0, y_1)$ with y_1 being an integer multiple of $2^{-(m-1)}$, and so on. This proves Claim II.

Claims I and II together imply that the discrepancy of the Van der Corput set P for corners is at most $\log_2 n + 2$. By Observation 1.3, the discrepancy for axis-parallel rectangles is also bounded by $O(\log n)$, and this is the end of the proof of Proposition 2.2. □

Let us remark that the consideration in the proof of Claim II about decomposing an interval $[0, k/2^m)$ into at most m canonical intervals is worth remembering very well. This trick recurs again and again, not only in discrepancy theory but also in combinatorics, computer science, and elsewhere. Later on, we will encounter more sophisticated examples of "canonical decompositions."

Next, we consider a higher-dimensional generalization of the Van der Corput construction.

2.3 Example (Halton–Hammersely set). We choose $d-1$ distinct primes $p_1, p_2, \ldots, p_{d-1}$ (say the first $d-1$ primes, $p_1 = 2$, $p_2 = 3, \ldots$). The ith point of the constructed set is

$$\left(\tfrac{i}{n}, r_{p_1}(i), r_{p_2}(i), \ldots, r_{p_{d-1}}(i)\right), \qquad i = 0, 1, \ldots, n-1.$$

Here the function $r_2(i) = r(i)$ is as in Example 2.1, and, in general, $r_p(i)$ is obtained by writing the digits of the p-ary notation for i in the reverse order: for $i = a_0 + pa_1 + p^2a_2 + p^3a_3 + \cdots$, where $a_j \in \{0, 1, \ldots, p-1\}$, we set

$$r_p(i) = \frac{a_0}{p} + \frac{a_1}{p^2} + \frac{a_2}{p^3} + \frac{a_3}{p^4} + \cdots.$$

2.4 Theorem. *For any fixed d and any fixed distinct primes $p_1, \ldots, p_{d-1}$, the discrepancy of the n-point Halton–Hammersely set for axis-parallel boxes is $O(\log^{d-1} n)$.*

Proof. This is a generalization of the proof given above for the Van der Corput set. We write it down for $d = 3$, $p_1 = 2$, and $p_2 = 3$ only, since the idea should be sufficiently apparent and the notation is much simpler.

For an integer $b \ge 2$, let us define a *b-ary canonical interval* as an interval

$$\left[\frac{k}{b^q}, \frac{k+1}{b^q}\right) \quad \text{for integers } q \ge 0 \text{ and } k \in \{0, 1, \ldots, b^q - 1\}.$$

(The canonical intervals in the proof of Proposition 2.2 are just binary canonical intervals.)

Claim I. *For any box R of the form $[0,a)\times I\times J$, where I is a binary canonical interval, J is a ternary canonical interval, and $a\in(0,1]$ is arbitrary, we have $|D(P,R)|\le 1$.*

Proof. Let $I=[k/2^q,(k+1)/2^q)$ and $J=[\ell/3^r,(\ell+1)/3^r)$, and let S denote the box $[0,1)\times I\times J$. This time, we obtain that the ith point of P falls into S if $i\equiv\bar{k}\,(\mathrm{mod}\ 2^q)$ and $i\equiv\bar{\ell}\,(\mathrm{mod}\ 3^r)$ for certain integers $\bar{k}$, $\bar{\ell}$. By the Chinese remainder theorem, the set $\{0,1,2,\ldots,2^q3^r-1\}$ contains exactly one number with remainder $\bar{k}$ modulo 2^q and with remainder $\bar{\ell}$ modulo 3^r. (Invoking the Chinese remainder theorem is the main new feature compared to the proof in the plane.) Therefore, the points of P lying in S are evenly spaced in the x-direction with step $\frac{2^q3^r}{n}$. The box S can again be divided into boxes of length $\frac{2^q3^r}{n}$ in the x-direction, and each of these boxes has zero discrepancy. Similar to the planar case, it follows that the discrepancy of the box $[0,a)\times I\times J$ considered in Claim I is at most 1.

Claim II. *Any corner $C_{(x,y,z)}$ can be expressed as a disjoint union of at most $\lceil\log_2 n\rceil\cdot 2\lceil\log_3 n\rceil$ boxes as in Claim I, plus a set $M\subset[0,1]^3$ with $|D(P,M)|\le 2$.*

Sketch of Proof. Let y_0 be the largest integer multiple of 2^{-m} not exceeding y, where $m=\lceil\log_2 n\rceil$, and let z_0 be the largest integer multiple of $3^{-m'}$ not exceeding z, with $m'=\lceil\log_3 n\rceil$. The corner $C_{(x,y_0,z_0)}$ can be sliced into at most $2mm'$ boxes as in Claim I (see Fig. 2.3). The remaining part of the corner $C_{(x,y,z)}$ is contained in the set $([0,1]\times[y_0,y]\times[0,1))\cup([0,1]\times[0,1]\times[z_0,z])$. This set has volume at most $(y-y_0)+(z-z_0)\le 2^{-m}+3^{-m'}\le\frac{2}{n}$, and it contains at most 2 points of P. We leave the details to the reader. This finishes the proof of Claim II and of Theorem 2.4 as well. □

Bibliography and Remarks. The Van der Corput construction with the $O(\log n)$ discrepancy bound is from [Cor35a], [Cor35b]. It was originally presented as an infinite one-dimensional sequence. For the sequences $\{n\alpha\}$, which will be discussed in Section 2.5, it was known much earlier that the discrepancy is at most $O(\log n)$ for suitable irrational numbers α. Actually, it seems difficult to decide whom this result should be attributed to, since the notion of discrepancy was not explicitly introduced at that time. In some form, it was proved by Ostrowski [Ost22] in 1922 (also see Behnke [Beh22],[Beh24]) and by Hardy and Littlewood [HL22a] in the same year. But, for example, already in 1904 Lerch probably possessed the ideas needed for the proof (in [Ler04], he proved that $\left|\sum_{i=1}^n(\{i\alpha\}-\frac{1}{2})\right|=O(\log n)$ for any α with bounded partial quotients of the continued fraction).

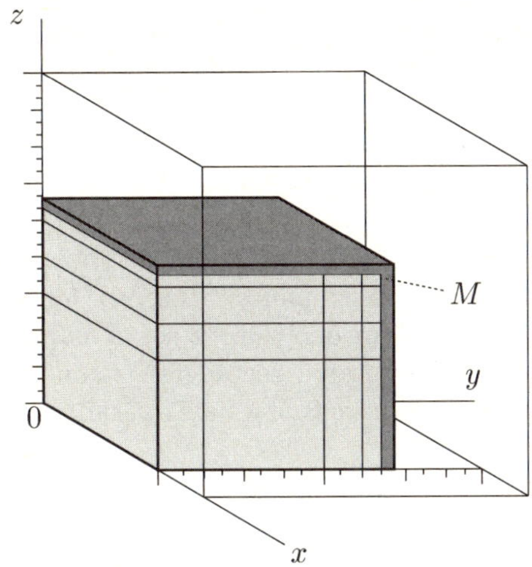

Fig. 2.3. Illustration for the proof of Claim II in dimension 3.

Hammersley [Ham60] proposed a generalization of the Van der Corput construction to higher dimensions as in Example 2.3 and asked for a discrepancy estimate. This was provided by Halton [Hal60], who also suggested that for practical purposes, infinite d-dimensional sequences as in Exercise 2 below may be more useful.

Detailed calculations of the constants of proportionality in the theoretical bounds for the discrepancy of the Halton–Hammersely construction can be found in Niederreiter [Nie92]. Although the results are not very favorable in comparison with other constructions, the Halton–Hammersely sets usually show a fairly good behavior in practice.

In the plane, it is known that the discrepancy functions for corners and for axis-parallel rectangles are bounded by multiples of $\log n$ both from above and from below, but the best possible constants of proportionality, let alone the precise values of the discrepancy functions, appear difficult to determine. According to Niederreiter [Nie92], the best known constructions in this respect in the plane, due to Faure [Fau81] [Fau92], give $D(n, \mathcal{R}_2) \leq 0.337 \ln n + o(\ln n)$ and $D(n, \mathcal{C}_2) \leq 0.224 \ln n + o(\ln n)$.

Exercises

1. (a) If you connect the 8 points in Fig. 2.2(a) by suitable segments, you get a picture of the 3-dimensional cube. Can you explain why?

(b) Show that the Van der Corput set with 2^m points is a projection of the vertex set of an m-dimensional cube, and write the projection down explicitly in coordinates.

2. (Halton–Hammersely sequence) Let $P_n \subset \mathbf{R}^d$ be the n-point set whose ith point is $(r_{p_1}(i), r_{p_2}(i), \ldots, r_{p_d}(i))$, $i = 0, 1, \ldots, n-1$, where the p_j and the r_{p_j} are as in Example 2.3. Show that $D(P_n, \mathcal{C}_d) = O(\log^d n)$. (So we have an infinite sequence of points in $[0,1]^d$ such that the n initial terms constitute a set of discrepancy $O(\log^d n)$, for all n.)

3.* Show that the n-point Van der Corput set P is bad as far as the discrepancy for arbitrarily rotated rectangles is concerned: there exists a rectangle R in the unit square (not an axis-parallel one) of area $\Omega(n^{-1/2})$ containing no point of P.
This result is due to Wernisch [Wer92].

2.2 Sets with Good Average Discrepancy

In the previous section, we have been considering the worst-case discrepancy for corners, and we have derived the upper bound $D(n, \mathcal{C}_d) = O(\log^{d-1} n)$, for any fixed d. Here we demonstrate the existence of sets for which most of the corners, although not all of them, have considerably smaller discrepancy.

2.5 Theorem. *For any fixed $d \geq 2$, the L_2-discrepancy for corners satisfies*

$$D_2(n, \mathcal{C}_d) = O(\log^{(d-1)/2} n).$$

Later on, we will see that this bound is best possible (Theorem 6.1).

Currently, several different proofs of Theorem 2.5 are known. Here we show one based on a suitable probabilistic modification ("randomization") of the Van der Corput and Halton–Hammersely sets. It may be instructive to first look why the Van der Corput set itself fails in this respect.

Why the Van der Corput Set Isn't Good Enough. Let $P \subset [0,1]^2$ be the n-point Van der Corput set as in Example 2.1, and let m be the smallest integer with $2^m \geq n$. Consider a corner $C = C_{(x,y)}$. We may assume that y is a multiple of 2^{-m}, since changing C by a strip of width smaller than 2^{-m} changes the discrepancy by at most 1.

Let y written in the binary notation have the form $0.a_1a_2\ldots a_m$, where the $a_i \in \{0,1\}$ are binary digits. As in the upper bound for the worst-case discrepancy of P (proof of Proposition 2.2), we decompose the interval $[0, y)$ into (binary) canonical intervals. It is easy to see that the canonical intervals used in this decomposition are, written in binary,

$$I_q = [0.a_1a_2\ldots a_{q-1}0, 0.a_1a_2\ldots a_{q-1}1), \quad q \in \{1, 2, \ldots, m\},\ a_q = 1.$$

(Note that I_q has length 2^{-q}.)

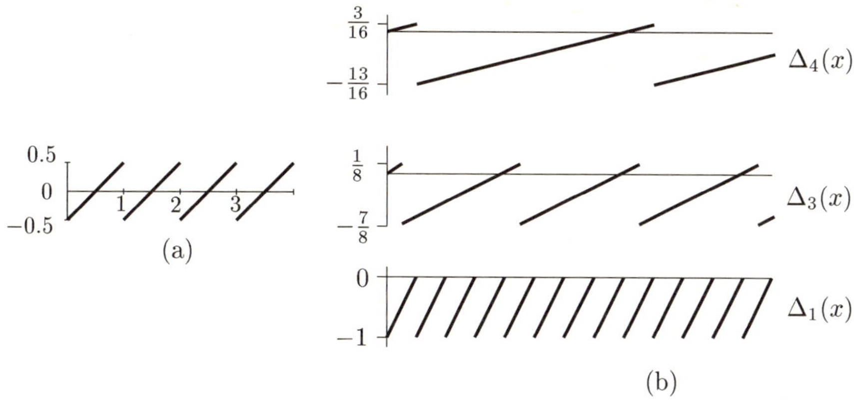

Fig. 2.4. (a) The sawtooth function $s(t) = t - \lceil t \rceil + \frac{1}{2}$; (b) the functions $\Delta_q(x)$ for $y = 0.1011$.

Let S_q be the horizontal strip $[0, x) \times I_q$, and let k_q denote the integer whose binary notation is $a_{q-1}a_{q-2}\ldots a_2a_1$. By the formula for the points of the Van der Corput set P, the strip S_q contains exactly the points of P whose first coordinate is $\frac{i}{n}$, where $i \equiv k_q \pmod{2^q}$ and $i < nx$. The number of such points is $\lceil \frac{nx-k_q}{2^q} \rceil$.

Let Δ_q denote the contribution of the strip S_q to the discrepancy of the corner $C_{x,y}$; that is,

$$\Delta_q = n\,\mathrm{vol}(S_q) - |P \cap S_q| = \frac{nx}{2^q} - \left\lceil \frac{nx - k_q}{2^q} \right\rceil.$$

For $y = 0.1011$, the quantities Δ_1, Δ_3, and Δ_4 as functions of x are drawn in Fig. 2.4(b); note that they are negative most of the time. They are all shifted and stretched copies of the "sawtooth function" $s(t) = t - \lceil t \rceil + \frac{1}{2}$ shown in Fig. 2.4(a). More precisely, denoting $\kappa_q = k_q/2^q$, we have

$$\Delta_q = s\left(\frac{nx}{2^q} - \kappa_q\right) + \kappa_q - \frac{1}{2}.$$

Assume for simplicity that $n = 2^m$. The expression $s\left(\frac{nx}{2^q} - \kappa_q\right)$, regarded as a function of x, has period $\frac{2^q}{n} = 2^{q-m}$. So the interval $[0, 1)$ contains an integral number of periods, and since the average of the sawtooth function s over each period is 0, we get that the average of Δ_q over $x \in [0, 1)$ is $\kappa_q - \frac{1}{2}$.

Recall that κ_q, written in binary, has the form $0.0a_{q-1}a_{q-2}\ldots a_1$. Thus, whenever $a_q = 1$ and $a_{q-1} = 0$, the contribution of Δ_q to the average is at most $-\frac{1}{4}$. Since "most" of binary digit sequences $a_1a_2\ldots a_m$ have at least $\frac{m}{8}$ alterations 01 in their binary notation, we see that even the L_1-average of $D(P, C_{(x,y)})$ over $(x, y) \in [0, 1)^2$ has the order $\Omega(m) = \Omega(\log n)$. This argument is admittedly somewhat informal but it can be made rigorous easily.

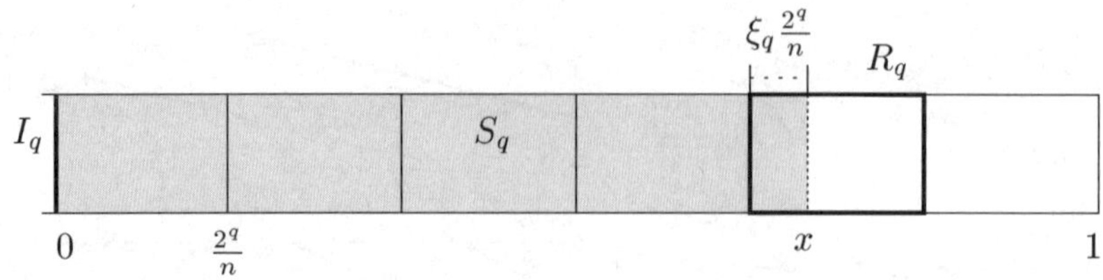

Fig. 2.5. Discrepancy of the strip $[0, 1) \times I_q$.

Solution: a Random Cyclic Shift. First we prove the two-dimensional case of Theorem 2.5. The basic idea is to shift the Van der Corput set cyclically by a random amount in the x-direction, and show that the expected square of the L_2-discrepancy is $O(\log n)$.

As usual, let m be the smallest integer with $2^m \geq n$, and set $N = 2^m$; this will be the period of the cyclic shift. For a real parameter $t \in [0, N)$, we define an n-point set $P_t \subset [0,1)^2$ by

$$P_t = \left\{ \left(\frac{(i+t) \pmod N}{n}, r(i) \right) : i = 0, 1, \ldots, N-1 \right\} \cap [0,1)^2 .$$

Here $r(.)$ denotes the "bit-reversal" function as in the definition of the Van der Corput set. So, expressed in words, we take the N-point Van der Corput set and re-scale it in the x-coordinate in such a way that exactly the first n points lie in the unit square. Then we shift it cyclically by $\frac{t}{n}$ within the interval $\left[0, \frac{N}{n}\right)$, and finally we intersect it with the unit square. (Note that the "unit" of the shift t is $\frac{1}{n}$, i.e. "one point.")

Let $C = C_{(x,y)}$ be a fixed corner. The plan is to prove that for t chosen at random from $[0, N)$, the expected squared discrepancy $D(P_t, C)^2$ is $O(\log n)$. If this is true for any fixed corner C, then there exists some specific t with $D_2(P_t, \mathcal{C}_2) = O(\sqrt{\log n})$ and we are done.

Let us adopt some of the notation introduced in the first part of this section. So we assume that $y = 0.a_1a_2\ldots a_m$ in binary, I_q are the binary canonical intervals in the decomposition of the interval $[0, y)$, $S_q = [0, x) \times I_q$ is the horizontal strip corresponding to I_q, and $\Delta_q = n \operatorname{vol}(S_q) - |P_t \cap S_q|$ is the contribution of the strip S_q to the discrepancy of the corner C.

As in the proof of Proposition 2.2, we decompose the horizontal strip $[0, 1) \times I_q$ (containing S_q) into rectangles of height 2^{-q} and width $\frac{2^q}{n}$ as in Fig. 2.5. Each of these rectangles contains exactly one point of P_t. Let R_q denote the rectangle that is only partially contained in the strip S_q (this is the only rectangle in this strip contributing to the discrepancy), and let ξ_q be the fraction of R_q contained in S_q. The value of Δ_q is ξ_q if the (single) point of P_t lying in R_q is to the right of S_q, and it is $\xi_q - 1$ if this point lies within S_q.

As the shift value t runs from 0 to N (think of t as time), the points of P_t move to the right. From the point of view of the rectangle R_q, a point enters it through the left boundary at some moment t_1, it moves through R_q to the right at uniform speed, and exits on the right at the moment $t_1 + 2^q$; at the same moment, another point is entering R_q on the left. Hence Δ_q, regarded as a function of t, can be expressed

$$\Delta_q(t) = f_{2^q,\xi_q,k_q}(t),$$

where $f_{a,\xi,-t_0}$ is the periodic function with period a whose each period consists of two constant portions of length $(1-\xi)a$ and ξa with values ξ and $\xi - 1$, respectively, and whose period starts at t_0:

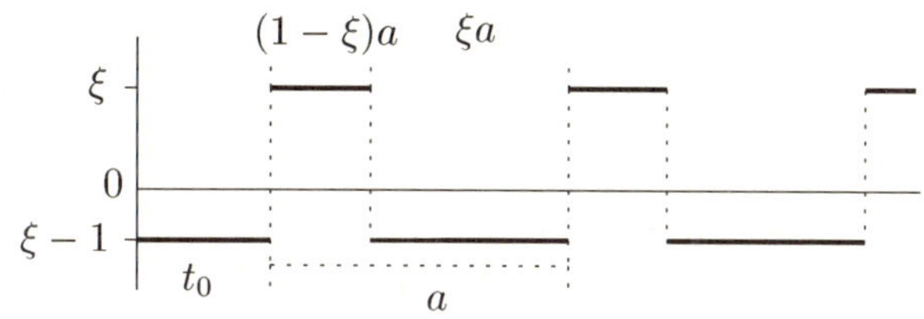

Note that the integral of $f_{a,\xi,-t_0}$ over any interval of length a is 0.

For our fixed corner C, let us now bound the expectation $\mathbf{E}\left[D(P_t,C)^2\right]$ with respect to a random choice of $t \in [0,N)$. We have

$$\mathbf{E}\left[D(P_t,C)^2\right] = \mathbf{E}\left[\left(\sum_q \Delta_q(t)\right)^2\right] = \sum_{q_1,q_2} \mathbf{E}\left[\Delta_{q_1}(t)\Delta_{q_2}(t)\right]$$

where the sum is over all q_1 and q_2 corresponding to the canonical intervals in the decomposition of $[0,y)$. We have

$$\mathbf{E}\left[\Delta_{q_1}(t)\Delta_{q_2}(t)\right] = \frac{1}{N}\int_0^N f_{2^{q_1},\xi_{q_1},k_{q_1}}(t) f_{2^{q_2},\xi_{q_2},k_{q_2}}(t)\,\mathrm{d}t.$$

For brevity, write $f_1 = f_{2^{q_1},\xi_{q_1},k_{q_1}}$ and $f_2 = f_{2^{q_2},\xi_{q_2},k_{q_2}}$; so f_1 has period 2^{q_1} and f_2 has period 2^{q_2}. Suppose that $q_1 \le q_2$, say. Each period of f_1 on which f_2 is constant contributes 0 to the integral $\int_0^N f_1(t)f_2(t)\,\mathrm{d}t$. Periods of f_1 containing jumps of f_2 may give nonzero contributions, but since $|f_1 f_2| \le 1$ the contribution of each such period is at most its length, i.e. 2^{q_1}. The function f_2 has $2N/2^{q_2}$ jumps on $[0,N)$, and hence

$$\left|\mathbf{E}\left[\Delta_{q_1}(t)\Delta_{q_2}(t)\right]\right| = \frac{1}{N}\left|\int_0^N f_1(t)f_2(t)\,\mathrm{d}t\right| \le 2^{q_1}\,\frac{2}{2^{q_2}} = O(2^{-(q_2-q_1)}).$$

Therefore

$$\mathbf{E}\left[D(C,P_t)^2\right] \le \sum_{q_1,q_2=1}^{m} O(2^{-|q_1-q_2|}) = O(m) = O(\log n)$$

and the planar case of Theorem 2.5 is proved. □

Higher Dimension. Instead of giving the proof for a general dimension d, we present it for the 3-dimensional case. Here the basis of the construction is the 3-dimensional Halton–Hammersely set from Example 2.3. We let m be the smallest integer with $2^m \geq n$, and m' the smallest integer with $3^{m'} \geq n$. This time the period of the cyclic shift is $N = 2^m 3^{m'}$. We thus define the shifted set

$$P_t = \left\{ \left(\frac{(i+t) \pmod N}{n}, r_2(i), r_3(i) \right) : i = 0, 1, \ldots, N-1 \right\} \cap [0,1)^3.$$

We again consider a fixed corner $C = C_{(x,y,z)}$, assuming that y is a multiple of 2^{-m} and z is a multiple of $3^{-m'}$. This time the rectangle $[0,y) \times [0,z)$ is decomposed into a collection $\mathcal{B}$ of canonical boxes (rectangles in this case) as in Fig. 2.3. The sides of each rectangle $B \in \mathcal{B}$ are 2^{-q} and 3^{-r} for some integers $q \leq m$ and $r \leq m'$. The 3-dimensional box $[0,1) \times B$ can be partitioned into boxes of length $\frac{2^q 3^r}{n}$ in the x-coordinate (a 3-dimensional analogue of Fig. 2.5), each of them containing exactly one point of each P_t. Setting $S_B = [0,x) \times B$ and $\Delta_B = n \operatorname{vol}(S_B) - |P_t \cap S_B|$, we find (precisely as in the planar case) that

$$\Delta_B(t) = f_{2^q 3^r, \xi_B, k_B}(t)$$

for a suitable $\xi_B \in [0,1)$ and an integer $k_B \in \{0, 1, \ldots, 2^q 3^r\}$.

This time we need to estimate

$$\mathbf{E}\left[D(C,P_t)^2\right] = \mathbf{E}\left[\left(\sum_{B \in \mathcal{B}} \Delta_B(t)\right)^2\right] = \sum_{B_1, B_2 \in \mathcal{B}} \mathbf{E}\left[\Delta_{B_1}(t)\Delta_{B_2}(t)\right].$$

Let $B_1 \in \mathcal{B}$ be a $2^{-q_1} \times 3^{-r_1}$ rectangle and let $B_2 \in \mathcal{B}$ be a $2^{-q_2} \times 3^{-r_2}$ rectangle. Define the *distance* of B_1 and B_2 as

$$\delta(B_1, B_2) = \max(|q_1 - q_2|, |r_1 - r_2|).$$

We want to show

$$\left|\mathbf{E}\left[\Delta_{B_1}(t)\Delta_{B_2}(t)\right]\right| \leq 2^{-\delta(B_1,B_2)}. \tag{2.2}$$

This will be sufficient, since then we obtain

$$\begin{aligned} \sum_{B_1, B_2 \in \mathcal{B}} \mathbf{E}\left[\Delta_{B_1}(t)\Delta_{B_2}(t)\right] &\leq \sum_{q_1,q_2=1}^{m} \sum_{r_1,r_2=1}^{m'} 2 \cdot 2^{-\max(|q_1-q_2|,|r_1-r_2|)} \\ &= O(mm') = O(\log^2 n), \end{aligned}$$

as is not difficult to calculate.

To prove (2.2), we again write

$$\mathbf{E}\left[\Delta_{B_1}(t)\Delta_{B_2}(t)\right] = \frac{1}{N}\int_0^N f_1(t)f_2(t)\,\mathrm{d}t,$$

where $f_1 = f_{2^{q_1}3^{r_1},\xi_{B_1},k_{B_1}}$ and similarly for f_2. This time, the simple-minded approach we have used to show the "near-orthogonality" of f_1 and f_2 in the 2-dimensional case fails (both f_1 and f_2 may have too many jumps), and we need a more clever argument. Here is the idea: If we reduce all the points of jumps of f_1 modulo the period of f_2, they are quite uniformly distributed within the period of f_2, and their contributions nearly cancel out. A detailed argument follows.

Let us suppose $r_1 \le r_2$ and $|r_1 - r_2| \ge |q_1 - q_2|$, say (the other cases are symmetric). Let $a_2 = 2^{q_2}3^{r_2}$ be the (smallest) period of f_2. Let $K > 1$ be an integer, and consider K equidistant points with spacing a_2/K, of the form t, $t + \frac{1}{K}a_2$, $t + \frac{2}{K}a_2, \ldots$, $t + \frac{K-1}{K}a_2$. We observe that the sum of the values of f_2 at these points is always at most 1 in absolute value:

$$\left|\sum_{j=0}^{K-1} f_2(t + \tfrac{j}{K}a_2)\right| \le 1. \tag{2.3}$$

The following picture, for $K = 9$, tries to indicate why:

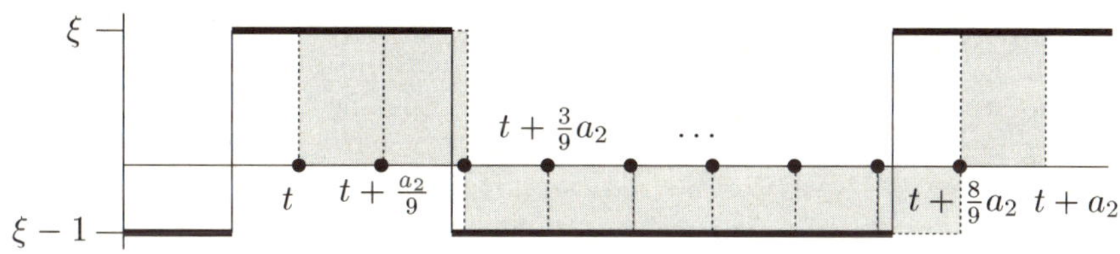

Set $K = 3^{r_2 - r_1}$ and $M = 2^m 3^{r_1}$. When i runs through $0, 1, \ldots, K-1$, the expression $iM \pmod{a_2}$ runs through the multiples of $\gcd(M, a_2) = 2^{q_2}3^{r_1}$, i.e. through the numbers $0, \frac{1}{K}a_2, \frac{2}{K}a_2, \ldots, \frac{K-1}{K}a_2$ (although typically in a different order)—this is a simple consequence of the Chinese remainder theorem. Therefore we have

$$\sum_{i=0}^{K-1} f_2(t + iM) = \sum_{j=0}^{K-1} f_2(t + \tfrac{j}{K}a_2). \tag{2.4}$$

As a final observation, we note that the integral $\int_0^N f_1(t+x)f_2(t+x)\,\mathrm{d}t$ is independent of x because the period of the function $f_1(t)f_2(t)$ divides N. We are ready to estimate $\int_0^N f_1(t)f_2(t)\,\mathrm{d}t$ by an averaging trick:

$$\begin{aligned}\int_0^N f_1(t)f_2(t)\,\mathrm{d}t &= \frac{1}{K}\sum_{i=0}^{K-1}\int_0^N f_1(t+iM)f_2(t+iM)\,\mathrm{d}t\\ &= \frac{1}{K}\int_0^N f_1(t)\left(\sum_{i=0}^{K} f_2(t+iM)\right)\mathrm{d}t\end{aligned}$$

because $f_1(t+M) = f_1(t)$. According to (2.3) and (2.4), the absolute value of the sum in parentheses is always at most 1, and hence the integral is at most N. From this equation (2.2) follows, and the proof of (the 3-dimensional case of) Theorem 2.5 is finished. □

Remark: L_p-Discrepancy. It is known that for any fixed $p \geq 1$, the L_p-discrepancy for corners is of the order $O(\log^{(d-1)/2} n)$ as well. This means that for suitable point sets, a great majority of corners have much smaller discrepancy than the best known worst-case bound. And, as we will see later, in the plane such a behavior is unavoidable, since the worst-case discrepancy really is of the order $\log n$, while the L_p-discrepancy for any fixed p is of the order $\sqrt{\log n}$.

Bibliography and Remarks. The two-dimensional case of Theorem 2.5 was proved by Davenport [Dav56]. He used (essentially) the explicitly given $2n$-point set $P \cup P^{\updownarrow}$, where $P = \{(\frac{i}{n}, \{i\alpha\}) \colon i = 0, 1, \ldots, n-1\}$ is a lattice set as in Example 2.19, with α irrational and having bounded partial quotients of its continued fraction, and $P^{\updownarrow} = \{(x, 1-y) \colon (x,y) \in P\}$. The proof employs harmonic analysis and can be admired, e.g., in [Cha98b]. (Interestingly, P itself is not good, as can be shown by an argument similar to the one at the beginning of this section.) As was shown by Chen and Skriganov (private communication from September 1998), if P is the 2^m-point Van der Corput set then $P \cup P^{\updownarrow}$ works too.

In [Rot79], Roth proved the 3-dimensional case of Theorem 2.5, and in [Rot80] he developed the method we have presented and proved the general case ([BC87] gives more information on the history of this result). At about the same time and independently of Roth, Frolov [Fro80] gave another construction with an optimal L_2-discrepancy bound (see Section 2.5). The L_p-analogue of Theorem 2.5 for every fixed p was obtained by Chen [Che81] (Chen [Che83] has an alternative proof, and another proof was given by Skriganov—see Section 2.5).

Exercises

1. Let $P \subset [0,1]^d$ denote the 2^m-point Van der Corput set.
   (a) Complete the proof sketched in the text of the fact that the L_1-discrepancy of P for corners is $\Omega(m)$.
   (b) Write down a specific corner $C \in \mathcal{C}_2$ such that $|D(P,C)| = \Omega(m)$.

2.3 More Constructions: b-ary Nets

The Halton–Hammersely set defined Section 2.1 shows the current best asymptotic upper bound for the discrepancy function of axis-parallel boxes: $D(n, \mathcal{R}_d) = O(\log^{d-1} n)$. But because of the great practical significance of uniformly distributed point sets, a large number of alternative constructions have been invented. Some of them behave better in practice than the Halton–Hammersley points, and they also have much better constants of proportionality in the asymptotic estimates of discrepancy.

In this section we will discuss low-discrepancy sets somewhat resembling the Halton–Hammersely set. First, let us recall the proof of the discrepancy upper bound (Theorem 2.4), again working with the specific case $d = 3$, $p_1 = 2$, and $p_2 = 3$. We observe that the second part, i.e. Claim II, works in the same way for any set P satisfying the following condition: any box $B = K \times I \times J$ contains exactly one point of P, where $I = [k/2^q, (k+1)/2^q)$ is a binary canonical interval, $J = [\ell/3^r, (\ell+1)/3^r)$ is a ternary canonical interval, and $K = [m2^q3^r/n, (m+1)2^q3^r/n)$ for some $m < n/2^q3^r$. This property can thus be taken as an axiom, and one can study ways of constructing various such sets. The next observation is that in this abstract setting, it is no longer important that p_1 and p_2 are primes. This was only needed to guarantee the just described property of the specific "bit reversal" sets, i.e. in order to apply the Chinese remainder theorem in the proof of Claim I. And, indeed, it turns out that one can construct such sets with all the p_i being the same, equal to some suitable number b. Moreover, if we assume that $n = b^m$ is a power of this b, then the special role of the first coordinate in the construction disappears.

Let $b \geq 2$ be an integer; mostly we will assume that it is a prime power so that the finite field $GF(b)$ with b elements exists. A b-ary canonical interval, as defined in Section 2.1, is an interval $[k/b^q, (k+1)/b^q)$ for an integer $q \geq 0$ and $k = 0, 1, \ldots, b^q - 1$. We define a *$b$-ary canonical box* in $[0, 1]^d$ as a Cartesian product of d b-ary canonical intervals.

2.6 Definition (b-ary nets). *Let $b, d, m, \lambda \geq 1$ be integer parameters. Call a set $P \subset [0, 1]^d$ a* b-ary net with λ points per box of volume b^{-m} *if*[1] *(what would you expect?) each b-ary canonical box of volume b^{-m} contains exactly λ points of P. Note that the size of P is already determined by this condition: we have $|P| = \lambda b^m$.*

Fig. 2.6 shows a simple example.

Proceeding in a way similar to the discrepancy estimate for the Halton–Hammersely sets, it is not difficult to show that if $P \subseteq [0, 1]^d$ is a b-ary net

---

[1] This concept appears under different names in the literature. In [BC87] it would be called something like a *λ-set of class m* with respect to $b, b, \ldots, b$. In [Nie92] and in many related recent works, it would be called a *$(t, m+t, d)$-net in base b*, provided that $\lambda = b^t$.

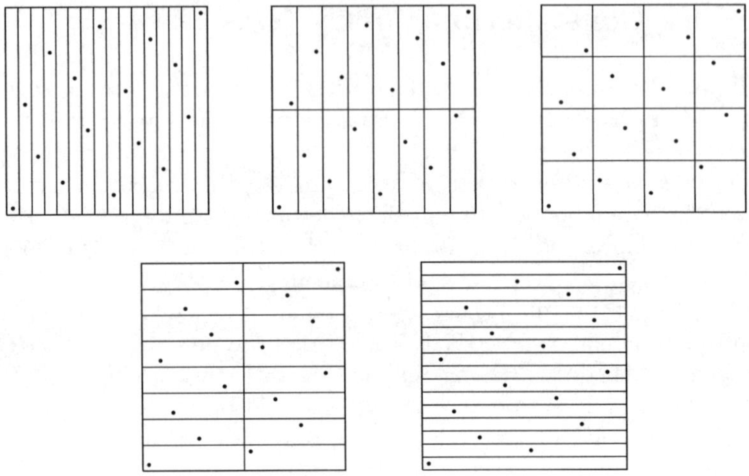

Fig. 2.6. A binary net with 1 point per box of volume 2^{-4} (a shifted copy of the Van der Corput set). Five copies are shown, with all the canonical binary boxes of volume 2^{-4}.

with λ points per box of volume b^{-m}, where d, b, λ are all considered fixed, then $D(P, \mathcal{R}_d) = O(\log^{d-1}|P|)$, i.e. P has asymptotically the smallest known discrepancy for axis-parallel boxes. The constant of proportionality can be bounded by an expression of the form $B(d, b)\lambda$, where $B(d, b)$ depends on b and d but not on λ. Hence it would be best to have $\lambda = 1$. The reason for introducing the λ parameter is that if b and λ are both too small then the appropriate b-ary nets do not exist. For instance, b-ary nets with 1 point per box can only exist if $b \geq d - 1$ (Exercises 3 and 4).

A Construction with Matrices Over $GF(b)$. The definition of b-ary nets captures an useful property of point sets, but it is an "empty form," so to speak—one has to come up with specific sets having this property, with favorable values of parameters. One construction, or rather a class of constructions, is based on a suitable collection $\mathcal{C} = \left(C^{(1)}, C^{(2)}, \ldots, C^{(d)}\right)$ of $m \times m$ matrices over $GF(b)$.

From a given collection $\mathcal{C}$ of d matrices, we are going to construct a set $P(\mathcal{C})$ of b^m points in $[0,1]^d$. The matrices $C^{(1)}$, $C^{(2)}, \ldots$, $C^{(d)}$ are called the *generator matrices* of the set $P(\mathcal{C})$. It is convenient to index the points of $P(\mathcal{C})$ by m-component vectors from $GF(b)^m$. For each vector $h \in GF(b)^m$, we define a point $x(h) \in [0,1)^d$. To obtain the kth component of $x(h)$, we first compute a vector $g_k = C^{(k)}h \in GF(b)^m$, with the matrix-vector multiplication in $GF(b)$. Then we read the components of g_k as b-ary digits of $x(h)_k$, i.e. we set $x(h)_k = \langle g_k, (b^{-1}, b^{-2}, \ldots, b^{-m})\rangle$, where $\langle .,. \rangle$ denotes the real scalar product, in which the entries of g_k are interpreted as integer numbers. That is, we fix a bijection between $GF(b)$ and the set $\{0, 1, \ldots, b-1\}$ of integers.

If b is a prime then $GF(b)$ is canonically identified with $\{0, 1, \ldots, b-1\}$; this is the case the reader may want to think of. For b being a prime power, some bijection has to be chosen.

Under suitable conditions on the matrix collection $\mathcal{C}$, the resulting set $P(\mathcal{C})$ turns out to be a b-ary net. To state the result, we define a number $\rho(\mathcal{C})$ characterizing the quality of $\mathcal{C}$. For nonnegative integers $m_1, m_2, \ldots, m_d$ with $m_k \leq m$ for all m, let $\mathcal{C}[\leq m_1, \leq m_2, \ldots, \leq m_d]$ be the matrix

row 1 of $C^{(1)}$ row 2 of $C^{(1)}$ …… row m_1 of $C^{(1)}$
row 1 of $C^{(2)}$ row 2 of $C^{(2)}$ …… row m_2 of $C^{(2)}$
$\vdots \quad \vdots$
row 1 of $C^{(d)}$ row 2 of $C^{(d)}$ …… row m_d of $C^{(d)}$

Then $\rho(\mathcal{C})$ is defined as the maximum number $\rho \leq m$ such that whenever $m_1 + m_2 + \cdots + m_d = \rho$, the matrix $\mathcal{C}[\leq m_1, \leq m_2, \ldots, \leq m_k]$ has the full rank ρ. In particular, $\rho(\mathcal{C}) = m$ means that whenever we piece together a square matrix from upper portions of the $C^{(k)}$'s, this matrix is nonsingular.

2.7 Proposition. *For any d-tuple $\mathcal{C}$ of $m \times m$ matrices as above, the set $P(\mathcal{C})$ is a b-ary net with $b^{m-\rho(\mathcal{C})}$ points per box of volume $b^{-\rho(\mathcal{C})}$.*

Proof. The connection between matrix rank and b-ary net properties may look mysterious at first sight, but actually it is quite natural once the definitions are unwrapped. Set $\rho = \rho(\mathcal{C})$. Let us fix a b-ary canonical box B of volume $b^{-\rho}$. Let the kth side of this box have length b^{-m_k}, so that $m_1 + m_2 + \cdots + m_d = \rho$. The condition that $x(h)$ lie in B means that for each k, the first m_k b-ary digits of $x(h)_k$ have some prescribed values. The jth digit of $x(h)_k$ is the scalar product of h with the jth row of $C^{(k)}$. Hence the values of h with $x(h) \in B$ are the solutions to the linear system $Ch = z$, where $z \in GF(b)^m$ is some fixed vector and $C = \mathcal{C}[\leq m_1, \leq m_2, \ldots, \leq m_k]$. Since we assume that C has full rank, the solution space of this system has dimension $m - \rho$, and hence the number of solutions is $b^{m-\rho}$ as claimed. □

2.8 Example (Faure's construction). Let $d \geq 1$ be an integer, let $b \geq d$ be a prime number, and let $m \geq 2$ be an integer. Define the entry at position (i, j) of the $m \times m$ matrix $C^{(k)}$ as 0 for $j < i$ and as $\binom{j-1}{i-1}(k-1)^{j-i}$ for $j \geq i$, $k = 1, 2, \ldots, d$, where the arithmetic is in $GF(b)$ and 0^0 means 1. Then the collection $\mathcal{C} = \left(C^{(1)}, C^{(2)}, \ldots, C^{(d)}\right)$ has $\rho(\mathcal{C}) = m$, and hence $P(\mathcal{C})$ is a b-ary net with 1 point per box of volume b^{-m}.

Sketch of Proof. The block of the first m_k rows of $C^{(k)}$ looks as follows:

$$\begin{pmatrix} \binom{0}{0} & \binom{1}{0} z_k & \binom{2}{0} z_k^2 & \binom{3}{0} z_k^3 & \cdots & \binom{m-1}{0} z_k^{m-1} \\ 0 & \binom{1}{1} & \binom{2}{1} z_k & \binom{3}{1} z_k^2 & \cdots & \binom{m-1}{1} z_k^{m-2} \\ 0 & 0 & \binom{2}{2} & \binom{3}{2} z_k & \cdots & \binom{m-1}{2} z_k^{m-3} \\ \vdots & \vdots & \vdots & \vdots & \vdots & \vdots \end{pmatrix},$$

with $z_k = k - 1$. A square matrix consisting of several blocks of this form is called a *generalized Vandermonde matrix*, and it can be shown that its determinant is

$$\prod_{1 \le i < j \le d} (z_j - z_i)^{m_i m_j}$$

(we omit a proof since the assertion of this example is a special case of a general result we prove later). In our case, each $z_j - z_i$ is an integer between 1 and $b - 1$, and since we assumed b is a prime, the determinant is nonzero over $GF(b)$ too. □

Adding an Extra Dimension. In the construction of the set $P(\mathcal{C})$, we haven't used the idea of the "bit-reversal sequence" employed in the Van der Corput and Halton–Hammersley constructions. By applying this idea suitably, one can add an extra dimension: from a collection $\mathcal{C}$ of $d - 1$ matrices (rather than d as before), we construct a d-dimensional set $P^{++}(\mathcal{C})$. If the matrices are $m \times m$, we again get b^m points. We put $P^{++}(\mathcal{C}) = \{x(h) \colon h \in GF(b)^m\}$, where $x(h)_1$ through $x(h)_{d-1}$ are defined using $C^{(1)}$ through $C^{(d-1)}$ exactly as in the construction of $P(\mathcal{C})$ above, and $x(h)_d = \langle h, (b^{-m}, b^{-m+1}, \ldots, b^{-1})\rangle$ (again scalar product over the reals!), i.e. the components of h are interpreted as b-ary digits in the reverse order. This time we not only need to assume that $\rho(\mathcal{C})$ is large, but we also have to consider $\rho(\mathcal{C}|_j)$ for $j = 1, 2, \ldots, m$, where $\mathcal{C}|_j$ is the collection of $j \times j$ matrices obtained by taking each of the matrices of $\mathcal{C}$ and removing all rows but the first j and all columns but the first j.

2.9 Theorem. *Let $\mathcal{C}$ be a $(d-1)$-tuple of $m \times m$ matrices over $GF(b)$, and suppose that $\rho(\mathcal{C}|_j) \ge j - q$ for all $j = 1, 2, \ldots, m$. Then the set $P^{++}(\mathcal{C}) \subset [0,1]^d$ is a b-ary net with b^q points per box of volume b^{m-q}.*

The proof of this theorem is not difficult, and we leave it to Exercise 2.

Note that in Faure's construction 2.8, we get that $\rho(\mathcal{C}|_j) = j$ with no extra effort, since $\mathcal{C}|_j$ is the same collection of matrices as the $\mathcal{C}$ obtained for $m = j$. (This will also be the case in the more general constructions of $\mathcal{C}$

considered below.) Therefore, we get that *if b is a prime and $d \leq b+1$ then b-ary nets with 1 point per box of volume b^{-m} exist for all m.* It can be shown that this remains true if we only assume that b is a prime power rather than a prime (Exercise 8). As was remarked above, $d \leq b+1$ is also necessary for the existence of b-ary nets with 1 point per box.

Constructions via Polynomials Over a Finite Field. Faure's construction can be seen as a special case of a more general class of constructions. A convenient way to express these constructions is using the so-called formal Laurent series over finite fields. We thus present the definition and a few properties of these series in a micro-course below.

One can perhaps start by recalling that a polynomial $p(x) = a_0 + a_1 x + \cdots + a_n x^n$ over a finite field F must be formally be regarded as an $(n+1)$-tuple of coefficients, and not as a function $F \to F$, because there are only finitely many functions but infinitely many polynomials. For a *power series* over F, $\sum_{i=0}^{\infty} a_i x^i$, it is not even clear what function it should represent (we would have to define some kind of convergence first), and so we consider it purely formally; it means nothing more or less than the sequence $(a_0, a_1, a_2, \ldots)$ of coefficients. Such series can still be added and multiplied together (formally but, of course, in a manner inspired by the "usual" power series over the real or complex numbers, where these operations correspond to addition and multiplication of analytic functions). Power series are multiplied in a way similar to polynomials: if we write $\left(\sum_{i=0}^{\infty} a_i x^i\right)\left(\sum_{j=0}^{\infty} b_j x^j\right) = \sum_{k=0}^{\infty} c_k x^k$ then the c_k are given by

$$c_k = \sum_{\substack{i,j \geq 0 \\ i+j=k}} a_i b_j. \tag{2.5}$$

It is not difficult to check that a formal power series $a(x) = \sum_{i=0}^{\infty} a_i x^i$ has a multiplicative inverse, i.e. a series $b(x)$ with $a(x)b(x) = 1$, if and only if $a_0 \neq 0$ (Exercise 7). Hence the set of all formal power series over F forms a ring, even an integrality domain (no zerodivisors), but not a field.

A suitable extension that embeds this ring into a field are the *formal Laurent series* over F. These are objects of the form

$$\sum_{i=i_0}^{\infty} a_i x^i$$

with i_0 being an arbitrary (possibly negative) integer. So we may have finitely many negative powers of x in such a series. In complex analysis, such series locally represent meromorphic functions, but here we again take them purely formally. Having agreed on this, we omit the adjective "formal" for the formal Laurent series from now on.

The Laurent series are added and multiplied analogously to the power series: the Laurent-series analogue of (2.5) is

$$c_k = \sum_{\substack{i \geq i_0,\, j \geq j_0 \\ i+j=k}} a_i b_j.$$

Having the above-mentioned result about multiplicative inverses of power series at disposal, it is straightforward to check that any Laurent series over F has a multiplicative inverse and all Laurent series over F form a field.

Another field we need to mention is that of *rational functions* over F. A rational function is a fraction $p(x)/q(x)$ of two polynomials with $q(x) \neq 0$; we again regard it formally. The rational functions are the quotient field of the ring of polynomials. Since the ring of polynomials is a subring of the field of the Laurent series (a polynomial over F can be regarded as a Laurent series with finitely many nonzero terms), it follows from simple results of algebra that there is a unique isomorphic embedding of the field of rational functions into the field of Laurent series. Let us denote this embedding by L_0, so $L_0(p(x)/q(x))$ denotes the Laurent series representing the rational function $p(x)/q(x)$. It is not difficult to give an algorithm for computing the first n terms of $L_0(p(x)/q(x))$ given the coefficients of $p(x)$ and $q(x)$.

The Laurent series we have introduced so far were Laurent series "at 0." For the subsequent construction of generator matrices, it is notationally more convenient to work with Laurent series "at ∞," which have the form

$$\sum_{i=i_0}^{\infty} a_i z^{-i}$$

(finitely many positive exponents and infinitely many negative ones). To this end, we simply substitute $z = x^{-1}$, and note that this substitution converts a rational function of x into a rational function of z. So we define

$$L_\infty(p(z)/q(z)) = \sum_{i=i_0}^{\infty} a_i z^{-i}$$

where $\sum_{i=i_0}^{\infty} a_i x^i = L_0(p(x^{-1})/q(x^{-1}))$.

Let us return to constructing generator matrices. Suppose that b is a prime power. The construction we are going to present requires d polynomials $p_1(z), p_2(z), \ldots, p_d(z)$ over the field $GF(b)$ as input data, such that no two of them have a nontrivial common divisor (of degree ≥ 1). Let δ_k denote the degree of p_k; we also require $\delta_k \geq 1$ for all k. In order to make the quality parameter, $\rho(\mathcal{C})$, of the resulting matrix collection $\mathcal{C}$ as large as possible, the degrees δ_k should be small. For example, Faure's construction corresponds (in the case of a prime b) to letting all the p_k be linear, namely $p_k(z) = z-(k-1)$. But this need not be the best way, since taking all the degrees small forces us to have the field size b sufficiently large, and this makes the resulting discrepancy estimate worse again. Hence a suitable compromise has to be found between the degrees of the p_k and the field size b.

Suppose the polynomials $p_k(z)$ as above have been fixed, and let $m \geq 1$ be an integer parameter. We define a collection $\mathcal{C}$ of d $m \times m$ matrices over $GF(b)$. The kth matrix $C^{(k)}$ is constructed from $p_k(z)$. Let

$$L_\infty\left(\frac{1}{p_k(z)}\right) = \sum_{i=i_0}^{\infty} a_i z^{-i};$$

then the first row of $C^{(k)}$ is set to $(a_1, a_2, \ldots, a_m)$. In general, each row of $C^{(k)}$ contains the coefficients of the powers z^{-1}, z^{-2},..., z^{-m} in the Laurent series of a suitable rational function. The rational functions for the first δ_k rows are

$$\frac{1}{p_k(z)}, \frac{z}{p_k(z)}, \ldots, \frac{z^{\delta_k - 1}}{p_k(z)},$$

for the next δ_k rows they are

$$\frac{1}{p_k(z)^2}, \frac{z}{p_k(z)^2}, \ldots, \frac{z^{\delta_k - 1}}{p_k(z)^2},$$

and so on; in the jth block by δ_k rows we have $p_k(z)^j$ in the denominator. We use $\lceil m/\delta_k \rceil$ blocks, the last one with possibly fewer than δ_k rows. We have

2.10 Theorem. *The collection* $\mathcal{C} = \left(C^{(1)}, C^{(2)}, \ldots, C^{(d)}\right)$ *constructed as above satisfies* $\rho(\mathcal{C}) \geq m - \sum_{k=1}^{d}(\delta_k - 1)$.

Let use remark that the same estimate for $\rho(\mathcal{C})$ also holds for a yet more general version of the construction. In the description above, we have used the polynomials $1, z, z^2, \ldots, z^{\delta_k - 1}$ in the numerators in each block; one can replace them by other suitable collections of polynomials. Namely, for the block of the rows in $C^{(k)}$ with $p_k(z)^j$ in the denominators, the numerators can be any δ_k polynomials $u_{k,j,1}(z)$, $u_{k,j,2}(z)$,..., $u_{k,j,\delta_k}(z)$, provided that they are pairwise relatively prime modulo $p_k(z)$ (for given k and j). The claim of Theorem 2.10 continues to hold, and the proof remains much the same as presented below, only the notation gets progressively more complicated.

Proof of Theorem 2.10. Let $\rho = m - \sum_{k=1}^{d}(\delta_k - 1)$. We should prove that for any partition $\rho = m_1 + m_2 + \cdots + m_d$, the matrix $C = \mathcal{C}[\leq m_1, \leq m_2, \ldots, \leq m_d]$ has rank ρ. Let c_i denote the ith row of C. We must show that if $\alpha = (\alpha_1, \alpha_2, \ldots, \alpha_\rho) \in GF(b)^\rho$ is a vector with $\sum_{i=1}^{\rho} \alpha_i c_i = 0$ then $\alpha = 0$.

The entries of c_i are the first m coefficients of the negative powers of the Laurent series $L_\infty(f_i(z))$ for a certain rational function $f_i(z)$. The equality $\sum_{i=1}^{\rho} \alpha_i c_i = 0$ means that $\sum_{i=1}^{\rho} \alpha_i L_\infty(f_i(z))$ has the coefficients of $z^{-1}, \ldots, z^{-m}$ all 0. Since each $f_i(z)$ has zero polynomial part (the degree of the numerator is smaller than the degree of the denominator), the coefficients of all nonnegative powers are also 0. We can write

$$\sum_{i=1}^{\rho} \alpha_i L_\infty(f_i(z)) = O(z^{-(m+1)}), \tag{2.6}$$

where the $O(z^{-j})$ symbol has the obvious formal meaning for Laurent series. We want to show that the left-hand side must in fact be the zero Laurent series.

Let $D(z)$ be the least common denominator of the $f_i(z)$, $i = 1, 2, \ldots, \rho$. By inspecting the construction of the matrices in $\mathcal{C}$, we see that the largest power of the polynomial $p_k(z)$ occurring in the denominators of the $f_i(z)$ is $\lceil m_k/\delta_k \rceil$. Therefore

$$\deg D(z) = \sum_{k=1}^{d} \delta_k \lceil m_k/\delta_k \rceil \leq \sum_{k=1}^{d} (m_k + \delta_k - 1) = m.$$

By multiplying both sides of (2.6) by $D(z)$, we obtain a polynomial on the left-hand side, while the right-hand side is $O(z^{-1})$, and so both sides must be 0 as Laurent series. Since the rational functions are a subfield of the Laurent series, we can infer

$$\sum_{i=1}^{\rho} \alpha_i f_i(z) = 0. \tag{2.7}$$

It remains to show that the rational functions $f_i(z)$ are linearly independent, which is a simple exercise in polynomial algebra. It is perhaps best to look at a particular example (the same argument works in general). Let $d = 3$, $\delta_1 = 4$, $\delta_2 = 3$, $\delta_3 = 1$, $m = 13$, $\rho = 8$, $m_1 = 2$, $m_2 = 5$, $m_3 = 1$. Then the rational functions in question are

$$\begin{aligned}
f_1(z) &= \frac{1}{p_1(z)}, & f_2(z) &= \frac{z}{p_1(z)}, & & \\
f_3(z) &= \frac{1}{p_2(z)}, & f_4(z) &= \frac{z}{p_2(z)}, & f_5(z) &= \frac{z^2}{p_2(z)}, \\
f_6(z) &= \frac{1}{p_2(z)^2}, & f_7(z) &= \frac{z}{p_2(z)^2}, & & \\
f_8(z) &= \frac{1}{p_3(z)}. & & & &
\end{aligned}$$

Multiplying (2.7) by the least common denominator $D = p_1 p_2^2 p_3$, we get

$$(\alpha_1 + \alpha_2 z) p_2^2 p_3 + (\alpha_3 + \alpha_4 z + \alpha_5 z^2) p_1 p_2 p_3 + (\alpha_6 + \alpha_7 z) p_1 p_3 + \alpha_8 p_1 p_2^2 = 0.$$

Since all terms but possibly the first one are divisible by p_1, the first term is divisible by p_1 as well, but this is only possible if $\alpha_1 = \alpha_2 = 0$ (since $\deg p_1 = 4$ and p_1 has no common factor with either p_2 or p_3). By a similar argument with divisibility by p_2, we derive that $\alpha_6 = \alpha_7 = 0$, and then divisibility by p_2^2 forces $\alpha_3 = \alpha_4 = \alpha_5 = 0$, etc. This concludes the proof of Theorem 2.10. □

Bibliography and Remarks. The constructions of b-ary nets originated by a paper of Sobol [Sob67], who gave essentially the construction explained above Theorem 2.10 for the case $b = 2$, but with a different presentation not using the formal Laurent series. Example 2.8 is from Faure's work [Fau82]. Theorem 2.10 is due to Niederreiter [Nie87] (see also [Nie92] for an account and references). He and his coworkers have analyzed various constructions of this type with the goal of obtaining the smallest possible constants in the leading term in the discrepancy bound (tables of numerical values are provided in Mullen et al. [MMN95]). A recent construction, with significantly better results in this direction than the previous ones, is by Niederreiter and Xing [NX96]; it uses advanced algebraic tools that are beyond the scope of the present book. A very recent survey of the results in this area is [Nie98]. The combined result of the various powerful constructions can be summarized, in our terminology, as follows: b-ary nets with b^t points per box of volume b^{-m} exist for all $m \geq 0$ provided that $t \geq C\frac{d-1}{\log \mathrm{ppmin}(b)} + 1$. Here C is an absolute constant and $\mathrm{ppmin}(b) = \min\{p_1^{\alpha_1}, \ldots, p_m^{\alpha_m}\}$, where $b = p_1^{\alpha_1} \ldots p_m^{\alpha_m}$ is the prime factorization of b (on the other hand, $t \geq \frac{d-1}{b} - \log_b \frac{(b-1)(d-1)+b+1}{2}$ is a necessary condition).

The relation of b-ary nets and projective planes in Exercise 3 below is from Niederreiter [Nie92]. The condition $b \geq d - 1$ for the existence of b-ary nets with 1 point per box following from that correspondence was already proved by Chen [Che83]. Exercise 5 (a fast algorithm for computing b-ary nets from generator matrices) is based on ideas of Antonov and Saleev [AS79], who suggested a similar implementation for Sobol's construction in base 2, and of Bratley and Fox [BF88]. Our presentation follows Tezuka [Tez95]. Properties of point sets generated from a random collection C of $m \times m$ matrices have been analyzed by Niederreiter (see [Nie92]), who has shown an $O(\log^d n)$ upper bound for the expected discrepancy for corners by a method resembling a discrete Fourier transform (Exercise 6 suggests a simple calculation establishing a weaker upper bound).

Exercises

1. (a) Prove the claim made after Definition 2.6, namely that for fixed b, d, and λ and for $m \to \infty$, a b-ary net $P \subset [0,1]^d$ with λ points per box of volume b^{-m} has discrepancy $D(P, \mathcal{R}_d) = O(\log^{d-1} n)$, where $n = |P| = \lambda b^m$. (Begin with the $d = 2$ case.)
(b) Show that if b and d are fixed and λ is arbitrary, then $D(P, \mathcal{R}_d) = O(\lambda \log^{d-1} n)$.

2. Prove Theorem 2.9. Very little needs to be added to the considerations in the proof of Proposition 2.7.
3. Consider the following statements for given integers $b, d \geq 2$:
   (i) A b-ary net with 1 point per box of volume b^{-2} exists in dimension d.
   (ii) There exist d mutually orthogonal b^2-tuples with entries $0, 1, \ldots, b-1$. Here two b^2-tuples $(x_1, x_2, \ldots, x_{b^2})$ and $(y_1, y_2, \ldots, y_{b^2})$ are called *orthogonal* if the b^2 ordered pairs (x_i, y_i), $i = 1, 2, \ldots, b^2$, are all distinct (and hence exhaust all pairs of elements of $\{0, 1, \ldots, b-1\}$).
   (a)* Prove that (i) implies (ii).
   (b)* Prove that (ii) implies (i).

   Remark. As is well-known in combinatorics, and easy to see, (ii) can only hold for $d \leq b+1$. Moreover, the validity of (ii) for $d = b+1$ is equivalent to the existence of a finite projective plane of order b.
4. Show that if a b-ary net with λ points per box of volume b^{-m} exists in dimension d, then also a b-ary net with λ points per box of volume $b^{-m'}$ exists in dimension d' for any $m' \leq m$ and $d' \leq d$. Using the previous exercise, derive that a b-ary net with one point per box of volume b^{-m} in dimension d does not exist unless $b \geq d-1$ (where $m \geq 2$).
5. (Efficient computation) An attractive feature of the set $P(\mathcal{C})$ constructed from a collection $\mathcal{C}$ of matrices is its very efficient computability. Consider the following algorithm.
   Initialize vectors $v_1, v_2, \ldots, v_d \in GF(b)^m$ to zeros. Then perform the following for $i = 0, 1, \ldots, b^m - 1$:
      Output the point x_i whose kth coordinate is $\langle v_k, (b^{-1}, b^{-2}, \ldots, b^{-m})\rangle$ (where the entries of v_k are interpreted as integers in range $0..b-1$).
      Let j be the minimum index with $a_j \neq b-1$, where $i = a_1 + a_2 b + a_3 b^2 + \cdots$ is the b-ary expansion of i (the a_j's are the b-ary digits).
      For $k = 1, 2, \ldots, d$, add the jth column of $C^{(k)}$ to v_k.
   (a)* Verify that if $C^{(1)}$ is the identity matrix then the vector v_1 runs through all vectors in $GF(b)^m$ in some order (for $b = 2$, the resulting order is the well-known *Gray code*).
   (b)* Prove that the algorithm correctly outputs all the points of $P(\mathcal{C})$ in some order, each of them exactly once.
6. (Random generator matrices)
   (a)* Find the probability that a $k \times m$ matrix over $GF(b)$, $k \leq m$, whose entries are chosen at random and independently of each other, has full rank (i.e. rank k).
   (b) Show that if a collection $\mathcal{C}$ of d $m \times m$ matrices over $GF(b)$ is chosen at random, with b, d regarded as constants and $m \to \infty$, then

$$\rho(\mathcal{C}) \geq m - (d-1)\log_b m - O(1)$$

   holds with probability at least $\frac{1}{2}$, say. Deduce that the point set $P(\mathcal{C})$ generated from $\mathcal{C}$ has discrepancy $D(P(\mathcal{C}), \mathcal{R}_d) = O(\log^{2d-2} n)$, where $n = b^m$.

This is a rough bound only; the best known estimate is $O(\log^d n)$ (see [Nie92]).

7. (a) Check that a formal power series $\sum_{i=0}^{\infty} a_i x^i$ over a field F has a multiplicative inverse if and only if $a_0 \neq 0$.
(b) Verify that the formal Laurent series over a field form a field.
8. (a) Let b be a prime, and put $p_k(z) = z - (k-1)$ in the construction described above Theorem 2.10. Show that the generator matrices obtained in this way are those from Faure's construction.
(b) Prove that if b is a prime power, $d \leq b+1$, and $m \geq 1$, then there exists a b-ary net with 1 point per box of volume b^{-m} in dimension d.

2.4 Scrambled Nets and Their Average Discrepancy

Scrambled b-ary Nets. Both theoretical results and empirical studies indicate that for numerical integration, it is desirable that the low-discrepancy point sets used have a some "randomness" in them. Here is one reason: if a multidimensional definite integral is approximated using a Monte-Carlo method, the error can be estimated statistically by repeating the computation several times because the result is a nicely enough distributed random variable. In contrast, deterministic constructions of low-discrepancy sets, such as the Halton–Hammersely set, provide only one set for a given size and dimension. Hence only various theoretical worst-case estimates are available for the error, and these are often unnecessarily pessimistic. By introducing randomness into the low-discrepancy constructions, one can hope to recover the possibility of estimating the error by repeated experiments.

We consider the following question: how can one modify a given b-ary net P with λ points per box of volume b^{-m} so that it is guaranteed to remain a b ary net with the same parameters? More specifically, we will consider "coordinate-wise" changes to P, i.e. mappings $\sigma: [0,1) \to [0,1)$ such that if the kth coordinate of each point $p = (p_1, p_2, \ldots, p_d) \in P$ is replaced by $\sigma(p_k)$ then the b-ary net property is preserved. For example, by inspecting the definition we see that if x is a coordinate of a point of P and I is the b-ary canonical interval of length b^{-m} containing x, we can replace x by any $x' \in I$ with no harm.

Here is a more general class of mappings preserving the b-ary net property. We say that a mapping $\sigma: [0,1) \to [0,1)$ is a *b-ary scrambling of depth m* if it satisfies the following condition: any b-ary canonical interval I of length b^{-i} with $0 \leq i \leq m$ is mapped by σ into a b-ary canonical interval I' of the same length, and distinct I are mapped into distinct I'. A short reflection about definitions reveals the truth of the following:

2.11 Observation. *Let $P \subset [0,1)^d$ be a b-ary net with λ points per box of volume b^{-m}, and let $\sigma = (\sigma_1, \sigma_2, \ldots, \sigma_d)$ be a d-tuple of b-ary scramblings of depth m. Then the set*

$$\sigma(P) = \{(\sigma_1(p_1), \sigma_2(p_2), \ldots, \sigma_d(p_d)) : p \in P\}$$

is a b-ary net with the same parameters.

First, let us give a simple example of a b-ary scrambling.

2.12 Example (Digit-scrambling). Let $\pi_1, \pi_2, \ldots, \pi_m$ be arbitrary permutations of the set $\{0, 1, \ldots, b-1\}$. Define a mapping $\sigma: [0,1) \to [0,1)$ as follows: if $x \in [0,1)$ has b-ary representation $0.a_1a_2a_3\ldots$, then $\sigma(x)$ has b-ary representation $0.b_1b_2b_3\ldots$, where $b_j = \pi_j(a_j)$ for $1 \le j \le m$, and $b_j = a_j$ for $j > m$. Such a σ can naturally be called a *b-ary digit-scrambling* (of depth m).

It is easy to see that such a σ is a b-ary scrambling of depth m (note that a b-ary canonical interval of length b^{-j} consists exactly of numbers having certain fixed sequence of the first j b-ary digits). Also, the choice of the digits $b_{m+1}, b_{m+2}, \ldots$ of $\sigma(x)$ does not really matter and it can be completely arbitrary.

Let us describe a general form of a b-ary scrambling of depth m. First, according to the definition, the b-ary canonical intervals of length b^{-1} are permuted in some way by σ. This means that if $0.a_1a_2a_3\ldots$ is the b-ary representation of a number $x \in [0,1)$, then the first b-ary digit of $\sigma(x)$ is $\pi(a_1)$, for some fixed permutation π of $\{0, 1, \ldots, b-1\}$. Similarly, the second b-ary digit of $\sigma(x)$ is given as $\pi_{a_1}(a_2)$, where π_{a_1} is a permutation of $\{0, 1, \ldots, b-1\}$, this time depending on the first digit a_1. In general, the jth digit of $\sigma(x)$ is given as $\pi_{a_1,a_2,\ldots,a_{j-1}}(a_j)$, $j = 1, 2, \ldots, m$, for some permutation $\pi_{a_1,a_2,\ldots,a_{j-1}}$ depending on the first $j-1$ digits of x.

Finally, each canonical interval I of length b^{-m} is mapped by σ into a canonical interval I' of the same length. According to the definition of a b-ary scrambling of depth m, the mapping of I into I' may be arbitrary. This means that the first m b-ary digits of $\sigma(x)$ are determined from the first m b-ary digits of x based on the various permutations $\pi_{a_1,a_2,\ldots,a_{j-1}}$, and the sequence of digits of $\sigma(x)$ from the $(m+1)$st on is chosen arbitrarily.

A b-ary scrambling σ of depth m can be visualized using a complete b-ary tree of depth m, as in the following drawing (for $b = 3$ and $m = 3$):

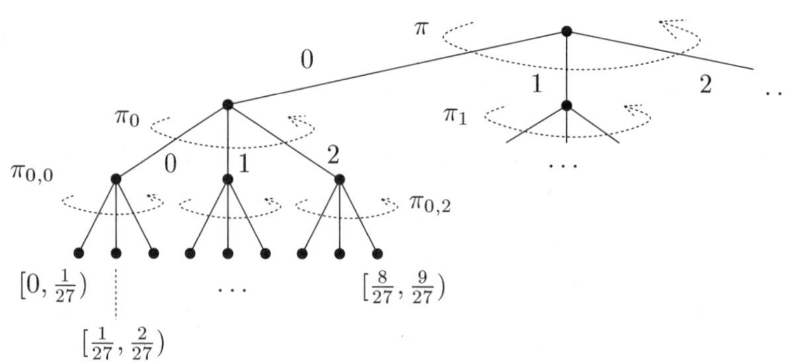

The leaves correspond to the b-ary canonical intervals of length b^{-m}. Initially, the leaves are labeled by these intervals in the natural order from left to right, and a b-ary scrambling of depth m is allowed to permute the left-to-right order of the subtrees of any node arbitrarily (and then do whatever it pleases within each interval in a leaf).

We can now proceed to define various types of random b-ary scramblings. In these definitions, we will not define the whole mapping σ; we will only say how to map a finite set of numbers (this is sufficient for the intended applications, since we always scramble finite point sets only).

For a *random b-ary digit-scrambling of depth m*, we pick m permutations of the set $\{0, 1, \ldots, b-1\}$ independently at random (for each π_j, all the $b!$ possible permutations have the same probability). Then, in order to map numbers of a finite set $F \subset [0,1)$, we proceed as follows. For $x \in F$ with b-ary representation $0.a_1a_2a_3\ldots$, we define $\sigma(x) = \sigma_0(x) + b^{-m}y$, where $\sigma_0(x)$ has b-ary representation $0.\pi_1(a_1)\pi_2(a_2)\ldots\pi_m(a_m)$ and y is chosen uniformly at random in $[0,1)$, these choices being independent for distinct $x \in F$.

Similarly, we define a *fully random b-ary scrambling of depth m*. We pick the permutations $\pi_{a_1,a_2,\ldots,a_j}$, $j = 0, 1, \ldots, m-1$, $a_1, a_2, \ldots \in \{0, 1, \ldots, b-1\}$ independently at random. This determines the first m b-ary digits of the image of each number, and the subsequent digits are chosen at random as in the previous definition.

Having defined these random scramblings, we can define the corresponding randomly scrambled b-ary nets. That is, if P is a b-ary net with λ points per box of volume b^{-m} and $\boldsymbol{\sigma}$ is a d-tuple of mutually independent random b-ary digit-scramblings of depth m, we call $\boldsymbol{\sigma}(P)$ a *randomly digit-scrambled version of P*. Similarly we define a fully randomly scrambled version of P. The following picture shows the 128-point Van der Corput set and a fully randomly scrambled version of it.

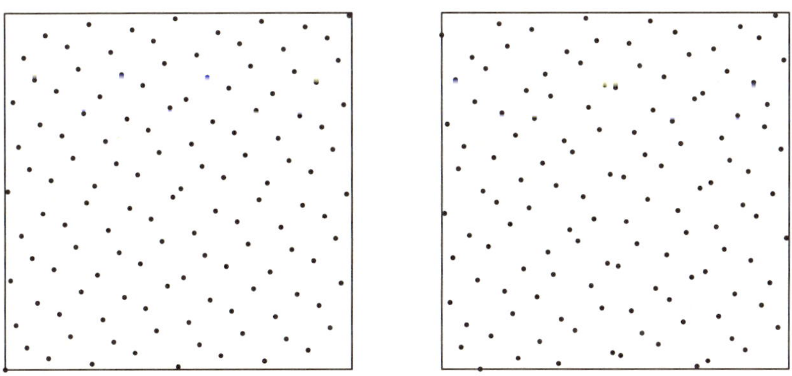

The following theorem gives fairly precise information about the expected L_2-discrepancy for corners of randomly scrambled b-ary nets. In particular, we obtain another proof of the asymptotically tight upper bound for the L_2-discrepancy for corners (as in Theorem 2.5).

2.13 Theorem. *Let $P \subset [0,1)^d$ be a b-ary net with 1 point per box of volume b^{-m}, and let $\boldsymbol{\sigma} = (\sigma_1, \sigma_2, \ldots, \sigma_d)$ be a d-tuple of independent random b-ary digit-scramblings of depth m. Then the expected squared L_2-discrepancy of $\boldsymbol{\sigma}(P)$ for corners, the expectation being with respect to the random choice of $\boldsymbol{\sigma}$, can be expressed solely in terms of d, b, and m (thus, it does not depend on the particular choice of P), and for b, d fixed and $m \to \infty$, we have*

$$\mathbf{E}\left[D_2(\boldsymbol{\sigma}(P), \mathcal{C}_d)^2\right] = O(m^{d-1}) = O(\log^{d-1}|P|).$$

Remarks. The expected squared L_2-discrepancy considered in the theorem can be expressed by a finite formula (to be presented below), and it can be quickly evaluated by a computer even for very large sizes of P. According to numerical studies, these values are very good compared to other constructions of low-discrepancy sets.

The theorem is formulated for randomly digit-scrambled b-ary nets, but the expected squared L_2-discrepancy remains exactly the same if we replace random digit-scrambling by fully random scrambling. In fact, it suffices that the random scrambling is drawn from some probability distribution satisfying certain simple conditions, which can be read off from the proof below.

Readers who dislike longer (although pretty) calculations should better skip the proof of Theorem 2.13. We begin the proof with deriving a formula for the L_2-discrepancy for corners of an arbitrary finite set.

2.14 Lemma (Warnock's formula). *Let $P \subset [0,1]^d$ be a finite set. Then we have*

$$D_2(P, \mathcal{C}_d)^2 = \frac{n^2}{3^d} - \frac{2n}{2^d}\sum_{p\in P}\prod_{k=1}^{d}\left(1-p_k^2\right) + \sum_{p,q\in P}\prod_{k=1}^{d}\left(1-\max(p_k,q_k)\right).$$

Note that this immediately gives an algorithm for evaluating the L_2-discrepancy for corners using $O(dn^2)$ arithmetic operations (an asymptotically even faster algorithm, with $O(n(\log n)^{d-1})$ operations for a fixed d, is presented in Exercise 11 below). In contrast, no similarly efficient algorithm is known for the worst-case discrepancy or for the L_p-discrepancy with $p \neq 2$.

Proof. This is a straightforward calculation. We have

$$\begin{aligned} D_2(P,\mathcal{C}_d)^2 &= \int_{[0,1]^d} \left(nx_1x_2\cdots x_d - |P\cap C_x|\right)^2 \,\mathrm{d}x \\ &= n^2\int_{[0,1]^d} x_1^2x_2^2\cdots x_d^2\,\mathrm{d}x - 2n\int_{[0,1]^d} x_1x_2\cdots x_d\cdot|P\cap C_x|\,\mathrm{d}x \\ &\quad + \int_{[0,1]^d} |P\cap C_x|^2\,\mathrm{d}x. \end{aligned} \tag{2.8}$$

The first integral is easily calculated, giving the $n^2/3^d$ term in the formula. To evaluate the second and third integrals, we write $|P \cap C_x| = \sum_{p \in P} I_p(x)$, where $I_p(x) = 1$ if $p \in C_x$ and $I_p(x) = 0$ otherwise. For the third integral, we thus get

$$\int_{[0,1]^d} |P \cap C_x|^2 \,\mathrm{d}x = \sum_{p,q \in P} \int_{[0,1]^d} I_p(x) I_q(x) \,\mathrm{d}x.$$

For given p and q, the integral inside the sum is the volume of the set

$$\{x \in [0,1]^d \colon p \in C_x,\ q \in C_x\}$$

$$= \{x \in [0,1]^d \colon x_k \geq \max(p_k, q_k), k = 1, 2, \ldots, d\} = \prod_{k=1}^{d} [\max(p_k, q_k), 1].$$

Hence $\int_{[0,1]^d} |P \cap C_x|^2 \,\mathrm{d}x = \sum_{p,q \in P} \prod_{k=1}^{d} (1 - \max(p_k, q_k))$, yielding the third term in the formula being proved. Checking that the second integral in (2.8) equals the middle term in Warnock's formula is left to the reader. □

Next, we derive some properties of random digit-scramblings.

2.15 Lemma. (i) *Let $x \in [0,1)$ be a fixed real number, and let σ stand for a random b-ary digit-scrambling (of any depth $m \geq 0$). Then the random variable $\sigma(x)$ is uniformly distributed in $[0,1)$.*

(ii) *Let $x, x' \in [0,1)$ be two distinct real numbers, and let $t = t(x,x')$ denote the number of b-ary digits shared by x and x'; that is, we assume that the b-ary representations of x and x' are $0.a_1 a_2 \ldots a_t a_{t+1} a_{t+2} \ldots$ and $0.a_1 a_2 \ldots a_t a'_{t+1} a'_{t+2} \ldots$ with $a_{t+1} \neq a'_{t+1}$. Then*

$$\mathbf{E}\big[\, |\sigma(x) - \sigma(x')| \,\big] = \frac{b+1}{3b}\, b^{-t},$$

where the expectation is with respect to a random b-ary digit-scrambling σ of depth $m \geq t+1$.

Proof. Write $B = \{0, 1, \ldots, b-1\}$. If $a \in B$ is fixed and π is a random permutation of B then $\pi(a)$ is random uniformly distributed in B. Hence for a random b-ary digit-scrambling σ, $\sigma(x) = \sum_{j=1}^{m} b_j b^{-j} + b^{-m} y$, where $(b_1, b_2, \ldots, b_m)$ is random uniformly distributed in B^m and y is uniformly distributed in $[0,1)$. This gives (i).

Concerning (ii), let us write b_j for $\pi_j(a_j)$. By the definition of a random b-ary digit-scrambling, we have

$$|\sigma(x) - \sigma(x')| = b^{-(t+1)} \left| b_{t+1} - b'_{t+1} \right| + \mathrm{sgn}^{\pm}(b'_{t+1} - b_{t+1}) \times \left(\sum_{j=t+2}^{m} b^{-j}(b_j - b'_j) + b^{-m}(y - y') \right),$$

where $\mathrm{sgn}^{\pm}(z) = 1$ for $z \geq 0$ and $\mathrm{sgn}^{\pm}(z) = -1$ for $z < 0$, and y, y' are independent uniformly distributed in $[0,1)$. For a random permutation π of B and two distinct elements $a, a' \in B$, one has $\mathbf{E}[\pi(a) - \pi(a')] = 0$ and $\mathbf{E}[|\pi(a) - \pi(a')|] = \frac{b+1}{3}$ (Exercise 7). Moreover, $\mathbf{E}[\pi(a) - \pi(a')] = 0$ obviously holds for $a = a'$ as well. By the mutual independence of the π_j's and by linearity of expectation, we obtain

$$\begin{aligned}\mathbf{E}[|\sigma(x) - \sigma(x')|] &= b^{-(t+1)}\mathbf{E}[|b_{t+1} - b'_{t+1}|] + \mathbf{E}\left[\mathrm{sgn}^{\pm}(b'_{t+1} - b_{t+1})\right] \\ &\quad \times \left(\sum_{j=t+2}^{m} b^{-j}\mathbf{E}\left[b_j - b'_j\right] + b^{-m}\mathbf{E}\left[y - y'\right]\right) \\ &= \frac{b+1}{3b} b^{-t}.\end{aligned}$$

□

As the next step, we re-express the expected squared L_2-discrepancy of $\boldsymbol{\sigma}(P)$ using Warnock's formula and Lemma 2.15(i). By Lemma 2.14, by linearity of expectation, and by the mutual independence of the scramblings $\sigma_1, \sigma_2, \ldots, \sigma_d$, we have

$$\begin{aligned}\mathbf{E}\left[D_2(\boldsymbol{\sigma}(P), \mathcal{C}_d)^2\right] &= \frac{n^2}{3^d} - \frac{2n}{2^d} \sum_{p \in P} \prod_{k=1}^{d} \mathbf{E}\left[1 - \sigma_k(p_k)^2\right] \\ &\quad + \sum_{p,q \in P} \prod_{k=1}^{d} \left(1 - \mathbf{E}\left[\max(\sigma_k(p_k), \sigma_k(q_k))\right]\right).\end{aligned} \tag{2.9}$$

(recall that the expectation of a product of independent random variables equals the product of expectations). Now for x uniformly distributed in $[0,1)$, we have $\mathbf{E}\left[x^2\right] = \frac{1}{3}$, and since $\sigma_k(p_k)$ is uniformly distributed in $[0,1)$ by Lemma 2.15(i), $\mathbf{E}\left[1 - \sigma_k(p_k)^2\right] = \frac{2}{3}$ and the middle addend in (2.9) equals $-2n^2/3^d$. Hence, we have $\mathbf{E}\left[D_2(\boldsymbol{\sigma}(P), \mathcal{C}_d)^2\right] = -n^2/3^d + R$ where R denotes the third addend in (2.9).

Using the equality $\max(x,y) = \frac{1}{2}(x + y + |x - y|)$, we rewrite

$$\begin{aligned}R &= \sum_{p,q \in P} \prod_{k=1}^{d} \left(1 - \frac{1}{2}\left(\mathbf{E}\left[\sigma_k(p_k)\right] + \mathbf{E}\left[\sigma_k(q_k)\right] + \mathbf{E}\left[|\sigma_k(p_k) - \sigma_k(q_k)|\right]\right)\right) \\ &= 2^{-d} \sum_{p,q \in P} \prod_{k=1}^{d} \left(1 - \mathbf{E}\left[|\sigma_k(p_k) - \sigma_k(q_k)|\right]\right).\end{aligned}$$

By multiplying out the product over k and moving the sum over p and q inside, we derive

$$R = 2^{-d} \sum_{S \subseteq \{1,2,\ldots,d\}} (-1)^{|S|} \sum_{p,q \in P} \prod_{k \in S} \mathbf{E}\left[|\sigma_k(p_k) - \sigma_k(q_k)|\right].$$

By Lemma 2.15(ii), we know that $\mathbf{E}\left[\,|\sigma_k(p_k)-\sigma_k(q_k)|\,\right]$ only depends on $t(p_k,q_k)$, the number of initial b-ary digits shared by p_k and q_k. Hence we can group the pairs (p,q) in the inner sum according to the values of the vector (indexed by S)

$$\mathbf{t}_S(p,q)=(t(p_k,q_k)\colon k\in S)\,.$$

Putting

$$F_S(\mathbf{t})=\{(p,q)\in P\times P\colon \mathbf{t}_S(p,q)=\mathbf{t}\}$$

and applying Lemma 2.15(ii) we have

$$\begin{aligned}\sum_{p,q\in P}\prod_{k\in S}\mathbf{E}\left[\,|\sigma_k(p_k)-\sigma_k(q_k)|\,\right] &= \sum_{\mathbf{t}}|F_S(\mathbf{t})|\prod_{k\in S}\frac{b+1}{3b}\,b^{-\mathbf{t}_k}\\ &= \left(\frac{b+1}{3b}\right)^{|S|}\sum_{\mathbf{t}}|F_S(\mathbf{t})|\,b^{-|\mathbf{t}|},\end{aligned}$$

where the sum is over all nonnegative $|S|$-component integer vectors $\mathbf{t}$ and $|\mathbf{t}|$ stands for the sum of the components of $\mathbf{t}$.

2.16 Lemma. *We have* $|F_S(\mathbf{t})|=f_{|S|,m}(|\mathbf{t}|)$, *where*

$$f_{s,m}(t)=b^m\sum_{r=0}^{s}\binom{s}{r}(-1)^r\lceil b^{m-t-r}\rceil.$$

Note that $|F_S(\mathbf{t})|$ does not depend on the specific choice of the b-ary net P; this already implies that the expected squared L_2-discrepancy of $\sigma(P)$ does not depend on the choice of P. We postpone the proof of Lemma 2.16 a little, and we finish the derivation of a formula for $\mathbf{E}\left[D_2(\sigma(P),\mathcal{C}_d)^2\right]$. There are $\binom{t+s-1}{s-1}$ different choices of an s-component vector $\mathbf{t}$ with $|\mathbf{t}|=t$, and $\binom{d}{s}$ choices of the set S of cardinality s. Therefore

$$\begin{aligned}\mathbf{E}\left[D_2(\sigma(P),\mathcal{C}_d)^2\right] &= -\frac{n^2}{3^d}+2^{-d}\sum_{s=0}^{d}\binom{d}{s}\left(-\frac{b+1}{3b}\right)^s\\ &\quad\times\sum_{t=0}^{\infty}\binom{t+s-1}{s-1}b^{-t}f_{s,m}(t),\end{aligned}\qquad(2.10)$$

with $f_{s,m}(t)$ as in Lemma 2.16.

We observe that for $t\geq m$, we have $f_{s,m}(t)=0$. This can be seen from the formula for $f_{s,m}(t)$ because $\sum_{r=0}^{s}\binom{s}{r}(-1)^r=(1-1)^s=0$, or it follows from the meaning of $f_{s,m}(t)$ and the fact that if two points of P share at least m initial b-ary digits (together in all coordinates) then they are necessarily equal. Therefore, the sum over t in (2.10) can actually be written with upper bound $m-1$, and (2.10) becomes the promised exact formula for the expected squared discrepancy of $\sigma(P)$.

It remains to prove Lemma 2.16, and to derive the asymptotics of (2.10) for d, b fixed and $m \to \infty$.

Proof of Lemma 2.16. Let $G_S(\mathbf{t})$ be the set of all pairs (p,q), $p \neq q$, of points that share *at least* $\mathbf{t}_k$ initial digits in the kth coordinate for all $k \in S$. Written formally, $G_S(\mathbf{t}) = \bigcup_{\mathbf{r}\geq\mathbf{t}} F_S(\mathbf{r})$, where the inequality among vectors means the inequality in all components simultaneously. For a fixed p, the set

$$\{x \in [0,1]^d : \mathbf{t}_S(p,x) \geq \mathbf{t}\},$$

i.e. the points sharing at least $\mathbf{t}_k$ initial digits with p in the kth coordinate for all $k \in S$, is a b-ary canonical box of volume $b^{-|\mathbf{t}|}$. Since P is a b-ary net with 1 point per box of volume b^{-m}, such a box contains $b^{m-|\mathbf{t}|}$ points of P for $|\mathbf{t}| \leq m$ and 1 point of P (namely p itself) for $|\mathbf{t}| > m$. We obtain

$$|G_S(\mathbf{t})| = b^m \left(\lceil b^{m-|\mathbf{t}|} \rceil - 1 \right).$$

Lemma 2.16 can be derived follows from the following general statement:

2.17 Lemma (A particular case of the Möbius inversion formula). *Let f be a real function defined on the set of all nonnegative s-component integer vectors, and put*

$$g(\mathbf{t}) = \sum_{\mathbf{r}\geq\mathbf{t}} f(\mathbf{r}).$$

Assuming that f is nonzero for finitely many values of $\mathbf{t}$ only, we have

$$f(\mathbf{t}) = \sum_{\mathbf{r}\in\{0,1\}^s} (-1)^{|\mathbf{r}|} g(\mathbf{t}+\mathbf{r}).$$

A proof is left as Exercise 9 (or, one can look up a general case of the Möbius inversion formula for partially ordered sets in a suitable book). Not surprisingly, the inversion formula also holds under more relaxed assumptions on f, but here we suffice with the simple finite case.

To prove Lemma 2.16, we apply Lemma 2.17 with $f(\mathbf{t}) = |F_S(\mathbf{t})|$ and $g(\mathbf{t}) = |G_S(\mathbf{t})|$. Lemma 2.16 follows by a simple formula manipulation, using the already mentioned fact that $\sum_{r=0}^{s} \binom{s}{r}(-1)^r = 0$. □

Asymptotics. The formula (2.10) expresses the expected squared L_2-discrepancy, which turns out to be of the order $\log^{d-1} n$, as a difference of two terms. The first term has the order of magnitude n^2, and hence these two terms have to nearly cancel each other. Thus, deriving the asymptotic behavior is a somewhat subtle matter. The trick is to write the first term, $-n^2/3^d$, as a sum with structure similar to the second term, in such a way that the corresponding terms nearly cancel each other.

Here is an intuitive explanation of the method. The point set P can be thought of as a discrete n-point approximation to the (continuous) uniform distribution of mass n in $[0,1]^d$. If $\tilde{P}$ stands for this continuous uniform distribution, then every canonical box of volume b^{-t} contains mass precisely b^{m-t} of $\tilde{P}$, for all $t \geq 0$ (while the discrete P achieves this for $t \leq m$ only). For such a $\tilde{P}$ replacing P, the analogue of the function $f_{s,m}(t)$ in Lemma 2.16 is

$$\tilde{f}_{s,m}(t) = b^m \sum_{r=0}^{s} \binom{s}{r} (-1)^r b^{m-t-r} = b^{2m-t} \left(1 - \frac{1}{b}\right)^s.$$

Using the binomial formula $(1-x)^{-s} = \sum_{t=0}^{\infty} \binom{t+s-1}{s-1} x^t$, we arrive at

$$\begin{aligned}\sum_{t=0}^{\infty} \binom{t+s-1}{s-1} b^{-t} \tilde{f}_{s,m}(t) &= b^{2m} \left(1 - \frac{1}{b}\right)^s \sum_{t=0}^{\infty} \binom{t+s-1}{s-1} b^{-2t} \\ &= b^{2m} \left(1 + \frac{1}{b}\right)^{-s}.\end{aligned}$$

Hence, after replacing f by $\tilde{f}$, the second term in (2.10) becomes

$$\frac{n^2}{2^d} \sum_{s=0}^{d} \binom{d}{s} \left(-\frac{b+1}{3b}\right)^s \left(1 + \frac{1}{b}\right)^{-s} = \frac{n^2}{2^d} \sum_{s=0}^{d} \binom{d}{s} \left(-\frac{1}{3}\right)^s = \frac{n^2}{3^d},$$

and so (2.10) with $\tilde{f}$ instead of f gives 0 in accordance with the intuition. Therefore,

$$\mathbf{E}\left[D_2(\sigma(P), \mathcal{C}_d)^2\right] = 2^{-d} \sum_{s=0}^{d} \binom{d}{s} \left(-\frac{b+1}{3b}\right)^s E(s),$$

where we put

$$\begin{aligned} E(s) &= \sum_{t=0}^{\infty} \binom{t+s-1}{s-1} b^{-t} \left(f_{s,m}(t) - \tilde{f}_{s,m}(t)\right) \\ &= \sum_{t=0}^{\infty} \binom{t+s-1}{s-1} b^{m-t} \sum_{r=0}^{s} \binom{s}{r} (-1)^r \left(\lceil b^{m-t-r} \rceil - b^{m-t-r}\right) \\ &= \sum_{r=0}^{s} \binom{s}{r} (-1)^r \sum_{t=0}^{\infty} \binom{t+s-1}{s-1} b^{m-t} \left(\lceil b^{m-t-r} \rceil - b^{m-t-r}\right). \end{aligned}$$

For $t \leq m - r$, b^{m-t-r} is an integer, and hence the inner summation can start from $t = \max(0, m-r+1)$. Since we are now interested in the behavior of the discrepancy for n (and hence m) large and d fixed, we may assume $m \geq d \geq r$. Then

$$|E(s)| \leq \sum_{r=0}^{s} \binom{s}{r} b^m \sum_{t=m-r+1}^{\infty} \binom{t+s-1}{s-1} b^{-t}.$$

We need one last technical lemma.

2.18 Lemma. *For any $b > 1$ and integers $s, t_0 \geq 0$, we have*

$$\sum_{t=t_0}^{\infty} \binom{t+s-1}{s-1} b^{-t} \leq b^{-t_0} \binom{t_0+s-1}{s-1} \left(1 - \frac{1}{b}\right)^{-s}.$$

Leaving the proof as yet another exercise, we thus obtain

$$|E(s)| \leq \sum_{r=0}^{s} \binom{s}{r} b^{r-1} \binom{m-r+s}{s-1} \left(1 - \frac{1}{b}\right)^{-s}.$$

Considering d and b as constants and $m \to \infty$, we have $|E(s)| = O(m^{s-1})$ and finally $\mathbf{E}\left[D_2(\sigma(P), \mathcal{C}_d)^2\right] = O(m^{d-1}) = O(\log^{d-1} n)$. This proves Theorem 2.13. □

Bibliography and Remarks. The idea of improving the properties of the constructed sets by adding randomness in some way was present in several constructions following the Halton–Hammersely sets.

Fully random b-ary scramblings in the above-defined sense were studied by Owen ([Owe97] and other papers), who also investigated the influence of random scrambling to error estimates in numerical integration. The terminology and the definition are slightly modified here compared to Owen's presentation. He essentially works with b-ary scramblings of infinite depth, given by countably many permutations $\pi_{a_1,\ldots,a_{j-1}}$; in such a situation, one has slight formal problems with numbers having all b-ary digits from some position on equal to $b-1$. The difference is unessential; our treatment emphasizes the finitary nature of random b-ary scramblings in applications to finite sets.

Theorem 2.13 for fully random scrambling was proved by Hickernell [Hic96]. The validity for random digit-scrambling (and for other types of random scrambling which are easier to implement than a fully random scrambling, as in Exercise 8(b)) was noted in [Mat98b]. Much earlier Chen [Che83] considered modifications of the Halton–Hammersely construction which, in our terminology, can be described as applying a digit-scrambling to each coordinate, where the kth coordinate associated with the prime p_k is scrambled by a p_k-ary digit-scrambling with $\pi_j(a_j) = a_j + c_j$, with $c_j \in GF(p_k)$ chosen independently at random. Chen proved that the expected squared L_2-discrepancy of the resulting set is asymptotically optimal, i.e. $O(\log^{d-1} n)$ (also in [BC87]).

Hickernell [Hic96] determined, by slightly more careful estimates, the constant in the leading term of the bound in Theorem 2.13: $\mathbf{E}\left[D_2(\sigma(P), \mathcal{C}_d)^2\right] = C(b,d)m^{d-1} + o(m^{d-1})$ for b and d fixed and $m \to \infty$, with

$$C(b,d) = \frac{\left(b - \frac{1}{b}\right)^{d-1}}{6^d (d-1)! \log^{d-1} b}.$$

But a full asymptotic analysis of the formula, for all of b, d, and m as parameters, is still missing.

The formula in Lemma 2.14 for the L_2-discrepancy for corners was first derived, to my knowledge, by Warnock [War72]. Morokoff and Caflisch [MC94] derive an analogous formula for the L_2-discrepancy for axis-parallel boxes, and Hickernell [Hic98] provides a formula for his generalized notion of L_2-discrepancy (see Exercise 6 for a simple instance). Another formula of this type, expressing the toroidal L_2-discrepancy for boxes, will be presented Exercise 7.1.8.

Exercises

1. (Generalized Faure sets) Let $\mathcal{C}$ be a collection of d $m \times m$ matrices over $GF(b)$ with $\rho(\mathcal{C}) = \rho$. Let $L^{(1)}, \ldots, L^{(d)}$ be nonsingular lower-triangular $m \times m$ matrices over $GF(b)$.
(a) Prove that the collection $\mathcal{C}_1 = \left(L^{(1)}C^{(1)}, L^{(2)}C^{(2)}, \ldots, L^{(d)}C^{(d)}\right)$ has $\rho(\mathcal{C}_1) \geq \rho$. (In particular, if $\mathcal{C}$ is the collection of matrices for the Faure set in Example 2.8, the sets generated by the resulting $\mathcal{C}_1$ are known as *generalized Faure sets*; they were suggested by Tezuka [Tez95].)
(b) Is it true that any generalized Faure set in the sense of (a) arises from a Faure set by an application of a b-ary scrambling to each coordinate?
2. (Scrambled Halton–Hammersely set) Let P be the n-point Halton–Hammersley set as in Example 2.3. Let σ_j be a p_j-ary scrambling of depth $m_j \geq \log_{p_j} n + 1$, $j = 1, 2, \ldots, d-1$, and let P' arise from P by applying σ_j to the $(j+1)$st coordinate, for $j = 1, 2, \ldots, d-1$. Check that P' has discrepancy for axis-parallel boxes of the order $O(\log^{d-1} n)$, similar to the original P.
3. Complete the proof of Warnock's formula (Lemma 2.14).
4. Find the expectation $\mathbf{E}\left[D_2(P, \mathcal{C}_d)^2\right]$ for a random n-point set $P \subset [0,1]^d$ (the points of P are chosen uniformly at random and independently).
5. (Limits of usefulness of the L_2-discrepancy for corners) Consider a pathological n-point set $P \subset \mathbf{R}^d$, whose all points lie at the point $(1, 1, \ldots, 1)$ (or very near to it).
(a) Compute $D_2(P, \mathcal{C}_d)^2$.
(b) For how large n (in terms of d) does this become larger than the expected squared discrepancy of a random point set (determined in Exercise 4)? Evaluate the bound for $d = 5, 10, 20, 50, 100$.
This indicates that if the dimension is large and the number of points not too large, then smaller L_2-discrepancy for corners does not necessarily mean a more uniform distribution in the intuitive sense. See [Mat98b] for more observations in this spirit.
6. (Modified L_2-discrepancy for corners)

From the remarks to Section 1.4, recall the definition of the discrepancy $D_{2,proj}(P)$ (below Eq. (1.8)). Derive an analogue of Warnock's formula for $D_{2,proj}(P)^2$, and show that it can be evaluated using $O(dn^2)$ arithmetic operations.

Remark. This kind of discrepancy, inspired by Zaremba's inequality (1.8), fixes the problems with the L_2-discrepancy for corners indicated in Exercise 5 to some extent, and it seems to be more suitable for high-dimensional settings than the usual L_2-discrepancy for corners.

7. Let $B = \{0, 1, \ldots, b-1\}$, let $a, a' \in B$ be two distinct elements of B, and let π be a random permutation of B. Prove $\mathbf{E}[\pi(a) - \pi(a')] = 0$ and $\mathbf{E}\big[\,|\pi(a) - \pi(a')|\,\big] = (b+1)/3$.
8. (a) Check that Lemma 2.15 remains valid if σ is chosen as a fully random b-ary scrambling of depth $m \geq t+1$.
(b)* Suppose that b is a prime. Modify the definition of a random b-ary digit-scrambling as follows. Each permutation π_j is given by the formula $\pi_j(a) = (ha + g_j) \bmod b$, where h is chosen uniformly at random from $\{1, 2, \ldots, b-1\}$ and the g_j are chosen uniformly at random from $\{0, 1, \ldots, b-1\}$, mutually independent and also independent of h. Show that Lemma 2.15 still holds with this kind of random scrambling.
9.* Prove Lemma 2.17 (the Möbius inversion formula).
10. Prove Lemma 2.18.
11. (Heinrich's algorithm for the L_2-discrepancy for corners [Hei96]) Let $P \subset [0,1]^d$ be an n-point set.
(a)** Design an algorithm for evaluating $\sum_{p,q \in P} \prod_{k=1}^{d} \min(p_k, q_k)$ in time $O(n \log^d n)$, for any fixed d. Begin with the $d = 1$ case, and work by induction on the dimension. For simplicity, assume that no two coordinates of points in P coincide.
(b) Explain how to use such an algorithm for evaluating $D_2(P, \mathcal{C}_d)$.
12.* Let $x_1, x_2, \ldots, x_n \in [0,1]$ be numbers in increasing order ($x_1 \leq x_2 \leq \cdots \leq x_n$). Give an algorithm for evaluating $\sum_{i,j=1}^{n} \min(x_i, x_j)$ in $O(n)$ time. (This gives a faster algorithm for the $d = 1$ case in Exercise 11, and allows one to reduce the running time in that exercise to $O(n \log^{d-1} n)$ for higher dimensions d.)
This result is from Frank and Heinrich [FH96].

2.5 More Constructions: Lattice Sets

Another interesting class of low-discrepancy sets for axis-parallel boxes is obtained by intersecting suitable *lattices* with the unit cube. Discrepancy estimates for lattices use number-theoretic methods, and they were studied long before discrepancy theory emerged and before discrepancy was defined for arbitrary point sets or sequences.

We begin with a simple planar example.

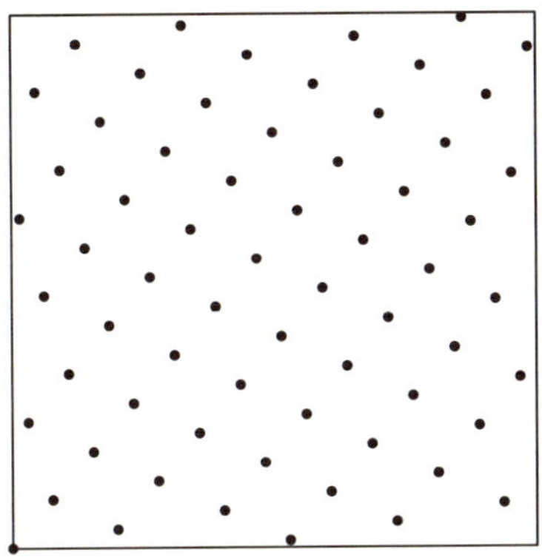

Fig. 2.7. The set $\{(\frac{i}{n}, \{i\alpha\}) \colon i = 0, 1, \ldots, n-1\}$ for $\alpha = \frac{1}{2}(\sqrt{5}+1)$ and $n = 64$.

2.19 Example. Let $\alpha > 0$ be a fixed irrational number. We construct the set

$$P = \left\{(\tfrac{i}{n}, \{i\alpha\}) \colon i = 0, 1, \ldots, n-1\right\},$$

where $\{x\}$ stands for the fractional part of x; see Fig. 2.7. For a suitable choice of the number α, the discrepancy of this set is of the order $\log n$.

The essential property of α used in estimating the discrepancy are the partial quotients of the *continued fraction* of α. These are the integers $a_0, a_1, \ldots$, defined recursively as follows: $\alpha_0 = \alpha$, $a_i = \lfloor \alpha_i \rfloor$, $\alpha_{i+1} = 1/(\alpha_i - a_i)$. It can be shown that if all the a_i are bounded above by some constant, then $D(P, \mathcal{R}_2) = O(\log n)$. The proof is not too difficult, and we leave it as Exercise 3. One class of numbers α with bounded partial quotients of their continued fraction are the *quadratic irrationalities*, which are numbers of the form $u + \sqrt{v}$, with u, v rational and $\sqrt{v}$ irrational. For example, the number $\frac{1}{2}(1+\sqrt{5})$, the famous golden section, has all the a_i equal to 1. For $n = 64$, the corresponding set is depicted in Fig. 2.7.

How should this construction be generalized to higher dimensions? It is very natural to choose $d-1$ real numbers $\alpha_1, \ldots, \alpha_{d-1}$, such that 1 and α_1, ..., α_{d-1} are linearly independent over the rationals (and they perhaps satisfy some additional conditions) and construct the d-dimensional n-point set

$$\left\{(\tfrac{i}{n}, \{i\alpha_1\}, \ldots, \{i\alpha_{d-1}\}) \colon i = 0, 1, \ldots, n-1\right\}. \tag{2.11}$$

These sets are well-known to be uniformly distributed (this can be shown elegantly by higher-dimensional Weyl's criterion, as was done in Weyl [Wey16]). It is even known that the discrepancy is close to $O(\log^{d-1} n \log\log n)$ for almost all vectors $(\alpha_1, \ldots, \alpha_{d-1}) \in [0,1)^{d-1}$, but no such good bounds seem to be known for any explicit $\alpha_1, \ldots, \alpha_{d-1}$. We describe, without proof, another recent construction of a similar type, for which the known discrepancy bounds match the best results obtained by other methods.

Lattices. At the beginning of this chapter, we noted that the "ordinary" integer lattice has a bad discrepancy. By looking at Fig. 2.7, the reader may suspect that the set from the above example is a lattice (the set of all integer linear combinations of two linearly independent vectors), or, more precisely, the part of such a lattice lying in $[0,1]^2$. It is indeed the case (Exercise 1). Let us recall the definition of a lattice in an arbitrary dimension. A *lattice* Λ in $\mathbf{R}^d$ is a set of the form

$$\Lambda = \Lambda(b_1, b_2, \ldots, b_d) = \left\{ \sum_{j=1}^{d} i_j b_j \colon i_1, \ldots, i_d \in \mathbf{Z} \right\},$$

where $b_1, b_2, \ldots, b_d \in \mathbf{R}^d$ are linearly independent vectors, called a *basis* of Λ. A lattice has many different bases.[2]

If B is the $d \times d$ matrix with $b_1, b_2, \ldots, b_d$ as the columns, we can also write $\Lambda = B\mathbf{Z}^d$. The *determinant* of a lattice $\Lambda = \Lambda(b_1, b_2, \ldots, b_d)$, denoted by $\det(\Lambda)$, is the volume of the parallelepiped spanned by the vectors $b_1, \ldots, b_d$. In other words, $\det(\Lambda) = |\det(B)|$. It is not hard to prove that the value of the determinant is a property of the lattice as a point set, and it does not depend on the particular choice of the basis.

We define the *norm* of Λ

$$\mathrm{Nm}(\Lambda) = \inf_{x \in \Lambda \setminus \{0\}} |x_1 x_2 \ldots x_d|$$

($x_1, \ldots, x_d$ are the coordinates of the point x). Geometrically, $\mathrm{Nm}(\Lambda) \geq \varepsilon$ means that the lattice points distinct from 0 avoid a region near the coordinate hyperplanes, delimited by the hyperbolic surfaces $x_1 x_2 \ldots x_d = \pm\varepsilon$. Here is a planar illustration:

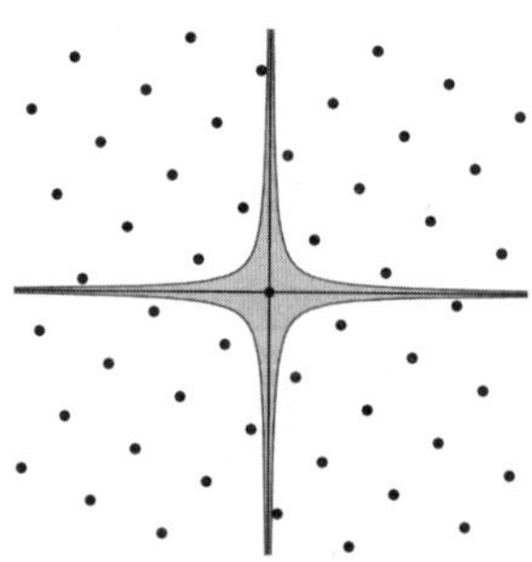

It turns out that lattices with $\mathrm{Nm}(\Lambda) > 0$ are good from the discrepancy point of view. Namely, if we re-scale such a lattice so that the unit cube contains n of its points, then this n-point set has discrepancy $O(\log^{d-1} n)$:

[2] In fact, a lattice can be defined in a seemingly more general way, as a full-dimensional discrete subgroup of $(\mathbf{R}^d, +)$. Then it is a moderately nontrivial theorem that every lattice has a basis.

2.20 Theorem. *If Λ is a lattice in $\mathbf{R}^d$ such that $\mathrm{Nm}(\Lambda) > 0$ and $\det(\Lambda) = 1$, and if we set $P_t = [0,1]^d \cap \frac{1}{t}\Lambda$ (where $t > 0$ is a real parameter), then $D(P_t, \mathcal{R}_d) = O(\log^{d-1}|P_t|)$ as $t \to \infty$. The constant of proportionality in this bound depends on d and on $\mathrm{Nm}(\Lambda)$ (tending to ∞ as $\mathrm{Nm}(\Lambda) \to 0$).*

The known proofs are too complicated to explain here, unfortunately. (The hard part is to get the exponent $d-1$ of the logarithm; proving a bound like $O(\log^d n)$ is less difficult.) But we at least describe some lattices Λ with nonzero norm.

For the planar lattice $\Lambda = \Lambda((1,1),(\sqrt{2},-\sqrt{2}))$, $\mathrm{Nm}(\Lambda) > 0$ is easy to see:

$$\mathrm{Nm}(\Lambda) = \inf_{(i,j)\neq(0,0)} \left|(i+j\sqrt{2})(i-j\sqrt{2})\right| = \inf_{(i,j)\neq(0,0)} \left|i^2 - 2j^2\right| = 1,$$

because $i^2 - 2j^2$ is integral and nonzero, since $i^2 = 2j^2$ would mean $\frac{i}{j} = \pm\sqrt{2}$. For a larger dimension d, we take a suitable degree d polynomial $p(x)$ with integer coefficients, with leading coefficient 1, irreducible over the rationals, and with d distinct real roots $\alpha_1, \ldots, \alpha_d$. (Or, more scientifically speaking, we consider some totally real number field of degree d over the rationals.) For $d = 3$, an example of such a polynomial is $x^3 - 3x + 1$, and Exercise 11 indicates one possible systematic method for producing such polynomials for higher dimensions. From the roots of such a $p(x)$, a lattice is produced as follows:

$$\Lambda = \Lambda((1,1,\ldots,1),(\alpha_1,\alpha_2,\ldots,\alpha_d),(\alpha_1^2,\ldots,\alpha_d^2),\ldots,(\alpha_1^{d-1},\ldots,\alpha_d^{d-1})).$$

Let us note that the above example for $d = 2$ was like that with $p(x) = x^2 - 2$.

2.21 Proposition. *If $p(x)$ is a monic irreducible polynomial of degree d with integer coefficients and with distinct real roots $\alpha_1, \alpha_2, \ldots, \alpha_d$ and if Λ is constructed from the α_i as above then $\mathrm{Nm}(\Lambda) > 0$.*

Proof. By definition, $\mathrm{Nm}(\Lambda)$ is the infimum, over all choices of nonzero integer vectors $(i_1, i_2, \ldots, i_d)$, of the absolute value of

$$\prod_{j=1}^{d} \left(i_1 + i_2\alpha_j + i_3\alpha_j^2 + \cdots + i_d\alpha_j^{d-1}\right). \tag{2.12}$$

First we note that the value of (2.12) is never 0 unless all the i_k are 0. This is because, by the irreducibility of $p(x)$, the α_j are not roots of any integer polynomial of degree smaller than d, and so none of the terms $i_1 + i_2\alpha_j + \cdots + i_d\alpha_j^{d-1}$ can be 0 unless all the i_k are 0.

Next, we claim that for integral $i_1, \ldots, i_d$, the value of (2.12) is always an integer. We note that (2.12) is a polynomial in $i_1, i_2, \ldots, i_d$, each of whose coefficients is a symmetric polynomial in $\alpha_1, \ldots, \alpha_d$ with integer coefficients.

We recall a theorem of Newton on symmetric polynomials, in the following form (see [Sti94]). Let $p(x) = x^d + a_{d-1}x^{d-1} + \cdots + a_0$ be a polynomial with roots $\alpha_1, \ldots, \alpha_d$. Then any symmetric polynomial in the α_j with integer coefficients can be written as a polynomial in the a_i, also with integer coefficients. In our case this means that the coefficient of each monomial $i_1^{\nu_1} i_2^{\nu_2} \ldots i_d^{\nu_d}$ in (2.12) is an integer. This concludes the proof of $\mathrm{Nm}(\Lambda) > 0$. □

According to an unproved conjecture (a generalization of Littlewood's conjecture), for $d > 2$, the lattices obtained by the above construction are essentially the only ones with nonzero norm (this is not true for $d = 2$; see Exercise 10).

Strong results (asymptotically tight ones, in fact) are also known about the L_p-discrepancy of lattices with $\mathrm{Nm}(\Lambda) > 0$:

2.22 Theorem. *Let Λ be a lattice in $\mathbf{R}^d$ with $\det(\Lambda) = 1$ and $\mathrm{Nm}(\Lambda) > 0$, let $p \geq 1$ be a fixed real number, and let $t > 0$ be a real parameter. Then there exists a vector $x \in \mathbf{R}^d$ such that the set*

$$P_{x,t} = [0,1]^d \cap \frac{1}{t}(\Lambda + x)$$

has L_p-discrepancy $O(\log^{(d-1)/2} |P_{x,t}|)$ as $t \to \infty$, where the constant of proportionality depends on $\mathrm{Nm}(\Lambda)$ and on p.

Again, we omit a proof. Note that this theorem provides a third proof of the optimal upper bound for the L_2-discrepancy (Theorem 2.5). As in the previous two proofs shown, we do not get an explicitly described set here, because the existence of a suitable translation x is proved nonconstructively.

Bibliography and Remarks. We have already encountered results about the distribution of the sets like the one in Example 2.19, or equivalently, of the sequences $(\{n\alpha\})$, in Section 1.1. This is a classical theme in uniform distribution theory and these sequences and their multidimensional versions (the *Kronecker sequences*) have been studied in great detail from various points of view. Here we quote just a small assortment of remarkable theorems in this area (for more information and references see, for instance, [DT97], [Sós83b]; or also [KN74] for older material).

It is known that the discrepancy of the $(\{n\alpha\})$ sequence achieves the asymptotically optimal $O(\log n)$ bound if and only if the sequence of the Cesàro means of the partial quotients of the continued fraction for α is bounded (Drmota and Tichy [DT97] attribute this result to Behnke [Beh22], [Beh24]). Kesten [Kes66] proved that, for each α, the only intervals for which the sequence $(\{n\alpha\})$ has bounded discrepancy are those of length $\{k\alpha\}$ for some integer k. This theorem has a generalization to arbitrary sequences as well as interesting generalizations in ergodic theory (see [DT97] or [BS95] for references).

Recently, many difficult and often surprising results on the detailed behavior of the discrepancy of the sequence $(\{n\alpha\})$ were discovered by Beck. Among others, he proved several "deterministic central limit theorems," showing that in many cases the discrepancy behaves as if it were a sum of $\log n$ independent random variables. For example, let $\alpha = \sqrt{2}$ and let $D(y) = \mathrm{vol}(T_y) - |T_y \cap (\mathbf{Z}^2 + (\frac{1}{2}, \frac{1}{2}))|$, where T_y is the gray right triangle in the picture:

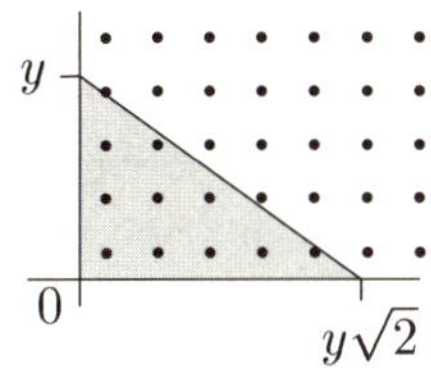

Then there is a constant c such that the distribution of $D(y)/c\sqrt{\log n}$, for y varying in $[0, n]$, approaches the standard normal distribution $N(0,1)$ as $n \to \infty$. At the time of finishing this book, these results remain unpublished; together with several other exciting discrepancy-theoretic results and problems, they are surveyed in [Bec].

For higher-dimensional Kronecker sequences, the theory is not so well-developed, mainly because a higher-dimensional analogue of the continued fractions is missing. As was proved by Khintchine [Khi23], [Khi24] for $d = 1$ and by Beck [Bec94] for higher dimensions, the discrepancy of the d-dimensional Kronecker sequence $(\{n\alpha_1\}, \ldots, \{n\alpha_d\})$ is between $\Omega(\log^d n (\log \log n)^{1-\varepsilon})$ and $O(\log^d n (\log \log n)^{1+\varepsilon})$ for almost all $(\alpha_1, \ldots, \alpha_d) \in [0,1]^d$, with an arbitrarily small constant $\varepsilon > 0$ (earlier higher-dimensional results of this type are due to Schmidt [Sch64]). Note that, by the correspondence of sequences in dimension d and sets in dimension $d+1$ mentioned in Section 1.1, this implies that the discrepancy of the n-point set (2.11) is almost always around $\log^{d-1} n \log \log n$. The use of certain Kronecker sequences for numerical integration was suggested by Richtmyer [Ric51], and they behave quite well in practice [JHK97].

Sets somewhat similar to initial segments of Kronecker sequences are obtained by taking suitable rational numbers for the α_j. One chooses a d-dimensional integer vector $a \in \{0, 1, \ldots, n-1\}^d$ and considers the n-point set

$$\left\{ \left(\left\{ i\frac{a_1}{n} \right\}, \ldots, \left\{ i\frac{a_d}{n} \right\} \right) : i = 0, 1, \ldots, n-1 \right\}. \tag{2.13}$$

For a randomly chosen vector a, the expected discrepancy for corners is $O(\log^d n)$. It is even known that a vector a achieving this low discrepancy can be chosen of the form $(1, g, g^2, \ldots, g^{d-1}) \bmod n$ for a suitable integer g [Kor59], which makes computer search for a good a more feasible. An advantage of sets of this type for numerical integration is that they achieve higher order convergence for sufficiently smooth and periodic functions. The error of approximating the integral of a function f by the average over the point set (2.13) can be simply expressed using the Fourier coefficients of f. Suppose that $f\colon [0,1)^d \to \mathbf{R}$ can be represented by its Fourier series: $f(x) = \sum_{k\in\mathbf{Z}^d} \hat{f}_k e^{2\pi \mathrm{i}\langle k,x\rangle}$. It is easy to derive that

$$\int_{[0,1)^d} f(x)\,\mathrm{d}x - \frac{1}{n}\sum_{j=0}^{n-1} f\left(\left\{\frac{a}{n}j\right\}\right) = -\sum_{k\in\mathbf{Z}^d\setminus\{0\},\,\langle k,a\rangle\equiv 0\,(\mathrm{mod}\,n)} \hat{f}_k.$$

The smoother f (or, rather, its periodic extension to $\mathbf{R}^d$) is, the faster the coefficients $\hat{f}_k$ tend to 0 as $\|k\| \to \infty$, and this is the reason for such point sets being advantageous for smooth functions. The choice of $a \in \mathbf{Z}^d$ should be such that $\langle k,a\rangle$ is not congruent to 0 modulo n for k with small norm, since then the Fourier coefficients of f with small indices are neutralized. This is just a very rough outline of the main idea.

These constructions are known under the name "good lattice points." Their investigation was initiated by Korobov [Kor59]. A quick introduction to the method can be found in [SM94], and the books [SJ94], [FW94], or [Nie92] offer more detailed treatments. Good lattice points are also related to another interesting field, namely to lattice packings—see the book of Conway and Sloane [CS88].

A pleasant introduction to the geometry of lattices is Siegel [Sie89].

Theorems 2.20 and 2.22 are due to Skriganov [Skr94]. For the L_2-discrepancy, a result analogous to Theorem 2.22 (but with specific lattices constructed in a way similar to the one shown in the text) was proved in an earlier work of Frolov [Fro80], and it is also contained in [Skr90]. The latter paper proves the weaker $O(\log^d n)$ worst-case discrepancy bound as well. The main ideas of that proof, which we very briefly outline below, are not too complicated (certainly much simpler than proofs of the $O(\log^{d-1} n)$ bound) but there are numerous technical details. For a fixed axis-parallel box R, let $\varphi(x) = \mathrm{vol}(R) - |(R+x)\cap\Lambda|$ be the discrepancy of R translated by x (here Λ is a lattice with $\mathrm{Nm}(\Lambda) > 0$ and $\det(\Lambda) = 1$). We have $\varphi(x+v) = \varphi(x)$ for any $v \in \Lambda$, and so φ can be regarded as a function on the compact Abelian group $G = \mathbf{R}^d/\Lambda$. In the proof, the Fourier series of φ is considered (see page 214 for the definition of the Fourier transform on an Abelian group), namely $\sum_{u\in\Lambda^*} \hat{\varphi}(u) e^{-2\pi\mathrm{i}\langle u,x\rangle}$, where Λ^* is the lattice dual to Λ (see Exercise 4 for definition) and where $\hat{\varphi}(u) = \int_G \varphi(x) e^{2\pi\mathrm{i}\langle u,x\rangle}\,\mathrm{d}x$. In order to get sufficiently nice Fourier series, the characteristic function of the box R is approximated from below and from above by suitable smooth functions. A key step in the proof, where the order of magnitude of the resulting bound appears, is showing that $\sum_{x\in\Lambda:\,0<\|x\|<\rho} \frac{1}{x_1 x_2 \cdots x_d} = O(\log^d(\rho+2))$. We refer to

[Skr90] for more information. The proofs of the L_2-discrepancy bound in [Skr90] and in [Fro80] are somewhat similar.

A new proof of Theorem 2.20 (on the worst-case discrepancy) was given by Skriganov in [Skr98], via ergodic theory and Fourier analysis. The paper also shows that for almost all lattices, the discrepancy is very close to $O(\log^{d-1} n)$. To state this precisely, we need to define the appropriate measure on lattices. Any lattice Λ with $\det(\Lambda) = 1$ can be written $\Lambda = B\mathbf{Z}^d$ with $B \in SL(\mathbf{R}, d)$. Here $SL(\mathbf{R}, d)$ is the *special linear group* of all real $d \times d$ matrices with determinant 1 and with matrix multiplication as the group operation. Moreover, two such matrices B and B' give the same lattice if and only if $B' = \pm TB$ for some $T \in SL(\mathbf{Z}, d)$ (these are integer matrices with determinant 1; see Exercise 5). Hence a lattice with determinant 1 can be regarded as an element of $\mathcal{L}_d = SL(\mathbf{R}, d)/SL(\mathbf{Z}, d)$. (Note that this is not a group!) This (noncompact) quotient space admits a unique probabilistic invariant measure. Lattices with nonzero norm form a null set in $\mathcal{L}_d$, but as is proved in [Skr98], for almost all $\Lambda \in \mathcal{L}_d$, the discrepancy of the intersection of Λ appropriately scaled with the unit cube is $O(\log^{d-1} n(\log\log n)^{1+\varepsilon})$ for an arbitrarily small constant $\varepsilon > 0$. Similar results hold for the family of translated and scaled copies of any fixed convex polytope (here the discrepancy bound is actually formulated in the "whole-space setting" explained in the remarks to Section 7.1). The behavior of the discrepancy for a lattice $\Lambda \in \mathcal{L}_d$ is shown to depend on the orbit of the dual lattice Λ^* in $\mathcal{L}_d$ under the action of the group of diagonal matrices. This orbit is bounded if and only if $\mathrm{Nm}(\Lambda) > 0$. The faster the orbit recedes to the "infinity" in $\mathcal{L}_d$, the worse the discrepancy becomes. Moreover, for any fixed lattice Λ, almost all rotations of Λ produce sets with discrepancy at most $O(\log^{2d-2} n(\log\log n)^{1+\varepsilon})$. A proof dealing with a similar phenomenon in a different and much simpler setting (only involving the standard planar lattice $\mathbf{Z}^2$) will be shown in Section 3.2. Let us stress that all the results mentioned in this paragraph concern the worst-case discrepancy, and the analogous questions for the L_p-discrepancies are open.

Skriganov [Skr94] also proves that the sets $P_{t,x}$ as in Theorem 2.22 have the asymptotically smallest possible error for integrating smooth functions from a certain class. More precisely, if $f\colon [0,1]^d \to \mathbf{R}$ is a function such that for some $k \geq 1$, the mixed derivative $\frac{\partial^{kd} f(x)}{\partial x_1^k \partial x_2^k \cdots \partial x_d^k}$ exists for all $x \in [0,1]^d$ and has a bounded L_p-norm for some fixed $p < \infty$, then the integration error converges as $O(n^{-k} \log^{(d-1)/2} n)$ for $n = |P_{t,x}| \to \infty$. This compares favorably with the known good lattice points methods, but I am aware of no numerical comparisons.

In higher dimensions, it may be difficult to generate the points of a given lattice lying in the unit cube efficiently, without considering many more lattice points outside of the cube. For example, the problem of finding, for a given d-dimensional lattice Λ, the smallest $a > 0$ such that the cube $[-a, a]^d$ contains a point of Λ distinct from 0 (in other words, computing the shortest vector of Λ in the L_∞-norm) is known to be NP-hard if d is a part of input (Lagarias [Lag85]; Lovász [Lov86] is a gentle introduction to algorithmic problems for lattices). This and some related algorithmic hardness results indicate that the difficulty with finding the points of a given lattice that lie in the unit cube might be essential.

Exercises

1. Show that the set in Example 2.19 equals $\Lambda \cap [0,1)^2$, where Λ is a lattice, i.e. $\Lambda = \{(ib_1 + jb_2)\colon i, j \in \mathbf{Z}\}$ for suitable vectors $b_1, b_2 \in \mathbf{R}^2$.
2. (a) Show that any real number is uniquely determined by its continued fraction.
   (b)* Show that the continued fraction of a quadratic irrationality is eventually periodic.
   (c)* Show that infinite periodic continued fractions determine quadratic irrationalities.
3.* Let α be as in Example 2.19, i.e. with bounded partial quotients of its continued fraction. Prove that the discrepancy for axis-parallel rectangles of the set constructed in that example is $O(\log n)$. Follow the basic scheme of the proof of Proposition 2.2, but replace the canonical intervals by intervals $[k/q_j, (k+1)/q_j)$, where the q_j are denominators of the convergents of α, defined by $q_0 = 1$, $q_1 = a_1$, and $q_n = a_n q_{n-1} + q_{n-2}$. (You may want to begin with the case $\alpha = \frac{1}{2}(\sqrt{5}+1)$.) A detailed proof can be found in several books, such as [Hla84].
4. Let Λ be a lattice in $\mathbf{R}^d$, $\Lambda = B\mathbf{Z}^d$. Show that the following two definitions of a set $\Lambda^* \subset \mathbf{R}^d$ are equivalent. This Λ^* is called the *dual lattice* to Λ.
   (i) $\Lambda^* = \{y \in \mathbf{R}^d\colon \langle x, y\rangle \in \mathbf{Z} \text{ for all } x \in \Lambda\}$.
   (ii) $\Lambda^* = (B^{-1})^T \mathbf{Z}^d$.
5. Let $\Lambda = B\mathbf{Z}^d$ be a lattice in $\mathbf{R}^d$, let T be a $d \times d$ matrix, and put $B' = TB$. Show that $B'\mathbf{Z}^d = \Lambda$ if and only if $T \in SL(\mathbf{Z}, d)$ or $-T \in SL(\mathbf{Z}, d)$, meaning that T is an integer matrix with determinant ± 1.
6. (Norm and the shortest vector) Show that the following are equivalent for a lattice Λ in $\mathbf{R}^d$:
   (i) $\mathrm{Nm}(\Lambda) > 0$.
   (ii) There exists an $\varepsilon > 0$ such that for any $d \times d$ diagonal matrix D with determinant 1, all nonzero vectors of the lattice $D\Lambda$ have length at least ε.
7. (Shortest vector in the dual lattice) Show that if a lattice Λ contains a linearly independent set $\{v_1, \ldots, v_d\}$ of d vectors, each of length at

most r, then all the nonzero vectors of the dual lattice Λ^* have length at least $\frac{1}{r}$.

8.* Prove that for any d and any $\varepsilon > 0$, there is a number $R = R(d, \varepsilon)$ with the following property. If Λ is a d-dimensional lattice in $\mathbf{R}^d$ with determinant 1 such that any nonzero $v \in \Lambda$ has length at least ε, then the ball $B(0, R)$ contains d linearly independent vectors $v_1, \ldots, v_d \in \Lambda$. You may want to use the following version of *Minkowski's First Theorem* from the geometry of numbers (see [Sie89]): If Λ is a lattice in $\mathbf{R}^d$ and $C \subset \mathbf{R}^d$ is a convex body centrally symmetric about 0 and with $\mathrm{vol}(C) > k2^d \det(\Lambda)$, for a natural number k, then $|\Lambda \cap C| \geq k + 1$. This result is usually stated with $k = 1$, but the generalization to an arbitrary k is easy. Let us remark that much more is known about the existence of "short" bases for lattices, and that such problems are studied in the theory of *basis reduction* (see [Lov86], [Sie89]).

9. Using Exercises 6, 7, and 8, show that a lattice Λ in $\mathbf{R}^d$ satisfies $\mathrm{Nm}(\Lambda) > 0$ if and only if $\mathrm{Nm}(\Lambda^*) > 0$. (For a more systematic approach, using Minkowski's Second Theorem, see [Skr94].)

10. (Planar lattices with nonzero norm) Show that a lattice $\Lambda((a, b), (c, d)) \subset \mathbf{R}^2$ has nonzero norm if and only if both the numbers $\frac{a}{b}$ and $\frac{c}{d}$ are *badly approximable*. A real number α is badly approximable if $|\frac{m}{q} - \alpha| > \frac{c}{q^2}$ for all integers $q > 0$ and m, $c > 0$ a constant.

11. (Polynomials suitable for constructing lattices with large norm) In this exercise, let $p \geq 5$ be a prime and let $\omega = e^{2\pi \mathrm{i}/p}$ be a primitive pth root of unity. Although completely elementary proofs are possible, a natural solution requires some basic field theory, such as can be found in Stillwell [Sti94], for instance.
(a)* Write $p = 2d + 1$ and put $\alpha_j = \omega^j + \omega^{-j}$, $j = 1, 2, \ldots, d$. Show that these α_j are real numbers and that the polynomial $q(x) = \prod_{j=1}^{d} (x - \alpha_j)$ is irreducible over the rationals and has integer coefficients.
(b)* More generally let $p = 2md + 1$ for some integer m, and let r be a primitive element modulo p, meaning that the powers $r^0, r^1, \ldots, r^{p-2}$ are all distinct modulo p (in other words, r is a generator of the multiplicative group of the field $\mathbf{Z}/p\mathbf{Z}$). Put $\alpha_j = \sum_{k=0}^{2m-1} \omega^{r^{kd+j}}$, $j = 1, 2, \ldots, d$. For these α_j, show the same claim as in (a).

3. Upper Bounds in the Lebesgue-Measure Setting

In this chapter we start proving upper bounds for the discrepancy for objects other than the axis-parallel boxes. We will encounter a substantially different behavior of the discrepancy function, already mentioned in Section 1.2. For axis-parallel boxes, the discrepancy is at most of a power of $\log n$, and similar results can be shown, for example, for homothets of a fixed convex polygon. The common feature is that the directions of the edges are fixed. It turns out that if we allow arbitrary rotation of the objects, or if we consider objects with a curved boundary, discrepancy grows as some fractional power of n. The simplest such class of objects is the set $\mathcal{H}_2$ of all (closed) halfplanes, for which the discrepancy function $D(n, \mathcal{H}_2)$ is of the order $n^{1/4}$. For halfspaces in higher dimensions, the discrepancy is of the order $n^{1/2-1/2d}$; so the exponent approaches $\frac{1}{2}$ as the dimension grows. Other classes of "reasonable" geometric objects, such as all balls in $\mathbf{R}^d$, all cubes (with rotation allowed), all ellipsoids, etc., exhibit a similar behavior. The discrepancy is again roughly $n^{1/2-1/2d}$, although there are certain subtle differences.

Since upper bounds for discrepancy in the axis-parallel case were obtained by various constructions of algebraic or number-theoretic nature, one might expect that similar sets will also work for balls, say. Interestingly, results in this spirit for other classes of objects are scarce and quite difficult, and they lead to suboptimal bounds. The methods used (so far) for obtaining strong upper bounds are quite different and essentially combinatorial. Combinatorial methods, in particular various semi-random constructions, currently yield the best known asymptotic upper bounds in most cases. The axis-parallel boxes in $\mathbf{R}^d$ are a (weak) exception, since there the combinatorial methods only provide an $O(\log^{d+1/2} n)$ bound while the best known explicit constructions give $O(\log^{d-1} n)$. The class of the convex sets in the plane can be regarded as another significant exception; for them, the best known construction combines number-theoretic and combinatorial methods (see Section 3.1).

In the first section of this chapter, we present an upper bound on the discrepancy for discs in the plane. The construction is random and the proof is a probabilistic argument, but we still remain in the "Lebesgue-measure" setting. Later on, mainly in Chapter 5, we will see purely combinatorial generalizations of this approach.

The methods in Section 3.2 are quite different. We consider the Lebesgue-measure L_1-discrepancy for halfplanes, and show that it is only $O(\log^2 n)$. This contrasts with the worst-case discrepancy, or even the L_2-discrepancy, for halfplanes, which are of the order $n^{1/4}$ (see Chapters 5 and 6). The set with this low L_1-discrepancy is the regular $\sqrt{n} \times \sqrt{n}$ grid. The proof is of a number-theoretic nature and it employs a suitable Fourier series expansion. The method is specific for the Lebesgue-measure setting, and it is quite possible that this result has no combinatorial analogue.

3.1 Circular Discs: a Probabilistic Construction

Here we exhibit the existence of an n-point set in the plane whose discrepancy for the family $\mathcal{B}_2$ of all circular discs is $O(n^{1/4}\sqrt{\log n})$. As we will see in Chapter 6, there is an $\Omega(n^{1/4})$ lower bound for this discrepancy, and so the upper bound is nearly the best possible. We consider discs for simplicity only; it will be apparent that a similar argument works for upper-bounding the discrepancy for other "reasonable" shapes as well. Moreover, a relatively straightforward generalization into higher dimensions gives a discrepancy bound of $O(n^{1/2-1/2d}\sqrt{\log n})$ for any fixed d for balls in $\mathbf{R}^d$ and for many other families.

3.1 Theorem. $D(n, \mathcal{B}_2) = O(n^{1/4}\sqrt{\log n})$.

We begin with a simple lemma, whose proof is left as Exercise 1.

3.2 Lemma. *Consider a $k \times k$ grid of squares (like a $k \times k$ chessboard). Any circle intersects the interiors of at most $4k$ squares of the grid.* □

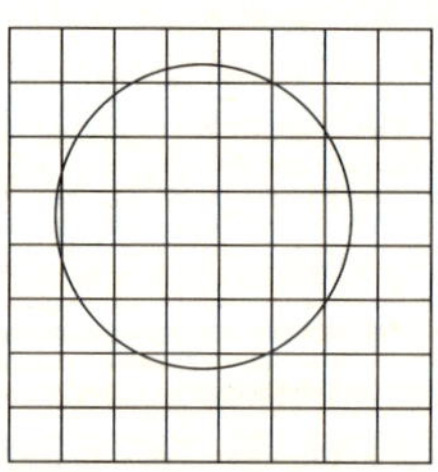

Proof of Theorem 3.1. As the first step, we are going to approximate the continuous Lebesgue measure in $[0,1]^2$ by a measure concentrated on a large but finite point set Q. A suitable choice for Q is a sufficiently fine grid, and for definiteness we can take the $n \times n$ grid:

$$Q = \left(\left(\tfrac{1}{2n}, \tfrac{1}{2n}\right) + \tfrac{1}{n}\mathbf{Z}^2\right) \cap [0,1]^2.$$

In Fig. 3.1, we have $n = 16$ and Q is the fine 16×16 grid.

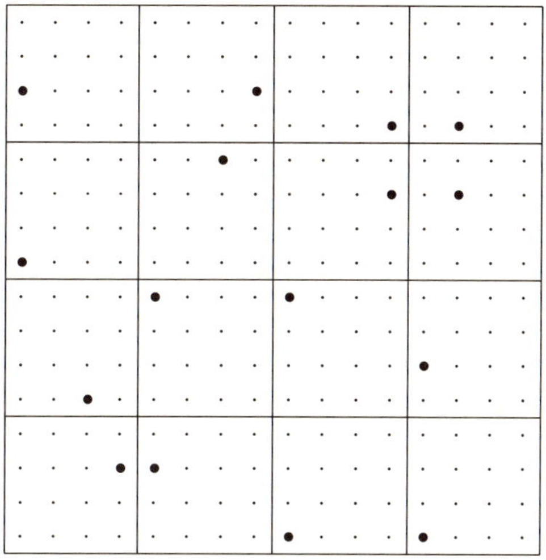

Fig. 3.1. A random construction of a low-discrepancy set for discs, with $n = 16$.

First we observe that Q has discrepancy at most $O(n)$ for discs, i.e. for any disc $B \in \mathcal{B}_2$, we have

$$|D(Q,B)| = \left| n^2 \cdot \mathrm{vol}_\square(B) - |Q \cap B| \right| = O(n), \tag{3.1}$$

which means, in the terminology of Section 1.3, that Q is a $\frac{1}{n}$-approximation for discs with respect to the measure $\mathrm{vol}_\square$. This is not a particularly good achievement for an n^2-point set, but it is good enough as a starting point for the subsequent proof.

To see that (3.1) holds, draw a little square of side $\frac{1}{n}$ centered at q for each point $q \in Q$, and make q "responsible" for the area of its square. Given a disc $B \in \mathcal{B}_2$, the points q whose little squares fall completely outside B or completely inside B fulfill their duty perfectly, contributing 0 to the difference $n^2 \cdot \mathrm{vol}_\square(B) - |Q \cap B|$, and the discrepancy can only be caused by the points whose little squares intersect the boundary of B. The number of such little squares is $O(n)$ by Lemma 3.2, and so $|D(Q,B)| = O(n)$ as claimed.

We are now going to select a random n-point subset P of Q in a suitable way and show that its discrepancy will typically be small. The finiteness of Q allows us to carry out a simple probabilistic argument. Suppose that n is a perfect square (this is no loss of generality; see Exercise 2). We divide the unit square into a $\sqrt{n} \times \sqrt{n}$ grid $\mathcal{G}$ of squares, as in Fig. 3.1. Each square G in this grid contains exactly n points of Q. Let us put $Q_G = Q \cap G$. For each $G \in \mathcal{G}$, we are going to pick one point, $q_G \in Q_G$, to "represent" the n points in Q_G. This q_G is chosen from Q_G uniformly at random, the choices being independent for distinct squares $G \in \mathcal{G}$ (a typical result of such random choice is illustrated in Fig. 3.1). Taking the points q_G for all $G \in \mathcal{G}$ yields a (random) n-point set P.

Let us put $\Delta = Cn^{1/4}\sqrt{\log n}$, where C is a sufficiently large constant (to be fixed later). We prove that with a positive probability we have, for all discs $B \in \mathcal{B}_2$ simultaneously,

$$\left|\frac{1}{n}|Q \cap B| - |P \cap B|\right| \leq \Delta. \tag{3.2}$$

An n-point set P satisfying this condition is a (Δ/n)-approximation for the set system induced by discs on Q, and by Observation 1.7 (iterated approximation), P is also a $(\frac{\Delta}{n} + \frac{1}{n})$-approximation for discs with respect to the measure $\mathrm{vol}_\square$. Hence $D(P, \mathcal{B}_2) = O(\Delta + 1)$, and so establishing (3.2) for all B is enough to prove Theorem 3.1.

Let us first consider the disc B arbitrary but fixed, and let us bound the probability that (3.2) is violated for this particular B if P is chosen at random in the above-described way. To this end, we consider the $\sqrt{n} \times \sqrt{n}$ grid $\mathcal{G}$ again. We recall that for each square $G \in \mathcal{G}$ we have one point $q_G \in P$ to represent the n points of Q_G in G; we can think of the points of Q_G as having weight $\frac{1}{n}$ each, while q_G has weight 1. If $G \subseteq B$ or $B \cap G = \emptyset$, the point q_G certainly represents its n points of Q_G perfectly, so we only need to worry about the set $\mathcal{G}_B \subseteq \mathcal{G}$ of squares intersecting the boundary of B, as in the following picture:

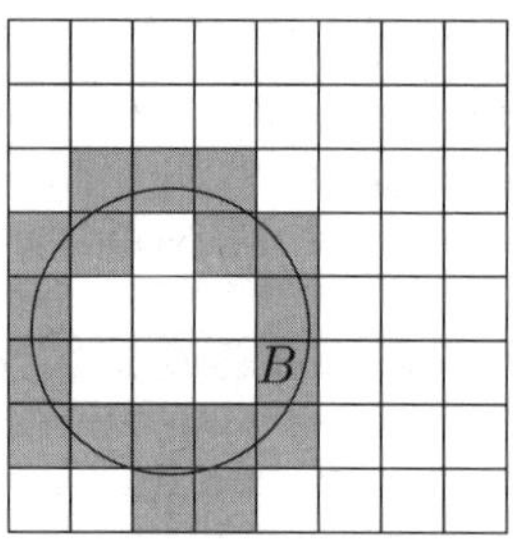

For $G \in \mathcal{G}_B$, let $k_G = |Q_G \cap B|$ be the number of points of Q_G falling into B. The total contribution of these points to the quantity $\frac{1}{n}|Q \cap B|$ is $\frac{k_G}{n}$. On the other hand, the point q_G contributes either 1 or 0 to $|P \cap B|$. Let X_G be the deviation of the contribution of q_G from the contribution of Q_G (i.e. by how much q_G deviates from representing its points perfectly):

$$X_G = \begin{cases} -\frac{k_G}{n} & \text{if } q_G \notin B \\ 1 - \frac{k_G}{n} & \text{if } q_G \in B. \end{cases}$$

Note that $X = \sum_{G \in \mathcal{G}_B} X_G = \frac{1}{n}|Q \cap B| - |P \cap B|$ is the left-hand side of (3.2), and so we need to show that $|X| \leq \Delta$ with high probability.

Since q_G was chosen from Q_G uniformly at random, the event $q_G \in B$ has probability $p_G = \frac{k_G}{n}$. Hence X_G is a random variable attaining value

$1-p_G$ with probability p_G and value $-p_G$ with probability $1-p_G$. We are in a situation fairly typical in combinatorial applications of probability. We have a random variable X that is the sum of $m=|\mathcal{G}_B|$ independent random variables, and Lemma 3.2 tells us that $m=O(\sqrt{n})$. The expectation of each X_G, and hence also of X, is 0, and we need to estimate the probability of X deviating from its expectation by more than Δ.

Here is a rough picture of the situation. Being a sum of independent random variables, X has an approximately normal distribution with mean 0. The variance of each X_G is clearly at most 1, so the variance of X is no more than m and the standard deviation of X is at most $\sqrt{m} \doteq O(n^{1/4})$ (this is where the mysterious exponent $\frac{1}{4}$ first appears). And for a normally distributed random variable X with zero mean, the probability of $|X|$ exceeding the standard deviation λ-times behaves roughly like $\exp(-\lambda^2/2)$, so we may hope that in our situation, with $\lambda = C\sqrt{\log n}$ and C large, $\Pr[|X|>\Delta]$ is bounded by something like n^{-c} for a large constant c. To get a rigorous bound on $\Pr[|X|>\Delta]$, we use an inequality of a Chernoff type, namely Theorem A.4 in Alon and Spencer [AS92]. This inequality says that if X is a sum of m independent random variables X_i, where X_i takes value $-p_i$ with probability $1-p_i$ and value $1-p_i$ with probability p_i (for some $p_i \in [0,1]$), then for any $\Delta>0$, we have $\Pr[|X|>\Delta] < 2\exp(-2\Delta^2/m)$.

For our carefully chosen value $\Delta = Cn^{1/4}\sqrt{\log n}$, this yields that for any fixed disc B, the probability that $\left|\frac{1}{n}|Q\cap B| - |P\cap B|\right| > \Delta$ is at most n^{-c}, where the constant c can be made as large as we wish by choosing C suffciently large. It remains to observe that although there are infinitely many discs, we only need to consider polynomially many (in n) of them. Namely, we have

3.3 Lemma. *Let Q be an arbitrary m-point set in the plane. Call two discs B and B' equivalent if $B\cap Q = B'\cap Q$. Then there are at most $O(m^3)$ equivalence classes. In other words, at most $O(m^3)$ distinct subsets of Q can be "cut off" by a circular disc.*

The proof of this lemma is simple but it deserves some attention, so we postpone it a little and we finish the proof of Theorem 3.1. If we have two discs B and B' with $B\cap Q = B'\cap Q$, then the required discrepancy estimate (3.2), i.e. $\left|\frac{1}{n}|Q\cap B| - |P\cap B|\right| \le \Delta$, either holds for both or for none of them, and so B and B' are equivalent for our purposes. By Lemma 3.3, there are at most $O(|Q|^3) = O(n^6)$ equivalence classes, and if $\mathcal{F}$ denotes a family of discs containing one representative from each equivalence class, we have

$$\Pr[(3.2) \text{ fails for some } B] = \Pr[(3.2) \text{ fails for some } B\in\mathcal{F}] \le$$

$$\sum_{B\in\mathcal{F}} \Pr[(3.2) \text{ fails for } B] = O(n^6)\cdot n^{-c} < 1$$

is c is large enough. Theorem 3.1 is proved, up to the proof of Lemma 3.3. □

Proof of Lemma 3.3. Somewhat informally, one can argue that any given disc can be moved and expanded or shrunk, while still defining the same subset of Q, until it has 3 points of Q on the boundary or 2 points determining a diameter. Since there are $O(m^3)$ triples and pairs of points of Q, there are no more than $O(m^3)$ equivalence classes. But this argument is somewhat dubious; for instance, we may sometimes have a point of Q "just hitting" the boundary from outside but we don't want it to lie in the considered disc.

To give a more rigorous proof, we first observe that Q can be assumed to be in general position (having no 3 collinear and no 4 cocircular points). Namely, if a (closed) disc B defines certain subset $S = B \cap Q$, we can replace B by a concentric disc B' with a little larger radius such that $B' \cap Q = S$. Consequently, if we perturb the points of Q by a sufficiently small amount, each subset that could have been cut off by a disc before the perturbation can still be cut off by some disc. (This perturbation argument is not strictly necessary here, since the proof can be done for a not necessarily general position as well, but assuming general position is convenient and it is useful to have this idea at hand.)

Next, it is helpful visualize the situation in the "space of discs." Each disc $B \in \mathcal{B}_2$ can be represented by the triple (x, y, r), where x and y are the coordinates of its center and r is its radius. For a point $p = (p_1, p_2) \in Q$, the set of all discs containing p is the cone $\{(x, y, r)\colon (x - p_1)^2 + (y - p_2)^2 \leq r^2\}$. The m points of Q define m (congruent) cones in the (x, y, r)-space with apexes in the $r = 0$ plane, and we want to bound the maximum possible number of regions defined by these cones. Note that it suffices to count the 3-dimensional connected regions arising by removing the surfaces of the m cones from the (x, y, r)-space (because for every subset $S \subseteq Q$ defined by a circular disc, the set of all discs B with $B \cap Q = S$ contains a small open ball, as is easy to check). The picture below illustrates the subdivision of space by 5 such conic surfaces:

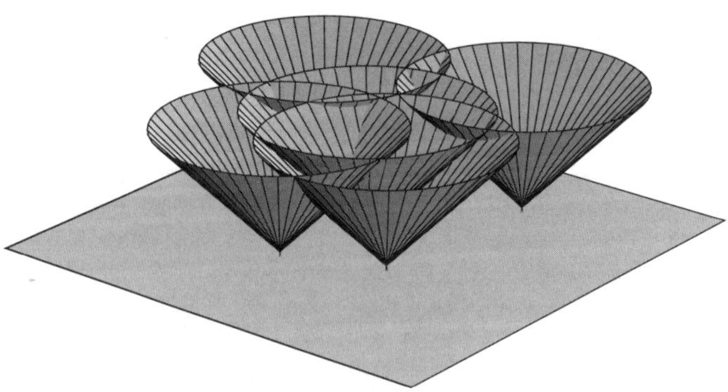

(of course, the cones should extend up to infinity, but then the reader wouldn't be able to admire the regions inside the cones). The general position assumption on the point set implies that no 4 of these conic surfaces have a point in common.

One way of counting the regions is using their lowest points. The closure of each region, except for the one outside of all the cones, has a unique lowest point. Each such lowest point is defined geometrically by at most 3 cones (it may be the apex of a cone, or the intersection of the surfaces of 3 cones, or the lowest point in the intersection of 2 cones), and one lowest point is adjacent to at most a constant-bounded number of regions (by the general position assumption). Finally, each subset of at most 3 cones defines a constant-bounded number of candidates for a lowest point, and so it follows that the total number of regions is $O(m^3)$. This type of argument could be phrased in terms of discs and points, but it would become much less intuitive.

Another, more geometric proof proceeds by induction on m, showing that by deleting one cone out of m, at most $O(m^2)$ pairs of regions are merged. This is because the intersections with the surfaces of the $m-1$ remaining cones subdivide the surface of the deleted cone into $O(m^2)$ regions. We omit further details here. We will return to similar considerations in more general context in Chapter 5. □

The family $\mathcal{F}$ constructed in the proof approximates the family of all discs "combinatorially" with respect to the finite point set Q. For any disc, there is a disc in $\mathcal{F}$ having the same intersection with Q. Alternatively, we could also use a family $\mathcal{F}_1$ of discs approximating the family of all discs "geometrically," in the sense of measure. This means that each disc B can approximated by a disc $B_{in} \in \mathcal{F}_1$ from inside and by a disc $B_{out} \in \mathcal{F}_1$ from outside, in such a way that the area of $B_{out} \setminus B_{in}$ is sufficiently small. Then, instead of sampling the points of P from the auxiliary finite set Q as in the proof above, we would sample each point q_G directly from the continuous uniform distribution in the little square G. The reader is invited to try this approach in Exercise 5. Both this method with continuous sampling and the one shown above have their advantages and possibilities of generalization, and none can be said to supersede the other.

Remark on Convex Sets. This is perhaps a suitable place to mention that the discrepancy has also been studied for the class of all convex sets in $[0,1]^d$. Here the behavior of the discrepancy function is considerably different from the case of balls, or rotated boxes or, more generally, geometric objects described by a bounded number of real parameters. The discrepancy for convex sets has order of magnitude roughly $n^{1-2/(d+1)}$ (see remarks below and Exercise 6).

Bibliography and Remarks. Historically, the correct bounds for discrepancy with rotation allowed were first approached by lower bounds, which will be discussed in Chapters 6 and 7. The first near-tight upper bounds were obtained by Beck (published in [BC87] and [Bec87]) by an argument based on continuous sampling from the grid squares and a geometric approximation argument, as was explained in the text following the proof.

Incidentally, random sampling similar to the one employed in Beck's proof is used in computer graphics in certain algorithms for improving image quality (*jittered sampling*).

In Chapter 5, we will consider a combinatorial generalization of the proof method shown in this section. For example, we will be able to deal with discrepancy with respect to arbitrary measures. But there are generalizations of Beck's original method (with geometric approximation arguments) which are not captured by the combinatorial approach. For instance, let C be a convex body in $\mathbf{R}^d$ that is not too small, meaning that it contains a ball of radius $n^{-1/d}$. Beck proved [Bec87], [BC87] that the discrepancy for the family of all translated, rotated, and scaled-down copies of C is bounded above by $O(n^{1/2-1/2d}\sqrt{S\log n})$, where S is the surface area of the set C. This result has no direct combinatorial counterpart—see Exercise 5.2.2.

We have derived an upper bound on the worst-case discrepancy (for discs, but the discussion again applies for various other shapes as well). This implies upper bounds for the L_p-discrepancy for all p, and for $p \geq 2$, this bound is nearly tight, i.e. $n^{1/4}$ is roughly the correct order of magnitude in the plane. In fact, one can even prove that for any fixed $p < \infty$, the L_p-discrepancy is bounded by $O(n^{1/4})$, without any logarithmic factors (which is tight for $p \geq 2$). For $p = 2$, this has been proved by Beck and Chen [BC90] by a method similar to Roth's one (Section 2.2), and it can also be established without much trouble using the construction presented in the current section. Exercise 5.4.3 shows a simple result on L_p-discrepancy in a combinatorial setting.

There are some explicit constructions of low-discrepancy sets in settings other than for the axis-parallel boxes. Perhaps most notably, Lubotzky et al. [LPS86], [LPS87] construct an n-point set on the two-dimensional sphere S^2 with discrepancy $O(n^{2/3})$ for spherical caps (intersections of S^2 with halfspaces) using very beautiful and advanced mathematics. Note that the asymptotic bound remains far behind the straightforward $O(\sqrt{n\log n})$ bound for a random point set, not speaking of the near-tight $O(n^{1/4}\sqrt{\log n})$ bound which can be obtained by the method of the present section. But an explicit construction has some advantages and certainly it provides new insights. Also, the construction is near-optimal in certain sense for numerical integration of functions on S^2.

The construction can be described quite concisely. Let ρ_x, ρ_y, and ρ_z denote the rotation in $\mathbf{R}^3$ by the angle $\arccos(-\frac{3}{5})$ around the x-axis, y-axis, and z-axis, respectively. By a k-step rotation, we mean any composition $\tau = \tau_1 \circ \tau_2 \circ \cdots \circ \tau_k$, where each τ_i is one of ρ_x, ρ_x^{-1}, ρ_y, ρ_y^{-1}, ρ_z, and ρ_z^{-1}, and where $\tau_i \neq \tau_{i+1}^{-1}$. Fix a starting point $p \in S^2$ lying on none of the coordinate axes, and form the set $\{\tau(p)\colon \tau \text{ is a } k\text{-step rotation}\}$. It can be proved that this set has ex-

Fig. 3.2. Lubotzky–Phillips–Sarnak sets in S^2 for $k = 2$ and 3.

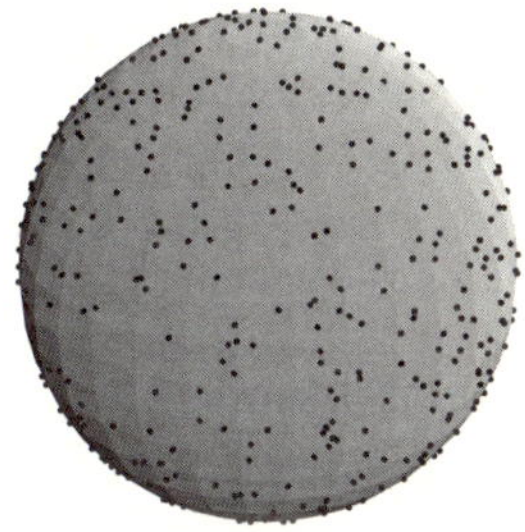

Fig. 3.3. A Lubotzky–Phillips–Sarnak set in S^2 for $k = 4$.

actly $n = \frac{3}{2}(5^k - 1)$ points and discrepancy $O(n^{2/3})$ for spherical caps. Figs. 3.2 and 3.3 show such sets for a randomly chosen initial point p. (Of course, the construction is just the simplest one from an infinite family; for each prime q congruent to 1 modulo 4, there is a construction involving $q + 1$ generator rotations, which correspond to ways of writing q as a sum of 4 squares of integer with the first addend being positive and odd.) In the proof, the operator $T: L_2(S^2) \to L_2(S^2)$ given by $Tf(x) = \sum_\tau f(\tau x)$ is analyzed (where τ in the summation runs through the six 1-step rotations). From Ramanujan's conjecture concerning the modular group, established by Deligne in the 1970s, Lubotzky et al. prove that the second largest eigenvalue of T is bounded away from the largest one (which is 6), and they derive the discrepancy bound and the quadrature error bound from this. The underlying mathematics resembles constructions of explicit expander graphs (by the same authors). The proof is also presented in Chazelle [Cha98b].

The indicated proof of Lemma 3.3 via the geometric approach is a small sample of investigating the number of certain geometric configurations using arrangements of suitable surfaces. An extensive background and recent results in this area can be found in Sharir and Agarwal [SA95].

The discrepancy for convex sets was introduced by Zaremba [Zar70] under the name *isotropic discrepancy.* Schmidt [Sch75] proved a lower bound of $\Omega(n^{1-2/(d+1)})$ for it by a method indicated in Exercise 6. Stute [Stu77] showed that for any fixed dimension $d \geq 3$, if n points are drawn from the uniform distribution in the unit cube independently at random, then the discrepancy matches this lower bound up to a small logarithmic factor with high probability. In contrast, in the plane a random point set cannot work, because the correct bound is only of the order $n^{1/3}$, while a random set gives discrepancy at least of the order $\sqrt{n}$. Beck [Bec88c] showed by an ingenious semi-random construction that the discrepancy for convex sets in the plane is $O(n^{1/3}\log^4 n)$, and so Schmidt's lower bound is nearly sharp in the plane as well.

The discrepancy of any n-point set $P \subset [0,1]^d$ for convex sets can be bounded in terms of the discrepancy of P for rectangles: it is at most $O(n^{1-1/d} D(P, \mathcal{R}_d)^{1/d})$, with the constant of proportionality depending on d. This was proved by Hlawka and further generalized by several researchers (see [DT97] or [KN74] for proofs and references, and Laczkovich [Lac95] for a recent result in this direction). Note, though, that even for the best sets for axis-parallel boxes, the resulting bound for convex sets is quite weak, and, on the other hand, random sets achieving almost tight bounds for convex sets behave poorly for axis-parallel boxes.

Exercises

1. Prove Lemma 3.2.
2. Show that if we know $D(n, \mathcal{A}) \leq f(n)$ for all n of the form 4^k (the number 4 is not important here, we could take any integer constant), where $\mathcal{A}$ is a class of sets and f is a nondecreasing function, we have $D(n, \mathcal{A}) = O(f(n) + f(\lfloor n/4 \rfloor) + f(\lfloor n/4^2 \rfloor) + \cdots)$ for all n.
3. Generalize the proof of Theorem 3.1 to dimension d, showing that the discrepancy for balls in $\mathbf{R}^d$ is bounded by $O(n^{1/2-1/2d}\sqrt{\log n})$. (For an analogoue of Lemma 3.3, you may use results of Section 5.1.)
4. The boundary of any convex set in the plane intersects at most $O(\sqrt{n})$ of the little squares used in the proof, which is the same bound as that for a circle, and yet the discrepancy for convex sets has a larger order

of magnitude than that for discs. Where does the upper bound proof for discs fail for convex sets?

5. (Proof of Theorem 3.1 without the discretization step)
(a) Let n be a natural number. Show that there exists a collection $\mathcal{F}_1$ of discs in the plane, with $|\mathcal{F}_1|$ bounded by a fixed polynomial function of n, such that for any disc $B \in \mathcal{B}_2$ there are $B_{in}, B_{out} \in \mathcal{F}_1 \cup \{\emptyset\}$ such that $B_{in} \cap [0,1]^2 \subseteq B \cap [0,1]^2 \subseteq B_{out} \cap [0,1]^2$ and $\mathrm{vol}_\square(B_{out} \setminus B_{in}) \leq \frac{1}{n}$.
(b)* Use (a) to prove Theorem 3.1 without introducing the auxiliary set Q; let each point q_G be sampled from the continuous uniform distribution in the grid square G.
6. (a)** Show that the discrepancy of any n-point set P in $[0,1]^2$ for the class of all convex sets is $\Omega(n^{1/3})$. To find a bad convex set, use a circular disc C inscribed into the unit square with suitable disjoint caps sliced off (depending on the point set).
(b)* Generalize (a) to any fixed dimension d, heading for the bound $\Omega(n^{1-2/(d+1)})$.

3.2 A Surprise for the L_1-Discrepancy for Halfplanes

Here we present a result showing that the L_1-discrepancy can behave in a manner completely different from the worst-case discrepancy or even the L_2-discrepancy. The results concerns the discrepancy for halfplanes. In order to speak about L_1-discrepancy or L_2-discrepancy for halfplanes, we have to introduce some probability measure on the halfplanes intersecting the unit square. There is a very natural measure on halfplanes which will be mentioned at the end of this section and discussed in more detail in Section 6.4. But here we use a slightly different (and a little unnatural) measure, which allows us to simplify the calculations significantly and to concentrate on the basic ideas.

Let us assume that n, the number of points, is of the form $n = m^2$ for an even integer m. Let P be the n-point grid set $\{(\frac{j}{m}, \frac{k}{m})\colon j,k = 0,1,\ldots,m-1\}$, and let S be the unit square shifted so that P lies symmetrically within it, namely $S = [0,1]^2 - (\frac{1}{2m}, \frac{1}{2m})$, as in Fig. 3.4. For the L_1-discrepancy, we consider only the lower halfplanes $h_{a,b} = \{y \leq ax + b\}$ with $0 \leq a \leq \frac{1}{3}$ and $0 \leq b \leq \frac{1}{2}$. Note that the bounding lines of all these halfplanes intersect both the vertical sides of the square S. The measure ν_0 on this set of halfplanes is given by the Lebesgue measure on the rectangle $\{(a,b)\colon 0 \leq a \leq \frac{1}{3}, 0 \leq b \leq \frac{1}{2}\}$ scaled by the factor 6 (so that we obtain a probability measure), and all other halfplanes receive measure 0. In this way, of course, we have omitted lots of interesting halfplanes from our considerations, and the reader may wonder if this omission is not critical for the L_1-discrepancy estimate. Well, it is not

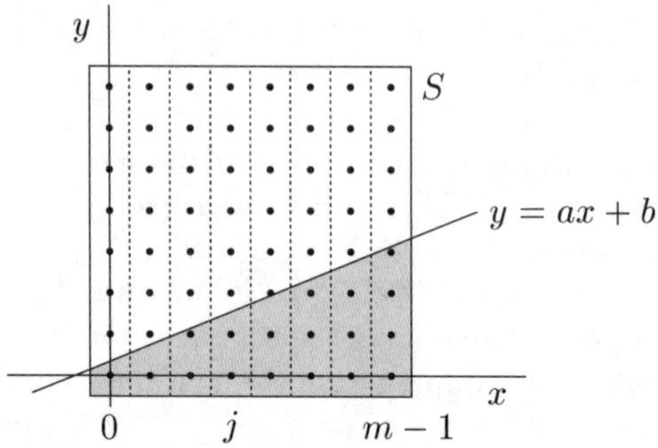

Fig. 3.4. The situation in Proposition 3.4.

(see Theorem 3.5 below) but it saves us some work in the proof. The result we are going to prove is

3.4 Proposition. *The L_1-discrepancy of the set P for halfplanes with the measure ν_0 defined above is at most $O(\log^2 n)$. Explicitly, this means that*

$$\int_0^{1/3}\int_0^{1/2} |D(P,h_{a,b})|\,\mathrm{d}b\,\mathrm{d}a = O(\log^2 n),$$

where[1] *$h_{a,b}$ denotes the halfplane $\{(x,y)\in\mathbf{R}^2\colon y\le ax+b\}$.*

Why should this result be interesting? Note that the set P in the proposition is fairly bad concerning the worst-case discrepancy, which is about $\sqrt{n}$. One can also calculate that the L_2-discrepancy of P is of the order $n^{1/4}$ (see Exercise 1), and using the methods of Section 6.6, it can even be shown that the L_2-discrepancy of *any* n-point set with respect to the measure ν_0 is at least of the order $n^{1/4}$. So there is a substantial difference between the L_1-discrepancy and L_2-discrepancy.

Proof of Proposition 3.4. Let us introduce the shorthand $D(a,b) = D(P,h_{a,b})$. To express $D(a,b)$ explicitly, divide the unit square into vertical strips of width $\frac{1}{m}$ (drawn by dashed lines in Fig. 3.4) and number these strips $0,1,\ldots,m-1$ from left to right. The middle line of the strip number j has x-coordinate $\frac{j}{m}$, and the line $\{y=ax+b\}$ intersects it at the point with y-coordinate $a\frac{j}{m}+b$. From this we get

$$\begin{aligned} D(a,b) &= m^2\,\mathrm{vol}(h_{a,b}\cap S) - |P\cap h_{a,b}| \\ &= \sum_{j=0}^{m-1}\left(aj+mb+\frac{1}{2}-\lceil aj+mb\rceil\right) \end{aligned}$$

[1] Here and in the proof of this proposition, the discrepancy D is meant with respect to the Lebesgue measure on the shifted unit square S, and not on the square $[0,1]^2$ as usual. The shifted coordinate system is more convenient for the subsequent calculation.

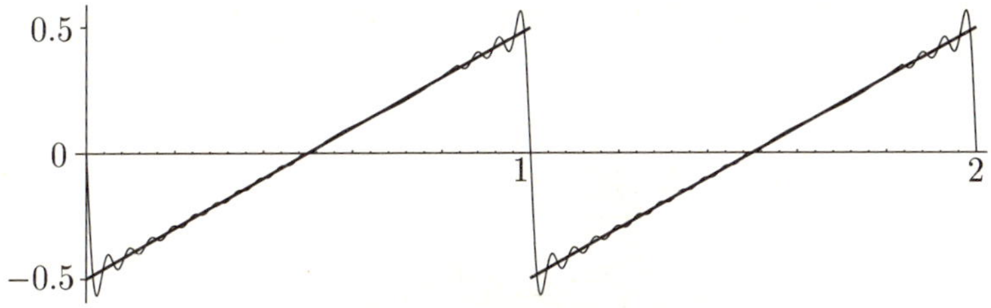

Fig. 3.5. The sawtooth function $s(x)$ and its approximation by the terms of the Fourier series for $|k| \leq 20$.

$$= \sum_{j=0}^{m-1} s(aj + mb),$$

where $s(x) = x - \lceil x \rceil + \frac{1}{2}$ is the sawtooth function (as in Section 2.2).

As a function of b, $D(a,b)$ is obviously periodic with period $\frac{1}{m}$. Introducing the substitution $\beta = mb$ and putting $g(a,\beta) = D(a, \frac{\beta}{m})$, the integral from the proposition is rewritten as

$$\int_0^{1/3}\int_0^{1/2} |D(a,b)| \,\mathrm{d}b\,\mathrm{d}a = \int_0^{1/3} \left(\frac{m}{2} \int_0^{1/m} |D(a,b)| \,\mathrm{d}b \right) \mathrm{d}a$$
$$= \frac{1}{2} \int_0^{1/3}\int_0^{1} |g(a,\beta)| \,\mathrm{d}\beta\,\mathrm{d}a.$$

The behavior of the function $g(a,\beta)$ looks fairly erratic. Fig. 3.6 shows the whole graph for $m = 10$, and Fig. 3.7 the dependence on a for a fixed value of β with $m = 30$. With these pictures in mind, one can perhaps better appreciate the subsequent application of the Fourier series.

We recall that any "reasonable" periodic function $f\colon \mathbf{R} \to \mathbf{R}$ with period 1 can be expressed by the Fourier series:

$$f(x) = \sum_{k \in \mathbf{Z}} c_k e^{2\pi i k x}.$$

Here "reasonable" may mean, for instance, piecewise continuous with a piecewise continuous derivative (this is fully sufficient for our purposes although much more refined sufficient conditions are known). For such an f, the Fourier series converges for all $x \in [0,1)$ and its sum equals $f(x)$ at all points of continuity of f (of course, if f is discontinuous, the convergence cannot be uniform; see Fig. 3.5). The (complex) coefficients c_k can be calculated as $c_k = \int_0^1 f(x) e^{-2\pi i k x}\,\mathrm{d}x$. Crucially, we will make use of the *Parseval equality*:

$$\int_0^1 f(x)^2 \,\mathrm{d}x = \sum_{k \in \mathbf{Z}} |c_k|^2.$$

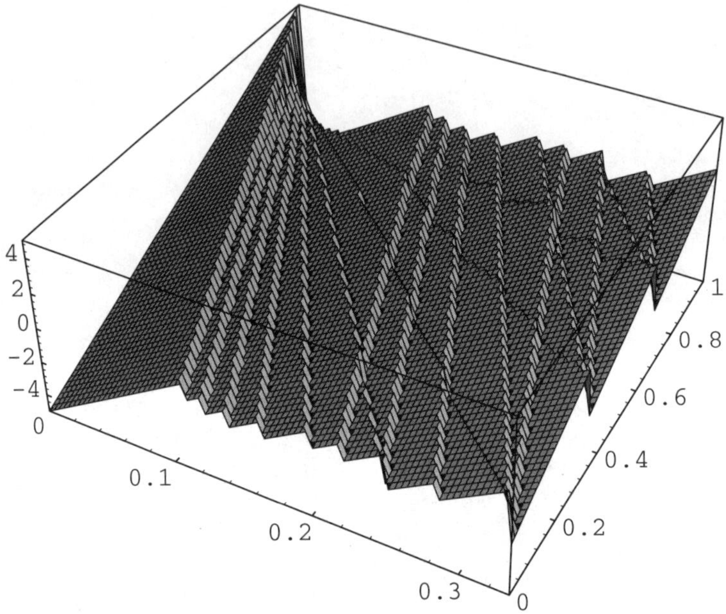

Fig. 3.6. The function $g(a, \beta)$ for $a \in [0, \frac{1}{3}]$ and $\beta \in [0, 1]$, with $m = 10$.

In our case, we need to expand $g(a, \beta)$ as a function of the variable β. By some integration per partes or by looking in a handbook of mathematical functions, we first find the Fourier series for the sawtooth function:[2]

$$s(x) = \sum_{k \in \mathbf{Z} \setminus \{0\}} -\frac{e^{2\pi i k x}}{2\pi i k}.$$

And from the expression $g(a, \beta) = \sum_{j=0}^{m-1} s(aj + \beta)$ we find the expansion $g(a, \beta) = \sum_{k \in \mathbf{Z}} c_k e^{2\pi i k \beta}$, where $c_0 = 0$ and

$$c_k = -\frac{1}{2\pi i k} \sum_{j=0}^{m-1} e^{2\pi i k a j} = -\frac{1}{2\pi i k} \cdot \frac{e^{2\pi i k a m} - 1}{e^{2\pi i k a} - 1}$$

[2] The Parseval equality for the sawtooth function happens to give a short and sweet proof of the well-known equality $\zeta(2) = \sum_{k=1}^{\infty} k^{-2} = \pi^2/6$, as the reader is invited to check.

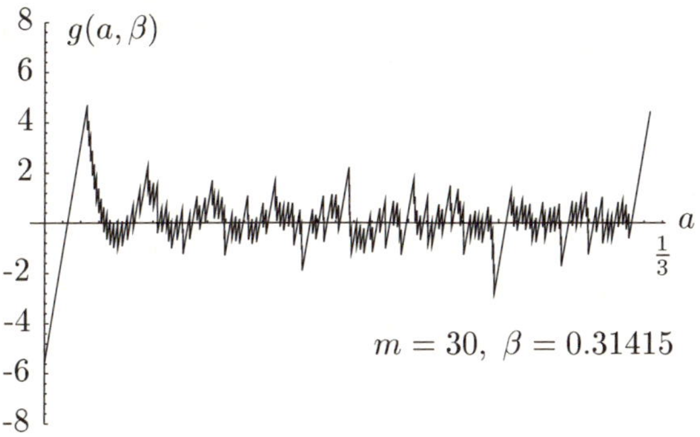

Fig. 3.7. The function $g(a, \beta)$: dependence on a with β fixed.

(note that the c_k are actually functions of a). From the first expression for c_k above, we see that $|c_k| \leq \frac{m}{2\pi k}$. On the other hand, a little calculation reveals that for all $x \in \mathbf{R}$, $\left|e^{2\pi \mathrm{i}x} - 1\right| = 2|\sin \pi x| \geq \delta(x)$, where $\delta(x)$ denotes the distance of x to the nearest integer. From the second expression for c_k, we thus obtain

$$|c_k| \leq \frac{1}{2\pi k} \cdot \frac{2}{|e^{2\pi \mathrm{i} ka} - 1|} = O\left(\frac{1}{k \cdot \delta(ka)}\right).$$

Now we estimate, using the inequality between L_1-norm and L_2-norm and the Parseval equality for the function $g(a, .)$,

$$\int_0^1 |g(a, \beta)| \,\mathrm{d}\beta \leq \left(\int_0^1 g(a, \beta)^2 \,\mathrm{d}\beta\right)^{1/2} = \left(\sum_{k \in \mathbf{Z}} |c_k|^2\right)^{1/2}.$$

From the estimate $|c_k| \leq \frac{m}{2\pi k}$, we get $\sum_{|k|>m^2} |c_k|^2 = O(1)$. Writing n instead of m^2, we have

$$\left(\sum_{k \in \mathbf{Z}} |c_k|^2\right)^{1/2} \leq \left(\sum_{|k| \leq n} |c_k|^2 + O(1)\right)^{1/2} \leq \sum_{|k| \leq n} |c_k| + O(1)$$

Now it is time to integrate over a. This gives

$$\begin{aligned} \int_0^{1/3} \int_0^1 |g(a, \beta)| \,\mathrm{d}\beta \,\mathrm{d}a &\leq O(1) + \int_0^{1/3} \sum_{|k| \leq n} |c_k| \,\mathrm{d}a \\ &\leq O(1) + O(1) \sum_{k=1}^{n} \frac{1}{k} \int_0^1 \min\left(m, \frac{1}{\delta(ka)}\right) \mathrm{d}a. \end{aligned} \quad (3.3)$$

By the substitution $ka = x$ and then by periodicity of the function δ, we calculate that the integral over a in (3.3) equals

$$\int_0^k \min\left(m, \frac{1}{\delta(x)}\right) \frac{1}{k}\,\mathrm{d}x = 2\int_0^{1/2} \min\left(m, \frac{1}{x}\right)\,\mathrm{d}x = 2(1 + \ln\frac{m}{2}).$$

Finally the summation over k in (3.3) gives the overall bound of $O(\ln n \ln m) = O(\log^2 n)$. Proposition 3.4 is proved. □

In conclusion, let us formulate the general result concerning the distribution of points in an arbitrary planar convex set U of unit area (instead of the unit square). For an angle $\varphi \in [0, 2\pi)$ and a radius $r > 0$, let $h(r,\varphi)$ be the halfplane $\{(x,y) \in \mathbf{R}^2\colon x\cos\varphi + y\sin\varphi \geq r\}$. Geometrically:

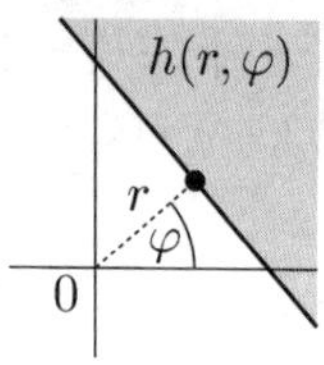

Moreover, suppose that $0 \in U$ and let $R_U(\varphi)$ be the largest r for which $h(r,\varphi) \cap U \neq \emptyset$.

3.5 Theorem. *For any convex set $U \subset \mathbf{R}^2$ of unit area and every $n \geq 2$, there exists an n-point set $P \subset U$ with*

$$\int_0^{2\pi} \int_0^{R_U(\varphi)} |D(P, h(r,\varphi))|\,\mathrm{d}r\,\mathrm{d}\varphi = O(\log^2 n),$$

where the constant of proportionality depends on U and where $D(P,h) = n\,\mathrm{vol}(h \cap U) - |P \cap h|$.

The underlying measure in the definition of L_1-discrepancy in this theorem is the motion-invariant measure on lines (appropriately scaled) we will discuss in Section 6.4. The point set P providing the upper bound is again essentially the regular square grid, but some adjustments must be made near the boundary of U. Namely, we first put in all the points p of the grid $\frac{1}{\sqrt{n}}\mathbf{Z}^2$ such that the square of side $\frac{1}{\sqrt{n}}$ centered at p is contained in U. Next, we remove all these squares from U, and we sweep the remaining region U_1 along the boundary by a semiline σ rotating around its origin, which is placed somewhere "near the center" of U:

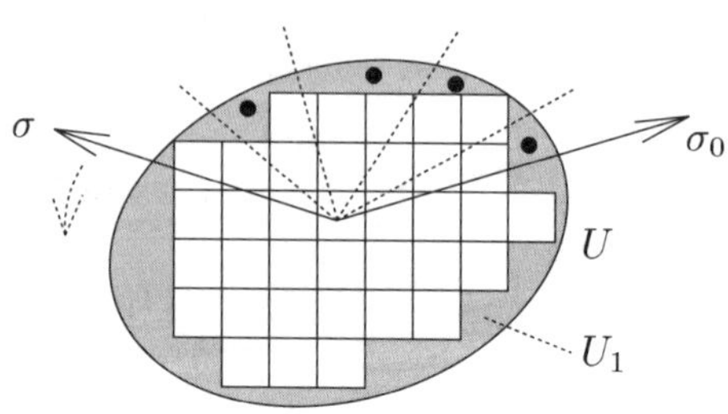

We start the sweep at some arbitrary position σ_0 of σ and we keep track the swept area of U_1. Whenever we sweep through an area $\frac{1}{n}$, we insert one point somewhere into the just swept part of U_1. The proof of Theorem 3.5 is considerably more complicated than in the simple case of Proposition 3.4, although the basic approach is similar.

Bibliography and Remarks. This section is based on Beck and Chen [BC93b], who established Theorem 3.5. The paper [BC93a] proves some other, similar results.

Earlier, Randol [Ran69] proved a somewhat related result. Namely, let C be a fixed polygon in the plane containing 0 in its interior, and let C_φ denote C rotated by the angle φ around 0. Then

$$\int_0^{2\pi} \left| t^2 \operatorname{vol}(C) - |(tC_\varphi) \cap \mathbf{Z}^2| \right| \mathrm{d}\varphi = O(\log^{3+\varepsilon} t)$$

as $t \to \infty$, with $\varepsilon > 0$ arbitrarily small (and the constant of proportionality depending on $\varepsilon > 0$). Also, instead of fixing the lattice and letting the polygon rotate, we could fix the polygon and let the lattice rotate; results in this direction are mentioned in the remarks to Section 2.5.

It is not known whether the upper bound in Theorem 3.5 (or even in Proposition 3.4) can be improved, for example, to $O(\log n)$. Also, it is not clear to what extent the phenomenon of the small L_1-discrepancy can be generalized beyond the case of the Lebesgue-measure for halfplanes. For instance, does anything like that happen for the combinatorial discrepancy? The paper [Mat97] gives a partial negative result concerning a possible combinatorial generalization.

Exercises

1. Let P be the n-point set as in Proposition 3.4.
(a) By modifying the proof of Proposition 3.4, show that the L_2-discrepancy of P for halfplanes, with respect to the measure ν_0, is $O(n^{1/4})$.
(b)* Show that the L_2-discrepancy as in (a), for the particular set P, is also at least $\Omega(n^{1/4})$.
(c) Generalizing (b), show that the L_p-discrepancy of P is $\Omega(n^{1/2-1/2p})$ for any fixed $p \in (1, \infty)$, with the constant of proportionality depending on p.
(d) Show that the bound in (c) is asymptotically tight for all $p \in (1, 2]$.

4. Combinatorial Discrepancy

In this chapter, we are going to investigate the combinatorial discrepancy, an exciting and significant subject in its own right. From Section 1.3, we recall the basic definition: If X is a finite set and $\mathcal{S} \subseteq 2^X$ is a family of sets on X, a *coloring* is any mapping $\chi: X \to \{-1, +1\}$, and we have $\mathrm{disc}(\mathcal{S}) = \min_\chi \max_{S \in \mathcal{S}} |\chi(S)|$, where $\chi(S) = \sum_{x \in S} \chi(x)$.

In Section 4.1, we prove some general upper bounds for $\mathrm{disc}(\mathcal{S})$ expressed in terms of the number and size of the sets in $\mathcal{S}$, and also a bound in terms of the maximum degree of $\mathcal{S}$. Section 4.2 discusses a technique for bounding discrepancy from below, related to matrix eigenvalues. Section 4.3 reviews variations on the notion of discrepancy, such as the linear discrepancy and the hereditary discrepancy, and it gives another general lower bound, in terms of determinants. The subsequent section considers set systems with discrepancy 0 and those with hereditary discrepancy at most 1. (The material in Sections 4.2 through 4.4 will not be used in the rest of this book.)

In Section 4.5, we introduce one of the most powerful techniques for upper bounds in discrepancy theory: the partial coloring method. Section 4.6 deals with a refinement of the partial coloring method, called the entropy method. With this approach, bounds obtained by the partial coloring method can often be improved by logarithmic factors. For several important problems, this it is the only known technique leading to asymptotically tight bounds.

4.1 Basic Upper Bounds for General Set Systems

We begin with the following question. Let X be an n-point set and let $\mathcal{S}$ be a set system on X having m sets. What is the maximum possible value, over all choices of $\mathcal{S}$, of $\mathrm{disc}(\mathcal{S})$? We will be most interested in the case when $n \leq m$ (more sets than points). This is what we usually have in geometric situations, and it also turns out that the $m < n$ case can essentially be reduced to the $m = n$ case (see Theorem 4.9).

A quite good upper bound for the discrepancy is obtained by using a random coloring.

4.1 Lemma (Random coloring lemma). *Let $\mathcal{S}$ be a set system on an n-point set X. For a random coloring $\chi: X \to \{+1, -1\}$, the inequalities*

$$|\chi(S)| \le \sqrt{2|S|\ln(4|\mathcal{S}|)}$$

hold for all sets $S \in \mathcal{S}$ simultaneously with probability at least $\frac{1}{2}$.

Note that if we know that $\mathcal{S}$ has at most m sets and have no information about their sizes, we get the upper bound $\mathrm{disc}(\mathcal{S}) = O(\sqrt{n \log m})$. Moreover, the above formulation shows that a random coloring gives better discrepancy for smaller sets, and this may be useful in some applications.

Proof. This is similar to considerations made in the proof of Theorem 3.1 (upper bound for the discrepancy for discs), and actually simpler. For any fixed set $S \subseteq X$, the quantity $\chi(S) = \sum_{x\in S} \chi(x)$ is a sum of $s = |S|$ independent random ± 1 variables. Such a sum has a binomial distribution, with standard deviation $\sqrt{s}$, and the simplest form of the Chernoff tail estimate (see Alon and Spencer [AS92]) gives

$$\Pr\left[|\chi(S)| > \lambda\sqrt{s}\right] < 2e^{-\lambda^2/2}.$$

Hence, if we set $\lambda = \sqrt{2\ln(4|\mathcal{S}|)}$, the above bound becomes $1/(2|\mathcal{S}|)$, and, with probability at least $\frac{1}{2}$, a random coloring satisfies $|\chi(S)| \le \sqrt{2|S|\ln(4|\mathcal{S}|)}$ for all $S \in \mathcal{S}$. □

The following theorem is a small improvement over the lemma just proved, at least if the set sizes are not much smaller than n:

4.2 Theorem (Spencer's upper bound). *Let $\mathcal{S}$ be a set system on an n-point set X with $|\mathcal{S}| = m \ge n$. Then*

$$\mathrm{disc}(\mathcal{S}) = O\left(\sqrt{n\log(2m/n)}\right).$$

In particular, if $m = O(n)$ then $\mathrm{disc}(\mathcal{S}) = O(\sqrt{n})$.

We will prove this result in Section 4.6. A probabilistic construction shows that this bound is tight in the worst case (see Exercise 1 or Alon and Spencer [AS92]). For $m = n$, there is a simple constructive lower bound based on Hadamard matrices, which we present in Section 4.2.

Another important upper bound, which we will not use but which is definitely worth mentioning, is this:

4.3 Theorem (Beck–Fiala theorem). *Let $\mathcal{S}$ be a set system on an arbitrary finite set X such that $\deg_{\mathcal{S}}(x) \le t$ for all $x \in X$, where $\deg_{\mathcal{S}}(x) = |\{S \in \mathcal{S}\colon x \in S\}|$. Then $\mathrm{disc}(\mathcal{S}) \le 2t - 1$.*

Proof. Let $X = \{1, 2, \ldots, n\}$. To each $j \in X$, assign a real variable $x_j \in [-1, 1]$ which will change as the proof progresses. Initially, all the x_j are 0. In the end, all x_j will be $+1$ or -1 and they will define the required coloring.

At each step of the proof, some of the variables x_j are "fixed" and the others are "floating;" initially all variables are floating. The fixed variables have values $+1$ or -1 and their value will not change anymore. The floating variables have values in $(-1, 1)$. At each step, at least one floating variable becomes fixed. Here is how this happens.

Call a set $S \in \mathcal{S}$ *dangerous* if it contains more than t elements j with x_j currently floating, and call S *safe* otherwise. The following invariant is always maintained:

$$\sum_{j \in S} x_j = 0 \quad \text{for all dangerous } S \in \mathcal{S}. \tag{4.1}$$

Let F be the current set of indices of the floating variables, and let us regard (4.1) as a system of linear equations whose unknowns are the floating variables. This system certainly has a solution, namely the current values of the floating variables. Since we assume $-1 < x_j < 1$ for all floating variables, this solution is an interior point of the cube $[-1, 1]^F$. We want to show that there also exists a solution lying on the boundary of this cube, i.e. such that at least one unknown has value $+1$ or -1. The crucial observation is that the number of dangerous sets at any given moment is smaller than the number of floating variables (this follows by a simple double counting of incidences of the floating indices j with the dangerous sets). Hence our system of linear equations has fewer equations than unknowns, and therefore the solution space contains a line. This line intersects the boundary of the cube $[-1, 1]^F$ at some point z. The coordinates of this point specify the new value of the floating variables for the next step; however, the variables x_j for which $z_j = \pm 1$ become fixed.

This step is iterated until all the x_j become fixed. We claim that their values specify a coloring with discrepancy at most $2t - 1$. Indeed, consider a set $S \in \mathcal{S}$. At the moment when it became safe, it had discrepancy 0 by (4.1). At this moment it contained at most t indices of floating variables. The value of each of these floating variables might have changed by less than 2 in the remaining steps (it might have been -0.999 and become $+1$, say). This concludes the proof. □

Remark. Beck and Fiala conjectured that in fact $\operatorname{disc}(\mathcal{S}) = O(\sqrt{t})$ holds under the assumptions of their theorem but no proof is known. The Beck–Fiala theorem remains the best known bound in terms of the maximum degree alone (except for a tiny improvement of the bound $2t - 1$ to $2t - 3$).

Remark on Algorithms. For statements establishing upper bounds for discrepancy of a set system $(X, \mathcal{S})$, it is interesting to learn whether they provide a polynomial-time algorithm (polynomial in $|X|$ and $|\mathcal{S}|$) for computing a coloring with the guaranteed discrepancy. For the Random coloring lemma, a randomized algorithm is obvious, and it can be made deterministic (slower but still polynomial) by the method of conditional probabilities; see [AS92]. The proof of the Beck–Fiala theorem 4.3 also provides a polynomial algorithm, but the proof of Spencer's upper bound 4.2 does not—it is a big challenge to find one.

Bibliography and Remarks. Spencer's theorem 4.2 is from Spencer [Spe85]; alternative proofs were given by Gluskin [Glu89] via Minkowski's lattice point theorem and by Giannopoulos [Gia97] using the Gaussian measure. The Beck–Fiala theorem is from [BF81] (and the improvement from $2t-1$ to $2t-3$ is in [BH97]). Exercise 1 is in the spirit of a lower-bound proof presented in [AS92]. For more bounds related to the Beck–Fiala theorem see the remarks to Section 4.3 and the end of Section 5.5.

In theoretical computer science, an intriguing question is an efficient computation of a coloring with small discrepancy for a given set system. In cases where randomized algorithms are easy, such as for the Random coloring lemma, the task is to find an efficient deterministic counterpart (i.e. to *derandomize* the algorithm). A related question is to parallelize the algorithms efficiently. Some such issues are treated in Spencer [Spe87] already; a sample of newer references are Berger and Rompel [BR91] and Srinivasan [Sri97].

Exercises

1. (a)* Let $S = \sum_{i=1}^{n} S_i$ be a sum of n independent random variables, each attaining values $+1$ and -1 with equal probability. Let $P(n,\Delta) = \Pr[S > \Delta]$. Prove that for $\Delta \le n/C$,
$$P(n,\Delta) \ge \tfrac{1}{C}\exp\left(-\tfrac{\Delta^2}{Cn}\right),$$
where C is a suitable constant. That is, the well-known Chernoff bound $P(n,\Delta) \le \exp(-\Delta^2/2n)$ is close to the truth. (For very precise lower bounds, proved by quite different methods, see Feller [Fel43].)
(b)* Let $X = \{1,2,\ldots,n\}$ be a ground set, let $\chi\colon X \to \{+1,-1\}$ be any fixed coloring of X, and let R be a random subset of X (a random subset means one where each i is included with probability $\frac{1}{2}$, the choices being independent for distinct i). Prove that for any $\Delta \ge 0$, $\Pr[|\chi(R)| \ge \Delta] \ge P(n, 2\Delta)$, where $P(.,.)$ is as in (a).
(c) Let $\mathcal{R}$ be a system of $m \ge n$ independently chosen random subsets of $\{1,2,\ldots,n\}$, and let $c_1 > 0$ be a sufficiently small constant. Use (a), (b) to show that $\operatorname{disc}(\mathcal{R}) > c_1\sqrt{n\log(2m/n)}$ holds with a positive probability, provided that $m \le 2^{c_1 n}$; that is, Theorem 4.2 is asymptotically tight.
2. (Discrepancy of the product of set systems) Let $\mathcal{S}$ and $\mathcal{T}$ be set systems (on finite sets). We let $\mathcal{S}\times\mathcal{T} = \{S\times T\colon S\in\mathcal{S}, T\in\mathcal{T}\}$.
(a) Show that $\operatorname{disc}(\mathcal{S}\times\mathcal{T}) \le \operatorname{disc}(\mathcal{S})\operatorname{disc}(\mathcal{T})$.
(b)* Find an example with $\operatorname{disc}(\mathcal{S}) > 0$ and $\operatorname{disc}(\mathcal{S}\times\mathcal{S}) = 0$.
These results are due to Doerr [Doe98a].

4.2 Matrices, Lower Bounds, and Eigenvalues

Let $(X, \mathcal{S})$ be a set system on a finite set. Enumerate the elements of X as $x_1, x_2, \ldots, x_n$ and the sets of $\mathcal{S}$ as $S_1, S_2, \ldots, S_m$, in some arbitrary order. The *incidence matrix* of $(X, \mathcal{S})$ is the $m \times n$ matrix A, with columns corresponding to points of X and rows corresponding to sets of $\mathcal{S}$, whose element a_{ij} is given by

$$a_{ij} = \begin{cases} 1 & \text{if } x_j \in S_i \\ 0 & \text{otherwise.} \end{cases}$$

As we will see, it is useful to reformulate the definition of discrepancy of $\mathcal{S}$ in terms of the incidence matrix. Let us regard a coloring $\chi: X \to \{-1, +1\}$ as the column vector $(\chi(x_1), \chi(x_2), \ldots, \chi(x_n))^T \in \mathbf{R}^n$. Then the product $A\chi$ is the row vector $(\chi(S_1), \chi(S_2), \ldots, \chi(S_m)) \in \mathbf{R}^m$. Therefore, the definition of the discrepancy of $\mathcal{S}$ can be written

$$\operatorname{disc}(\mathcal{S}) = \min_{x \in \{-1,1\}^n} \|Ax\|_\infty,$$

where the norm $\|.\|_\infty$ of a vector $y = (y_1, y_2, \ldots, y_m)$ is defined by $\|y\|_\infty = \max_i |y_i|$. The right-hand side of the above equation can be used as a definition of discrepancy for an arbitrary real matrix A.

Expressing discrepancy via incidence matrices helps in obtaining lower bounds. For many lower bound techniques, it is preferable to consider the L_2-discrepancy instead of the worst-case discrepancy. In our case, this means replacing the max-norm $\|.\|_\infty$ by the usual Euclidean norm $\|.\|$, which is usually easier to deal with. Namely, we have

$$\operatorname{disc}(\mathcal{S}) \geq \operatorname{disc}_2(\mathcal{S}) = \min_\chi \left(\frac{1}{m} \sum_{i=1}^{m} \chi(S_i)^2 \right)^{1/2} = \frac{1}{\sqrt{m}} \cdot \min_{x \in \{-1,1\}^n} \|Ax\|.$$

Since $\|Ax\|^2 = (Ax)^T (Ax) = x^T (A^T A) x$, we can further rewrite

$$\operatorname{disc}_2(\mathcal{S}) = \left(\frac{1}{m} \min_{x \in \{-1,1\}^n} x^T (A^T A) x \right)^{1/2}. \tag{4.2}$$

Now we present the example of n sets on n points with discrepancy about $\sqrt{n}$ promised in Section 4.1. To this end, we first recall the notion of an *Hadamard matrix*. This is an $n \times n$ matrix H with entries $+1$ and -1 such that any two distinct columns are orthogonal; in other words, we have $H^T H = nI$, where I stands for the $n \times n$ identity matrix. Since changing all signs in a row or in a column preserves this property, one usually assumes that the first row and the first column of the considered Hadamard matrix consist of all 1's. We will also use this convention.

Hadamard matrices do not exist for every n. For example, it is clear that for $n \geq 2$, n has to be even if an $n \times n$ Hadamard matrix should exist. The existence problem for Hadamard matrices is not yet fully solved, but various constructions are known. We recall only one simple recursive construction, providing a $2^k \times 2^k$ Hadamard matrix for all natural numbers k. Begin with the 1×1 matrix $H_0 = (1)$, and, having defined a $2^{k-1} \times 2^{k-1}$ matrix H_{k-1}, construct H_k from four blocks as follows:

$$\begin{pmatrix} H_{k-1} & H_{k-1} \\ H_{k-1} & -H_{k-1} \end{pmatrix}.$$

Thus, we have

$$H_1 = \begin{pmatrix} 1 & 1 \\ 1 & -1 \end{pmatrix}, \quad H_2 = \begin{pmatrix} 1 & 1 & 1 & 1 \\ 1 & -1 & 1 & -1 \\ 1 & 1 & -1 & -1 \\ 1 & -1 & -1 & 1 \end{pmatrix}.$$

The orthogonality is easy to verify by induction.

4.4 Proposition (Hadamard set system). *Let H be an $n \times n$ Hadamard matrix, and let $\mathcal{S}$ be the set system with incidence matrix $A = \frac{1}{2}(H + J)$, where J denotes the $n \times n$ matrix of all 1's (in other words, A arises from H by changing the -1's to 0's). Then* $\mathrm{disc}(\mathcal{S}) \geq \mathrm{disc}_2(\mathcal{S}) \geq \frac{\sqrt{n-1}}{2}$.

Proof of Proposition 4.4. We have $A^T A = \frac{1}{4}(H+J)^T(H+J) = \frac{1}{4}(H^T H + H^T J + J^T H + J^T J) = \frac{1}{4}(nI + nR + nR^T + nJ)$, where $R = \frac{1}{n} H^T J$ is the matrix with 1's in the first row and 0's in the other rows. Therefore, for any $x \in \{-1,1\}^n$, we get

$$\begin{aligned} \frac{1}{n} \cdot x^T(A^T A)x &= \frac{1}{4}\left(\sum_{i=1}^n x_i^2 + 2x_1\left(\sum_{i=1}^n x_i\right) + \left(\sum_{i=1}^n x_i\right)^2\right) \\ &= \frac{1}{4}\left(\sum_{i=2}^n x_i^2 + (2x_1 + x_2 + \cdots + x_n)^2\right) \\ &\geq \frac{1}{4}\left(\sum_{i=2}^n x_i^2\right) = \frac{n-1}{4}, \end{aligned}$$

and so $\mathrm{disc}(\mathcal{S}) \geq \frac{\sqrt{n-1}}{2}$ follows from (4.2). □

A slightly different treatment of this result is outlined in Exercise 3. The proof just given used the estimate $\mathrm{disc}(\mathcal{S}) \geq \left(\frac{1}{m} \min_{x \in \{-1,1\}^n} x^T(A^T A)x\right)^{1/2}$. Often it is useful to take the minimum on the right-hand side over a larger set of vectors: instead of the discrete set $\{-1,1\}^n$, we minimize over all real vectors with Euclidean norm $\sqrt{n}$. (Combinatorially, this means that we allow "generalized colorings" assigning real numbers to the points of X, and we only require that the sum of squares of these numbers is the same as if we used ± 1's.) So we have

$$\operatorname{disc}_2(\mathcal{S}) \geq \left(\frac{1}{m} \cdot \min_{\|x\|=\sqrt{n}} x^T(A^T A)x\right)^{1/2}$$
$$= \left(\frac{n}{m} \cdot \min_{\|x\|=1} x^T(A^T A)x\right)^{1/2}. \qquad (4.3)$$

The right-hand side of this inequality can naturally be called the *eigenvalue bound* for the discrepancy of $\mathcal{S}$. This is because for any real $m \times n$ matrix A, the matrix $B = A^T A$ is positive semidefinite, it has n nonnegative eigenvalues $\lambda_1 \geq \lambda_2 \geq \cdots \geq \lambda_n \geq 0$, and for the smallest eigenvalue λ_n we have $\lambda_n = \min_{\|x\|=1} x^T Bx$. (The eigenvalues of $A^T A$ are often called the *singular values* of the matrix A.) All this is well-known in linear algebra, and not too difficult to see, but perhaps it is useful to recall a geometric view of the situation. For simplicity, suppose that $m = n$. Then the mapping $x \mapsto Ax$ is a linear mapping of $\mathbf{R}^n$ into $\mathbf{R}^n$, and it maps the unit sphere onto the surface of an ellipsoid (possibly a flat one if A is singular). The quantity $\min_{\|x\|=1} \|Ax\|$ is the length of the shortest semiaxis of this ellipsoid. At the same time, the lengths of the semiaxes are $\sqrt{\lambda_1}, \sqrt{\lambda_2}, \ldots, \sqrt{\lambda_n}$, where the λ_i are eigenvalues of $A^T A$ as above. Here is an illustration for $n = 2$:

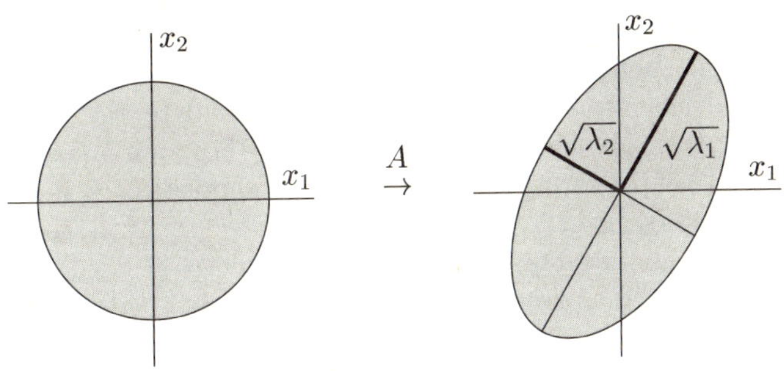

For $m > n$, the mapping $x \mapsto Ax$ maps $\mathbf{R}^n$ to an n-dimensional linear subspace of $\mathbf{R}^m$, and the image of the unit ball is an ellipsoid in this subspace.

The eigenvalue bound can be smaller than the L_2-discrepancy, but the eigenvalues of a matrix are efficiently computable and there are various techniques for estimating them. The following theorem summarizes our discussion:

4.5 Theorem (Eigenvalue bound for discrepancy). *Let $(\mathcal{S}, X)$ be a system of m sets on an n-point set, and let A denote its incidence matrix. Then we have*

$$\operatorname{disc}(\mathcal{S}) \geq \operatorname{disc}_2(\mathcal{S}) \geq \sqrt{\frac{n}{m} \cdot \lambda_{min}},$$

where λ_{min} denotes the smallest eigenvalue of the $n \times n$ matrix $A^T A$.

A very neat application of the eigenvalue bound concerns the discrepancy of a finite projective plane (Exercise 5.1.5). A more advanced example is the lower bound for the discrepancy of arithmetic progressions (a version due to Lovász; see Exercise 5). Moreover, numerous lower bounds in the Lebesgue-measure setting are in fact continuous analogues of the eigenvalue bound, in the sense that they hold for the L_2-discrepancy and for point sets with arbitrary weights, although they usually are not stated in this form (see Chapters 6 and 7). On the other hand, we should remark that for the Hadamard set system from Proposition 4.4, the eigenvalue bound fails miserably (Exercise 4). This can be fixed by deleting one set and one point (Exercise 3).

Bibliography and Remarks. An early result in combinatorial discrepancy was Roth's $\Omega(n^{1/4})$ lower bound on the discrepancy of arithmetic progressions [Rot64]. This beautiful proof uses harmonic analysis. Lovász suggested a version based on the eigenvalue bound, which is outlined in Exercise 5. Roth thought that the bound $n^{1/4}$ was too small and that the discrepancy was actually close to $n^{1/2}$. This was disproved by Sárközy (see [ES74]), who established an $O(n^{1/3+\varepsilon})$ upper bound. Beck [Bec81b] obtained the near-tight upper bound of $O(n^{1/4}\log^{5/2})$, inventing the powerful partial coloring method (see Section 4.5) for that purpose. The asymptotically tight upper bound $O(n^{1/4})$ was finally proved by Matoušek and Spencer [MS96]. (Proofs of slightly weaker upper bounds are indicated in Exercises 4.5.7 and 5.5.4.) Knieper [Kni98] generalized Roth's lower bound to the set system of *multidimensional arithmetic progressions*; these are the subsets of $\{1,2,\ldots,n\}^d$ of the form $A_1\times A_2\times\cdots\times A_d$, where each A_i is an arithmetic progression in $\{1,2,\ldots,n\}$ (note that the size of the ground set is n^d rather than n). The lower bound is $\Omega(n^{d/4})$ (d fixed), and this is easily seen to be asymptotically tight, using the $O(n^{1/4})$ bound for the case $d=1$ and the observation on product hypergraphs in Exercise 4.1.2 .

Proposition 4.4, with the approach as in Exercise 3 below, is due to Olson and Spencer [OS78]. The eigenvalue bound (Theorem 4.5) is attributed to Lovász and Sós in [BS95]. For another convenient formulation of the eigenvalue bound see Exercise 1.

Exercises

1. Let $(X,\mathcal{S})$ and A be as in Theorem 4.5. Show that if D is a diagonal matrix such that A^TA-D is positive semidefinite, then $\mathrm{disc}_2(\mathcal{S})\geq\sqrt{\mathrm{trace}(D)}$, where trace$(M)$ denotes the sum of the diagonal elements of a square matrix M.
2. Let the rows of a $2^k\times 2^k$ matrix H be indexed by the k-component 0/1 vectors, and let the columns be indexed similarly. Let the entry of H corresponding to vectors x and y be $+1$ if the scalar product $\langle x,y\rangle$ is

even and -1 if $\langle x, y\rangle$ is odd. Check that H is a Hadamard matrix (in fact, the same one as constructed recursively in the text).

3. Let H be a $4n \times 4n$ Hadamard matrix, and let A be the $(4n-1)\times(4n-1)$ matrix arising from H by deleting the first row and first column and changing the -1's to 0's.
(a) Verify that A is the incidence matrix of a 2-$(4n-1, 2n-1, n-1)$-design, meaning that there are $4n-1$ points, all sets have size $2n-1$, and any pair of distinct points is contained in exactly $n-1$ sets.
(b) Show that the eigenvalue bound (Theorem 4.5) gives at least $\sqrt{n}$ for the matrix A.
4.* Let $A = \frac{1}{2}(H+J)$ be the incidence matrix of the set system as in Proposition 4.4. Show that the eigenvalue bound (Theorem 4.5) is quite weak for A, namely that the smallest eigenvalue of A^TA is $O(1)$. (Note that in the proof of Proposition 4.4, we used the fact that all components of x are ± 1.)
5. (Discrepancy of arithmetic progressions) Let k be an integer; let $n = 6k^2$, and define the set system $\mathcal{A}_0$ on $\{0, 1, \ldots, n-1\}$ ("wrapped arithmetic progressions of length k with difference $\leq 6k$") as follows:

$$\begin{aligned}\mathcal{A}_0 = \{&\{i, (i+d) \bmod n, (i+2d) \bmod n, \ldots,\\ &(i+(k-1)d) \bmod n\}\colon d = 1, 2, \ldots, 6k,\ i = 0, 1, \ldots, n-1\}\end{aligned}$$

(a)** Use Theorem 4.5 to prove that $\operatorname{disc}(\mathcal{A}_0) = \Omega(n^{1/4})$.
(b) Deduce that the system of all arithmetic progressions (without wrapping) on $\{0, 1, \ldots, n-1\}$ has discrepancy $\Omega(n^{1/4})$.
6. Call a mapping $\chi\colon X \to \{+1, -1, 0\}$ *perfectly balanced* on a set system $(X, \mathcal{S})$ if $\chi(S) = 0$ for all $S \in \mathcal{S}$. Define $g(m)$ as the maximum possible size of X such that there exists a system $\mathcal{S}$ of m sets on X for which any perfectly balanced mapping χ is identically 0.
(a)* Show that $g(m)$ equals the maximum number n of columns of an $m \times n$ zero-one matrix A such that $\sum_{i\in I} a_i \neq \sum_{j\in J} a_j$ whenever I, J are distinct subsets of $\{1, 2, \ldots, n\}$, where a_i denotes the ith column of A.
(b) Deduce the bound $g(m) = O(m \log m)$ from (a).
(c) Prove that the discrepancy of an arbitrary system of m sets is always bounded by the maximum possible discrepancy of m sets on $g(m)$ points.

These results are from Olson and Spencer [OS78]. They also show that the bound in (b) is asymptotically tight.

4.3 Linear Discrepancy and More Lower Bounds

The discrepancy of a set system $\mathcal{S}$ can be thought of as a certain measure of "complexity" of $\mathcal{S}$. But from this point of view, it is not a very good measure, since disc($\mathcal{S}$) may happen to be small just "by chance." For example, let X be a disjoint union of two n-point sets Y and Z, and let $\mathcal{S}$ consist of all sets $S \subseteq X$ with $|S \cap Y| = |S \cap Z|$. Then disc($\mathcal{S}$) = 0 although we feel that $\mathcal{S}$ is nearly as complicated as the system of all subsets of X. A better concept from this point of view is the *hereditary discrepancy* of $\mathcal{S}$, defined as

$$\mathrm{herdisc}(\mathcal{S}) = \max_{Y \subseteq X} \mathrm{disc}(\mathcal{S}|_Y).$$

(Or, using our previously introduced notation, we can also write herdisc($\mathcal{S}$) = $\max_{0 \le m \le n} \mathrm{disc}(m, \mathcal{S})$.) As the just mentioned example shows, the hereditary discrepancy can be much larger than the discrepancy.

Another useful concept is the *linear discrepancy* of $\mathcal{S}$. It arises in the in the following "rounding" problem. Each point $x \in X$ is assigned a weight $w(x) \in [-1, 1]$. We want to color the points by $+1$'s and -1's in such a way that the sum of the colors in each set $S \in \mathcal{S}$ is close to the total weight of its points. The discrepancy of $\mathcal{S}$ with respect to the given weights is thus

$$\min_{\chi: X \to \{-1,1\}} \max_{S \in \mathcal{S}} |\chi(S) - w(S)|,$$

and the linear discrepancy of $\mathcal{S}$ is the supremum of this quantity over all choices of the weight function $w: X \to [-1, 1]$. (The usual discrepancy corresponds to the case $w(x) = 0$ for all x.) The linear discrepancy can again be defined for an arbitrary matrix A. Namely, we have

$$\mathrm{lindisc}(A) = \max_{w \in [-1,1]^n} \min_{x \in \{-1,1\}^n} \|A(x - w)\|_\infty.$$

Clearly lindisc(A) $\geq$ disc(A) for any matrix A, and one cannot expect to bound lindisc in terms of disc. But, perhaps surprisingly, the following bound in terms of the hereditary discrepancy holds:

4.6 Theorem (lindisc $\le$ 2·herdisc). *For any set system $\mathcal{S}$, we have*

$$\mathrm{lindisc}(\mathcal{S}) \le 2\,\mathrm{herdisc}(\mathcal{S}).$$

The same inequality holds between the linear and hereditary discrepancies for an arbitrary real matrix A.

This result can be proved in way somewhat similar to the proof of the transference lemma (Proposition 1.8). A proof in this spirit can be found in Spencer [Spe87], but here we give another proof using a geometric interpretation of the various notions of discrepancy. Unlike to the geometric discrepancy mostly treated in this book, the geometry here is introduced into the picture somewhat artificially, but once it is there it constitutes a powerful tool.

Let A be an $m \times n$ matrix, and let us define the set

$$U_A = \{x \in \mathbf{R}^n : \|Ax\|_\infty \le 1\}.$$

This U_A is a convex polyhedron symmetric about the origin, as is illustrated in the following picture for $n = 2$:

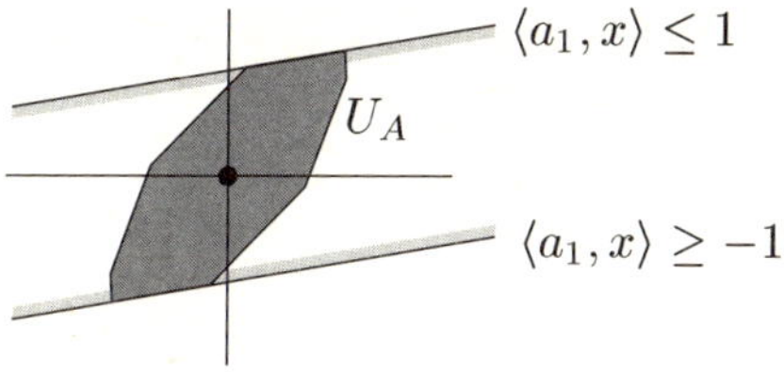

If a_i denotes the ith row of A, U_A is the intersection of the $2m$ halfspaces $\langle a_i, x\rangle \le 1$ and $\langle a_i, x\rangle \ge -1$, as is marked in the picture for $i = 1$. For any vector $x \in \mathbf{R}^n$, we have $\|Ax\|_\infty = \min\{t \ge 0\colon x \in tU_A\}$, and so disc$(A)$ is the smallest t such that the convex body tU_A contains a vertex of the cube $[-1,1]^n$. In other words, disc(A) is the smallest t such that for some vertex $a \in \{-1,1\}^n$, the translated body $tU_A + a$ contains the origin. This geometric interpretation of disc(A) allows perhaps the best comparison with the other notions of discrepancy introduced above. Their geometric interpretation is indicated in the following picture:

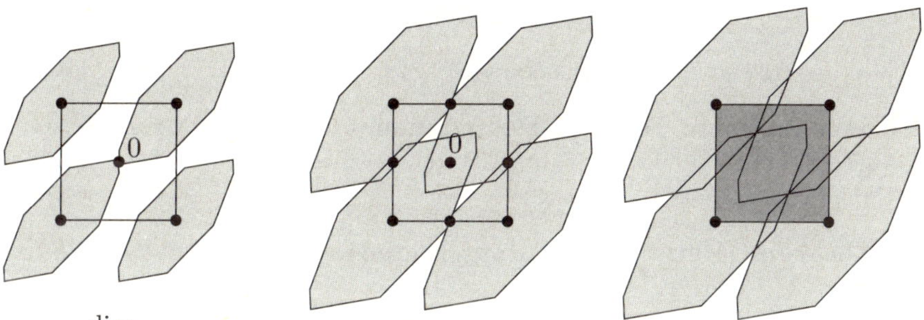

disc herdisc lindisc

We can imagine that at time $t = 0$, we start growing a similar copy of U_A from each vertex of the cube $[-1,1]^n$, in such a way that at time t, we have a congruent copy of tU_A centered at each vertex. The reader is invited (and recommended) to check that

- disc(A) is the first moment when some of the growing bodies swallows the origin,
- herdisc(A) is the first moment such that for each face F of the cube (of each possible dimension), the center of F is covered by some of the bodies centered at the vertices of F, and
- lindisc(A) is the first moment such that the whole cube is covered.

Proof of Theorem 4.6 (lindisc $\le$ 2·herdisc). In view of the above geometric interpretation, it suffices to prove the following statement.

> *If U is a closed convex body such that $\bigcup_{a\in\{-1,1\}^n}(U+a)$ covers all the points of $\{-1,0,1\}^n$ then the set $C = \bigcup_{a\in\{-1,1\}^n}(2U+a)$ covers the whole cube $[-1,1]^n$.*

Indeed, if $\mathrm{herdisc}(A) \le t$ then the body $U = tU_A$ satisfies even a stronger assumption, namely that each point $v \in \{-1,0,1\}^n$ is covered by the copy of U centered at one of the vertices closest to v.

Since U is closed, it is enough to prove that C covers all dyadic rational points in $[-1,1]^n$, i.e. all points $v = \frac{z}{2^k} \in [-1,1]^n$ for some integer vector $z \in \mathbf{Z}^n$. We proceed by induction on k, where the case $k=0$ follows immediately from the assumption. Consider some $v = \frac{z}{2^k} \in [-1,1]^n$. Since all components of $2v$ are in the interval $[-2,2]$, there is a vector $b \in \{-1,1\}^n$ such that $2v - b \in [-1,1]^n$. Since $2v - b = \frac{z+2^{k-1}b}{2^{k-1}}$, the inductive hypothesis for $k-1$ provides a vector $a \in \{-1,1\}^n$ such that $2v - b \in 2U + a$. Therefore, we obtain

$$v \in U + \frac{a+b}{2}.$$

The vector $\frac{a+b}{2}$ has all entries in $\{-1,0,1\}$, and so by the assumption on U, it is covered by some $U + c$ for $c \in \{-1,1\}^n$. Hence

$$v \in U + (U + c) = 2U + c,$$

where the last equality uses the convexity of U. This proves Theorem 4.6. □

A Lower Bound in Terms of Determinants. The hereditary discrepancy of a set system can be lower-bounded in terms of determinants of submatrices of the incidence matrix.

4.7 Theorem (Determinant lower bound). *For any set system $\mathcal{S}$, we have*

$$\mathrm{herdisc}(\mathcal{S}) \ge \frac{1}{2} \max_k \max_B |\det(B)|^{1/k},$$

where B ranges over all $k \times k$ submatrices of the incidence matrix of $\mathcal{S}$. An analogous bound also holds for the hereditary discrepancy of an arbitrary $m \times n$ real matrix A.

This is a consequence of the bound "lindisc $\le 2\cdot$herdisc" (Theorem 4.6) and of the following lemma:

4.8 Lemma. *Let A be an $n \times n$ matrix. Then* $\mathrm{lindisc}(A) \ge |\det(A)|^{1/n}$.

Proof. Let $t = \mathrm{lindisc}(A)$ and set $U = tU_A$. By the above geometric interpretation of the linear discrepancy, the sets $U + a$ for $a \in \{-1,1\}^n$ cover the whole cube $[-1,1]^n$, and therefore the sets $U + a$ for $a \in 2\mathbf{Z}^n$ cover the whole space. Hence

$$\mathrm{vol}(U) = t^n \,\mathrm{vol}(U_A) \ge \mathrm{vol}([-1,1]^n) = 2^n.$$

On the other hand, the linear mapping $x \mapsto Ax$ changes the volume by the factor $|\det(A)|$ (since it maps the unit cube to a parallelepiped of volume $|\det(A)|$), and since U_A is the inverse image of the cube $[-1,1]^n$, we get

$\mathrm{vol}(U_A) = |\det(A)|^{-1}2^n$. Together with the previous inequality for $\mathrm{vol}(U_A)$, this gives $t \geq |\det(A)|^{1/n}$. □

It is instructive to compare the determinant lower bound and the eigenvalue lower bound (Theorem 4.5). For simplicity, let us consider the case of a square matrix A first, in which case the eigenvalue bound becomes $\sqrt{\lambda_{min}}$. We recall the geometric interpretation of the eigenvalue bound: $\sqrt{\lambda_{min}}$ is the length of the shortest semiaxis of the ellipsoid E that is the image of the unit ball $B(0,1)$ under the linear mapping $x \mapsto Ax$. The ratio $\mathrm{vol}(E)/\mathrm{vol}(B(0,1))$ equals, on the one hand, the product of the semiaxes of E, i.e. $\sqrt{\lambda_1\lambda_2\cdots\lambda_n}$, and on the other hand, it is equal to $|\det A|$. Therefore, since λ_{min} is the smallest among the n eigenvalues of A^TA, we get $\sqrt{\lambda_{min}} \leq |\det A|^{1/n}$. Thus, for a square matrix, the determinant lower bound for discrepancy is never weaker than the eigenvalue lower bound (and it can be much stronger if the ellipsoid E happens to be very flat). Also for non-square matrices A, the determinant lower bound is never smaller than the eigenvalue bound, possibly up to a small constant multiplicative factor; see Exercise 7. But one should not forget that the eigenvalue bound estimates discrepancy, while the determinant bound only applies to hereditary discrepancy.

Few Sets on Many Points. Set systems coming from geometric settings typically have more sets than points, so we are mainly interested in this case. For studying discrepancy of set systems with few sets and many points, the following result is important:

4.9 Theorem. *Let $(X,\mathcal{S})$ be a set system such that $\mathrm{disc}(\mathcal{S}|_Y) \leq K$ for all $Y \subseteq X$ with $|Y| \leq |\mathcal{S}|$. Then $\mathrm{disc}(\mathcal{S}) \leq 2K$.*

Proof. This is a nice application of the concept of linear discrepancy. We note that if w and w_0 are two weight functions on X such that $w(S) = w_0(S)$ for all $S \in \mathcal{S}$ then the discrepancy of any coloring χ for w is the same as that for w_0. We also have

4.10 Lemma. *Let $(X,\mathcal{S})$ be a set system, $|X| = n \geq |\mathcal{S}| = m$, and let $w: X \to [-1,1]$ be a weight function. Then there exist an n-point set $Y \subseteq X$ and a weight function $w_0: X \to [-1,1]$ such that $w_0(S) = w(S)$ for all $S \in \mathcal{S}$ and $w_0(x) = \pm 1$ for all $x \in X \setminus Y$.*

The proof of this lemma is quite similar to the proof of the Beck–Fiala theorem 4.3 and we leave it as Exercise 1. From the lemma and the observation above it, we get that $\mathrm{lindisc}(\mathcal{S}) \leq \sup_{Y\subseteq X,|Y|=n} \mathrm{lindisc}(\mathcal{S}|_Y)$. The left-hand side of this inequality is at least $\mathrm{disc}(\mathcal{S})$ while the right-hand side is at most $2\max_{Y\subseteq X,|Y|\leq n} \mathrm{disc}(\mathcal{S}|_Y)$ by the bound "lindisc $\leq$ 2·herdisc" (Theorem 4.6). This proves Theorem 4.9. □

The theorem just proved plus Spencer's upper bound (Theorem 4.2) imply that an arbitrary set system with m sets has discrepancy $O(\sqrt{m})$.

Bibliography and Remarks. Hereditary discrepancy and linear discrepancy were introduced by Lovász et al. [LSV86]. (In [BS95], linear discrepancy is called the *inhomogeneous discrepancy.*) Lovász et al. [LSV86] also established Theorem 4.6 and Theorem 4.7, and our presentation mostly follows their proofs.

We should remark that they use definitions of discrepancy giving exactly half of the quantities we consider. They work with 0/1 vectors instead of $-1/1$ vectors, which is probably somewhat more natural in the context of linear discrepancy.

Another potentially useful lower bound for the discrepancy of an $m \times n$ matrix A is

$$\operatorname{lindisc}(A) \geq \frac{2\operatorname{vol}(\operatorname{conv}(\pm A))^{1/n}}{c_n^{2/n}}. \tag{4.4}$$

Here $c_n = \pi^{n/2}/\Gamma(\frac{n}{2}+1)$ is the volume of the n-dimensional unit ball and $\operatorname{conv}(\pm A)$ denotes the convex hull of the $2m$ vectors a_1, $-a_1$, a_2, $-a_2$, ..., a_m, $-a_m$, where $a_i \in \mathbf{R}^n$ stands for the ith row of A. This lower bound is obtained from $\operatorname{lindisc}(A) \geq 2\operatorname{vol}(U_A)^{-1/n}$ (see the proof of Lemma 4.8) using so-called *Blaschke's inequality*, stating that $\operatorname{vol}(K)\operatorname{vol}(K^*) \leq c_n^2$ holds for any centrally symmetric convex body K in $\mathbf{R}^n$. Here

$$K^* = \{x \in \mathbf{R}^n \colon \langle x, y\rangle \leq 1 \text{ for all } y \in K\}$$

is the *polar body* of K. For $K = U_A$, it turns out that $U_A^* = \operatorname{conv}(\pm A)$. The inequality (4.4) is due to Lovász and Vesztergombi [LV89], who used it to estimate the maximum possible number $m(n,d)$ of distinct rows of an integral $m \times n$ matrix A with $\operatorname{herdisc}(A) \leq d$. They proved that this $m(n,d)$ is between $\binom{n+d}{n}$ and $\binom{n+2\pi d}{n}$. If one asks a similar question for a set system, i.e. what is the maximum possible number of distinct sets in a set system $\mathcal{S}$ on n points with $\operatorname{herdisc}(\mathcal{S}) \leq d$, then a precise answer can be given—see Exercise 5.2.5.

The next few remarks concern the relationship of the linear and hereditary discrepancies. In the inequality $\operatorname{lindisc}(\mathcal{S}) \leq 2\operatorname{herdisc}(\mathcal{S})$ (Theorem 4.6), the constant 2 cannot be improved in general; see Exercise 3. On the other hand, the inequality is always strict, and in fact, $\operatorname{lindisc}(\mathcal{S}) \leq 2(1-\frac{1}{2m})\operatorname{herdisc}(\mathcal{S})$ holds for any set system $\mathcal{S}$ with m sets (Doerr [Doe98b]).

There is a simple example showing that the hereditary discrepancy of a matrix cannot be bounded in terms of the linear discrepancy [LSV86], namely the single-row matrix $(1, 2, 4, \ldots, 2^{n-1})$ (Exercise 4). The question whether the hereditary discrepancy of a set system can be estimated by a function of the linear discrepancy seems to be open. On the one hand, there is a set system such that any system containing it as an induced subsystem has linear discrepancy at least 2 (unpublished result of the author). On the other hand, the hereditary discrepancy can be strictly bigger than the linear discrepancy (Exercise 6), and so the situation cannot be too simple.

The fact that the determinant lower bound for the hereditary discrepancy in Theorem 4.7 is never asymptotically smaller than the eigenvalue lower bound for disc_2 in Theorem 4.5 (Exercise 7) is a simple consequence of observations of Chazelle [Cha98a] (I haven't seen it explicitly mentioned anywhere). Chazelle's result actually says that if the eigenvalue lower bound for some system $\mathcal{S}$ on n points equals some number Δ then there is a subsystem $\mathcal{S}_0 \subseteq \mathcal{S}$ of at most n sets with $\mathrm{herdisc}(\mathcal{S}_0) = \Omega(\Delta)$. So, in some sense, the eigenvalue bound is always "witnessed" by the hereditary discrepancy of at most n sets, no matter how many sets the original set system may have. Little seems to be known about possible strengthenings and analogues of this result. One related observation is that for a set system $\mathcal{S}$ on an n-point set with a large eigenvalue bound, all the systems $\mathcal{S}' \subseteq \mathcal{S}$ consisting of n sets may have the eigenvalue bound very small (Exercise 8).

A result somewhat weaker than Theorem 4.9, namely that the discrepancy of a system of m sets is always bounded by the maximum possible discrepancy of m sets on $O(m \log m)$ points, was first obtained by Olson and Spencer [OS78] (see Exercise 4.2.6). The fact that any m sets have discrepancy $O(\sqrt{m})$ was proved by Spencer [Spe85].

We have mentioned a natural generalization of the notion of discrepancy from incidence matrices of set systems to arbitrary real matrices. A different and interesting notion of matrix discrepancy arises from *vector sum* problems. Having n vectors $v_1, \ldots, v_n \in \mathbf{R}^m$ of norm at most 1, we ask for a choice of n signs $\varepsilon_1, \ldots, \varepsilon_n$ so that the vector $w = \sum_{i=1}^n \varepsilon_i v_i$ is as short as possible. We have a whole class of problems, since the vectors v_i can be measured by one norm in $\mathbf{R}^m$ (supremum norm, L_1-norm, Euclidean norm, etc.) and the vector w can be measured by another, possibly different, norm. Let us mention that for the case when both the norms are the supremum norm, an extension of Spencer's theorem shows that the norm of w can be made $O(\sqrt{m})$.

A famous conjecture of Komlós asserts that if all the v_i have Euclidean length at most 1 then the supremum norm of w can be bounded by an absolute constant (this is a generalization of the Beck–Fiala conjecture mentioned in Section 4.1; see Exercise 9). The current best result on Komlós' conjecture is due to Banaszczyk [Ban98]. He proves the following more general result: There is an absolute constant c such that if K is a convex body in $\mathbf{R}^m$ with $\gamma_m(K) \geq \frac{1}{2}$, then for any vectors $v_1, v_2, \ldots, v_n \in \mathbf{R}^m$ of Euclidean norm at most 1 there exist signs $\varepsilon_1, \varepsilon_2, \ldots, \varepsilon_n \in \{-1, +1\}$ such that $v_1\varepsilon_1 + v_2\varepsilon_2 + \cdots + v_n\varepsilon_n \in cK$. Here γ_m denotes the m-dimensional

Gaussian measure whose density at a point x is $(2\pi)^{-m/2}e^{-\|x\|^2/2}$; this is the density of the normalized normal distribution. This theorem improves an earlier result of Giannopoulos [Gia97], where the conclusion was that $v_1\varepsilon_1+v_2\varepsilon_2+\cdots+v_n\varepsilon_n \in c(\log n)K$ (this was already sufficient for proving Spencer's upper bound 4.2). Banaszczyk's result easily implies that in the situation of Komlós' conjecture, $\|w\|_\infty = O(\sqrt{\log n})$ can be achieved. Further, for the Beck–Fiala conjecture, this yields that the discrepancy of a set system of maximum degree t on n points is $O(\sqrt{t}\log n)$, which is the best known bound in a wide range of the parameters n and t. (We will prove a weaker bound in Section 5.5.)

More about these and related subjects can be found, for instance, in Beck and Sós [BS95], Alon and Spencer [AS92], Bárány and Grinberg [BG81], and Spencer [Spe87].

Exercises

1. Prove Lemma 4.10.
2. Find a set system $(X, \mathcal{S})$ and a set $A \subseteq X$ such that $\mathrm{disc}(\mathcal{S}) = 0$ but $\mathrm{disc}(\mathcal{S} \cup \{A\})$ is arbitrarily large.

   Remark. It is not known whether an example exists with $\mathrm{herdisc}(\mathcal{S}) \le 1$ and with $\mathrm{disc}(\mathcal{S} \cup \{A\})$ large.
3. Show that the the set system $\{\{1\}, \{2\}, \ldots, \{n\}, \{1, 2, \ldots, n\}\}$ has hereditary discrepancy 1 and linear discrepancy at least $2 - \frac{2}{n+1}$.
4. Show that the $1 \times n$ matrix $(2^0, 2^1, 2^2, \ldots, 2^{n-1})$ has hereditary discrepancy at least 2^{n-1} and linear discrepancy at most 2.
5. Let A be an $m \times n$ real matrix, and set

$$\Delta = \max_{w \in \{-1,0,1\}} \min_{x \in \{-1,1\}} \|A(x - w)\|_\infty$$

   ("linear discrepancy with weights $-1, 0, 1$"). Prove that $\mathrm{lindisc}(A) \le 2\Delta$.
6. Show that the set system $\{\{1,2\}, \{1,3\}, \{2,3,4\}\}$ has hereditary discrepancy 2 and linear discrepancy strictly smaller than 2.
7. (Relation of the determinant and eigenvalue bounds) Let $(X, \mathcal{S})$ be a system of m sets on an n-point set, $m \ge n$, and let A be the incidence matrix of $\mathcal{S}$.

   (a)* Put $\Delta = \left(\frac{n}{m} \det(A^T A)^{1/n}\right)^{1/2}$. Prove the existence of a subsystem $\mathcal{S}_0 \subseteq \mathcal{S}$ consisting of n sets with $\mathrm{herdisc}(\mathcal{S}_0) = \Omega(\Delta)$. Use the Binet–Cauchy theorem from linear algebra, asserting that for any $m \times n$ real matrix A, $m \ge n$, we have $\det(A^T A) = \sum_B \det(B)^2$, where B ranges over all $n \times n$ submatrices of A.

   (b)* Prove that if Δ_{eig} is the eigenvalue lower bound from Theorem 4.5 and Δ_{det} is the determinant lower bound from Theorem 4.7 then $\Delta_{eig} = O(\Delta_{det})$.

(c) Show that the lower bound in Theorem 4.7 for herdisc($\mathcal{S}$), and consequently also the eigenvalue bound in Theorem 4.5, is never bigger than $O(\sqrt{n})$.

8. Let A be an $(n+1) \times n$ zero-one matrix obtained from an $(n+2) \times (n+2)$ Hadamard matrix by deleting the first row and the first two columns and changing -1's to 0's. Show that the eigenvalue lower bound for A is $\Omega(\sqrt{n})$ (this is similar to Exercise 4.2.3), and that for any $n \times n$ submatrix B of A, the eigenvalue bound is only $O(1)$. Therefore, unlike the determinant lower bound, the eigenvalue lower bound for $n+1$ sets on n points need not be "witnessed" by n sets on these points.
9. Verify that Komlós' conjecture implies the Beck–Fiala conjecture. Komlós' conjecture says that there is a constant K such that for any vectors $v_1, v_2, \ldots, v_n$ of unit Euclidean norm, there exist signs $\varepsilon_1, \varepsilon_2, \ldots, \varepsilon_n$ such that $\|\varepsilon_1 v_1 + \varepsilon_2 v_2 + \cdots + \varepsilon_n v_n\|_\infty \le K$. The Beck–Fiala conjecture states that $\mathrm{disc}(\mathcal{S}) \le C\sqrt{t}$ for any set system $\mathcal{S}$ of maximum degree t.
10. Let $A = \frac{1}{2}(H+J)$ be the $n \times n$ incidence matrix of a set system as in Proposition 4.4. Derive an $\Omega(\sqrt{n})$ lower bound for herdisc(A) using the determinant lower bound (Theorem 4.7); use the specific Hadamard matrices H_k of size $2^k \times 2^k$ whose construction was indicated above Proposition 4.4.

4.4 On Set Systems with Very Small Discrepancy

A very important class of set systems are those with hereditary discrepancy at most 1 (note that requiring hereditary discrepancy to be 0 leads to rather trivial set systems). Such set systems are called *totally unimodular*. They are of interest in polyhedral combinatorics, theory of integer programming, etc., and there is an extensive theory about them, which has been developing more or less independently of discrepancy theory. Here we only touch this subject very briefly, but it is useful to be aware of its existence.

An Example Destroying Several Conjectures. The following question has been open for some time: if $\mathcal{S}_1$ and $\mathcal{S}_2$ are two set systems on the same ground set, can $\mathrm{disc}(\mathcal{S}_1 \cup \mathcal{S}_2)$ be upper-bounded by some function of $\mathrm{disc}(\mathcal{S}_1)$ and $\mathrm{disc}(\mathcal{S}_2)$? The following important example shows that this is not the case. Even the union of two set systems with the best possible behavior in terms of discrepancy, namely with herdisc $= 1$, can have arbitrarily large discrepancy.

4.11 Proposition (Hoffman's example). *For an arbitrarily large number K, there exist set systems $\mathcal{S}_1, \mathcal{S}_2$ such that* $\mathrm{herdisc}(\mathcal{S}_1) \le 1$, $\mathrm{herdisc}(\mathcal{S}_2) \le 1$, *and* $\mathrm{disc}(\mathcal{S}_1 \cup \mathcal{S}_2) \ge K$.

Proof. The ground set of both set systems is the set of edges of the complete K-ary tree T of depth K (a picture shows the case $K=3$).

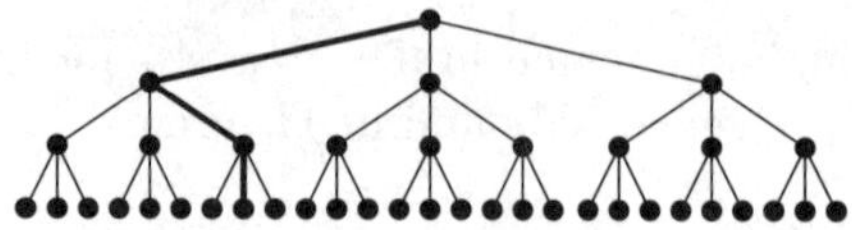

The sets in $\mathcal{S}_1$ are the edge sets of all root-to-leaf paths (the picture shows one of them drawn thick). The sets of $\mathcal{S}_2$ are the "fans" in the tree: for each non-leaf vertex v, we put the set of the K edges connecting v to its successors into $\mathcal{S}_2$. The bound $\text{herdisc}(\mathcal{S}_2) \leq 1$ is obvious, and $\text{herdisc}(\mathcal{S}_1) \leq 1$ is simple and it is left as Exercise 1. Finally $\text{disc}(\mathcal{S}_1 \cup \mathcal{S}_2) \geq K$ follows from a Ramsey-type observation: whenever the edges of T are each colored red or blue, there is either a red root-to-leaf path or a vertex with all successor edges blue. □

Based on this example, one can refute several other plausible-looking conjectures about discrepancy (see Exercises 2 and 3).

An Etude in Discrepancy Zero. We consider the following function: let $f(n)$ denote the smallest number of sets of size n each that constitute a system with nonzero discrepancy. For instance, we have $f(n) = 1$ for every odd n. The question is whether $f(n)$ can be bounded by some universal constant K for all n. The answer is negative:

4.12 Theorem. *We have* $\limsup_{n\to\infty} f(n) = \infty$.

This theorem is included mainly because of the beauty of the following proof.

Proof. For contradiction, suppose that $f(n) \leq K$ for all n. This means that for every n there is a set system $\mathcal{S}^{(n)} = \{S_1^{(n)}, S_2^{(n)}, \ldots, S_K^{(n)}\}$ consisting of K n-element sets such that $\text{disc}(\mathcal{S}^{(n)}) \geq 1$. Let us fix one such $\mathcal{S}^{(n)}$ for each n.

A system of 3 sets, say, can be described by giving the number of elements in each field of the Venn diagram, as an example illustrates:

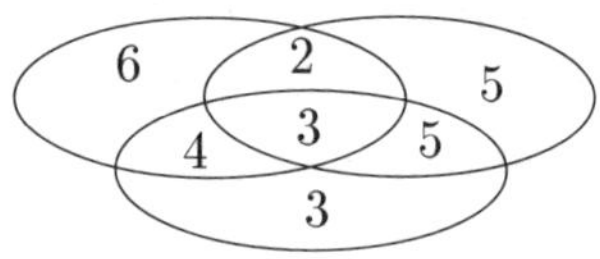

Similarly, the system $\mathcal{S}^{(n)} = \{S_1^{(n)}, S_2^{(n)}, \ldots, S_K^{(n)}\}$ is determined by an integer vector indexed by nonempty subsets of $\{1, 2, \ldots, K\}$. Namely, for each nonempty index set $I \subseteq \{1, 2, \ldots, K\}$, we let $s_I^{(n)}$ be the number of elements that belong to all $S_i^{(n)}$ with $i \in I$ and to no $S_j^{(n)}$ with $j \notin I$. In this way, the set system $\mathcal{S}^{(n)}$ determines an integer vector $s^{(n)} \in \mathbf{R}^{2^K-1}$. The condition that all sets of $\mathcal{S}^{(n)}$ have size n implies that $\sum_{I:\, j\in I} s_I^{(n)} = n$ for all $j = 1, 2, \ldots, K$.

Similarly, a red-blue coloring χ of the ground set $\mathcal{S}^{(n)}$ can be described by an integer vector $c \in \mathbf{R}^{2^K-1}$, where this time the component c_I tells us how many elements colored red lie in all the sets $S_i^{(n)}$ with $i \in I$ and in none of the sets $S_j^{(n)}$ with $j \notin I$.

Let us put $\sigma^{(n)} = \frac{1}{n} s^{(n)}$, and let us consider the following system of linear equations and inequalities for an unknown vector $\gamma \in \mathbf{R}^{2^K-1}$:

$$\begin{array}{ll} 0 \le \gamma_I \le \sigma_I^{(n)} & \text{for all nonempty } I \subseteq \{1,2,\ldots,K\} \\ \sum_{I:\, j\in I} \gamma_I = \frac{1}{2} & \text{for } j = 1,2,\ldots,K. \end{array} \tag{4.5}$$

Let $\Gamma^{(n)} \subseteq \mathbf{R}^{2^K-1}$ denote the set of all real vectors γ satisfying the system (4.5) for a given n. If c were an integer vector encoding a coloring of the ground set of $\mathcal{S}^{(n)}$ with zero discrepancy, then we would conclude that $\frac{1}{n}c \in \Gamma^{(n)}$. But we assume that no such c exists, and so $\Gamma^{(n)}$ contains no vector γ with $n\gamma$ integral.

To arrive at a contradiction, we will look at the values of n of the form $q!$ for $q = 1, 2, \ldots$; the important thing about these values is that all the numbers up to q divide $q!$. So let us consider the vectors $\sigma^{(q!)}$, $q = 1, 2, 3, \ldots$. This is an infinite and bounded sequence of vectors in $\mathbf{R}^{2^K-1}$, and hence it has a cluster point; call it σ. Let $(n_1, n_2, \ldots)$ be a subsequence of the sequence $(1!, 2!, 3!, \ldots)$ such that the $\sigma^{(n_k)}$ converge to σ.

Let us choose a rational vector $\bar\sigma$ with $\frac{1}{2}\sigma \le \bar\sigma \le \frac{2}{3}\sigma$ (the inequalities should hold in each component), and let $\bar\Gamma$ denote the solution set of the system (4.5) with $\bar\sigma$ replacing $\sigma^{(n)}$. We have $\frac{1}{2}\sigma \in \bar\Gamma$ and so $\bar\Gamma \ne \emptyset$. At the same time, the inequalities and equations defining $\bar\Gamma$ have all coefficients rational, and hence $\bar\Gamma$ contains a rational vector $\bar\gamma$. This $\bar\gamma$ satisfies $\bar\gamma_I < \sigma_I$ (strict inequalities) for all I with $\sigma_I \ne 0$, and hence also $\bar\gamma \in \Gamma^{(n_k)}$ for all large enough k. But since the n_k were selected among the numbers $1!, 2!, 3!, \ldots$, we get that for sufficiently large k, $n_k\bar\gamma$ is a vector of integers and hence it encodes a zero-discrepancy coloring of $\mathcal{S}^{(n_k)}$. This contradiction finishes the proof of Theorem 4.12. □

Bibliography and Remarks. For a basic overview and references concerning the theory of total unimodularity, the reader may consult Schrijver [Sch95]. A matrix A is called totally unimodular if the determinant of each square submatrix of A is 0, 1 or -1 (this implies, in particular, that the entries of A are 0's and ± 1's). A famous theorem of Ghouila-Houri [GH62] asserts, in our terminology, that a matrix consisting of 0's and ± 1's is totally unimodular if and only if its hereditary discrepancy is at most 1; see Exercise 4. On the other hand, the linear discrepancy of a totally unimodular matrix can be arbitrarily close to 2; see Exercise 4.3.3.

The question about bounding $\operatorname{disc}(\mathcal{S}_1 \cup \mathcal{S}_2)$ in terms of bounding $\operatorname{disc}(\mathcal{S}_1)$ and $\operatorname{disc}(\mathcal{S}_2)$ was raised by Sós. Hoffmann's example is cited as an oral communication from 1987 in Beck and Sós [BS95]. The conjectures in Exercises 2 and 3 were stated in Lovász et al. [LSV86].

Theorem 4.12 is due to Alon et al. [AKP+87], who also give fairly precise quantitative bounds for $f(n)$ in terms of the number-theoretic structure of n, more precisely in terms of the smallest number not dividing n.

Exercises

1. Verify the assertion $\mathrm{herdisc}(\mathcal{S}_1) \leq 1$ in the proof of Proposition 4.11.
2. Let $(X, \mathcal{S})$ be a set system and let $(\mathcal{S}, \mathcal{S}^*)$ be the set system dual to $\mathcal{S}$; explicitly $\mathcal{S}^* = \{\{S \in \mathcal{S}: x \in S\}: x \in X\}$. Using Proposition 4.11, show that $\mathrm{herdisc}(\mathcal{S}^*)$ cannot in general be bounded by any function of $\mathrm{herdisc}(\mathcal{S})$.
3. Using Proposition 4.11, show that $\mathrm{herdisc}(\mathcal{S})$ cannot be bounded from above by any function of $\max_k \max_B |\det(B)|^{1/k}$, i.e. of the right-hand side of the inequality in Theorem 4.7, where B is a $k \times k$ submatrix of the incidence matrix of $\mathcal{S}$.
4. (On Ghouila-Houri's theorem)
   (a) Show that if A is a nonsingular $n \times n$ totally unimodular matrix (the definition was given above the exercises), then the mapping $x \mapsto Ax$ maps $\mathbf{Z}^n$ bijectively onto $\mathbf{Z}^n$.
   (b)* Show that if A is an $m \times n$ totally unimodular matrix and b is an m-dimensional integer vector such that the system $Ax = b$ has a real solution x, then it has an integral solution as well.
   (c)* (Kruskal–Hoffmann theorem—one implication) Let A be an $m \times n$ totally unimodular matrix and let $u, v \in \mathbf{Z}^n$ and $w, z \in \mathbf{Z}^m$ be integer vectors. Show that if the system of inequalities $u \leq x \leq v$, $w \leq Ax \leq z$ (the inequalities should hold in each component) has a real solution then it has an integer solution as well. Geometrically speaking, all the vertices of the polytope in $\mathbf{R}^n$ determined by the considered system are integral.
   (d)* Prove that the discrepancy of a totally unimodular set system with all sets of even size is 0.
   (e) Prove that the hereditary discrepancy of a totally unimodular set system is at most 1 (this is one of the implications in Ghouila-Houri's theorem for 0/1 matrices).

4.5 The Partial Coloring Method

Here we introduce one of the most powerful methods for producing low-discrepancy colorings.

Let X be a set. A *partial coloring* of X is any mapping $\chi: X \to \{-1, 0, +1\}$. For a point $x \in X$ with $\chi(x) = 1$ or $\chi(x) = -1$, we say that x is *colored by* χ, while for $\chi(x) = 0$ we say that x is *uncolored.*

4.13 Lemma (Partial coloring lemma). *Let $\mathcal{F}$ and $\mathcal{M}$ be set systems*[1] *on an n-point set X, $|\mathcal{M}| > 1$, such that $|M| \le s$ for every $M \in \mathcal{M}$ and*

$$\prod_{F \in \mathcal{F}} (|F| + 1) \le 2^{(n-1)/5}. \tag{4.6}$$

Then there exists a partial coloring $\chi\colon X \to \{-1, 0, +1\}$, such that at least $\frac{n}{10}$ elements of X are colored, $\chi(F) = 0$ for every $F \in \mathcal{F}$, and $|\chi(M)| \le \sqrt{2s \ln(4|\mathcal{M}|)}$ for every $M \in \mathcal{M}$.

For brevity, let us call a partial coloring that colors at least 10% of the points a *no-nonsense partial coloring.*

Intuitively, the situation is as follows. We have the "few" sets of $\mathcal{F}$, for which we insist that the discrepancy of χ be 0. Each such $F \in \mathcal{F}$ thus puts one condition on χ. It seems plausible that if we do not put too many conditions then a coloring χ randomly selected among those satisfying the conditions will still be "random enough" to behave as a true random coloring on the sets of $\mathcal{M}$. In the lemma, we claim something weaker, however: instead of a "true" coloring $\chi\colon X \to \{+1, -1\}$ we obtain a no-nonsense partial coloring χ, which is only guaranteed to be nonzero at a constant fraction of points. (And, indeed, under the assumptions of the Partial coloring lemma, one cannot hope for a full coloring with the discrepancy stated. For example, although every system of $\frac{n}{10 \log n}$ sets on n points has a no-nonsense partial coloring with zero discrepancy, there are such systems with discrepancy about $\sqrt{n/\log n}$.)

Proof of Lemma 4.13. Let $\mathcal{C}_0$ be the set of all colorings $\chi\colon X \to \{-1, +1\}$, and let $\mathcal{C}_1$ be the subcollection of colorings χ with $|\chi(M)| \le \sqrt{2s \ln(4|\mathcal{M}|)}$ for all $M \in \mathcal{M}$. We have $|\mathcal{C}_1| \ge \frac{1}{2}|\mathcal{C}_0| = 2^{n-1}$ by the Random coloring lemma 4.1.

Now let us define a mapping $b\colon \mathcal{C}_1 \to \mathbf{Z}^{|\mathcal{F}|}$, assigning to a coloring χ the $|\mathcal{F}|$-component integer vector $b(\chi) = (\chi(F)\colon F \in \mathcal{F})$ (where the sets of $\mathcal{F}$ are taken in some arbitrary but fixed order). Since $|\chi(F)| \le |F|$ and $\chi(F) - |F|$ is even for each F, the image of b contains at most

$$\prod_{F \in \mathcal{F}} (|F| + 1) \le 2^{(n-1)/5}$$

distinct vectors. Hence there is a vector $b_0 = b(\chi_0)$ such that b maps at least $2^{4(n-1)/5}$ elements of $\mathcal{C}_1$ to b_0 (the pigeonhole principle!). Put $\mathcal{C}_2 = \{\chi \in \mathcal{C}_1\colon b(\chi) = b_0\}$. Let us fix an arbitrary $\chi_1 \in \mathcal{C}_2$ and for every $\chi_2 \in \mathcal{C}_2$, let us define a new mapping $\chi'\colon X \to \{-1, 0, 1\}$ by $\chi'(x) = \frac{1}{2}(\chi_2(x) - \chi_1(x))$. Then $\chi'(F) = 0$ for all $F \in \mathcal{F}$, and also $\chi'(M) \le \sqrt{2s \ln(4|\mathcal{M}|)}$ for all $M \in \mathcal{M}$. Let $\mathcal{C}_2'$ be the collection of the χ' for all $\chi_2 \in \mathcal{C}_2$.

To prove the lemma, it remains to show that there is a partial coloring $\chi' \in \mathcal{C}_2'$ that colors at least $\frac{n}{10}$ points of X. The number of mappings $X \to \{-1, 0, +1\}$ with fewer than $\frac{n}{10}$ nonzero elements is bounded by

[1] $\mathcal{F}$ for "few" sets, $\mathcal{M}$ for "minute" (or also "many") sets.

$$N = \sum_{0 \le q < n/10} \binom{n}{q} 2^q;$$

we will show that $N < |\mathcal{C}_2'|$. We may use the estimate

$$\sum_{0 \le i \le z} \binom{n}{i} a^i \le \left(\frac{ean}{z}\right)^z \tag{4.7}$$

(valid for any $n \ge z > 0$ and any real $a \ge 1$),[2] which in our case yields that

$$N < \left(\frac{2en}{n/10}\right)^{n/10} < 60^{n/10} < 2^{6n/10} < 2^{4(n-1)/5} \le |\mathcal{C}_2'|.$$

Hence there exists a partial coloring $\chi' \in \mathcal{C}_2'$ with at least $\frac{n}{10}$ points colored. □

Suppose that we want a low-discrepancy coloring of a set system $(X, \mathcal{S})$. How do we apply the Partial coloring lemma? Usually we look for an auxiliary set system $\mathcal{F}$ such that

- $\mathcal{F}$ has sufficiently few sets. More exactly, it satisfies the condition $\prod_{F \in \mathcal{F}}(|F|+1) \le 2^{(n-1)/5}$, where $n = |X|$.
- Each set $S \in \mathcal{S}$ can be written as a disjoint union of some sets from $\mathcal{F}$, plus some extra set M_S which is small (smaller than some parameter s, for all $S \in \mathcal{S}$).

We then define $\mathcal{M} = \{M_S \colon S \in \mathcal{S}\}$. The Partial coloring lemma yields a partial coloring χ which has zero discrepancy on all sets of $\mathcal{F}$, and so $|\chi(S)| = |\chi(M_S)| = O(\sqrt{s \log |\mathcal{S}|})$. In this way, some 10% of points of X are colored. We then look at the set of yet uncolored points, restrict the system $\mathcal{S}$ on these points, and repeat the construction of a partial coloring. In $O(\log n)$ stages, everything will be colored. This scheme has many variations, of course.

Here we describe an application in an upper bound for the so-called *Tusnády's problem*: What is the combinatorial discrepancy for axis-parallel rectangles in the plane, i.e. $\operatorname{disc}(n, \mathcal{R}_2)$? By the transference lemma (Proposition 1.8), this discrepancy is asymptotically at least as large (for infinitely many n) as the Lebesgue-measure discrepancy $D(n, \mathcal{R}_2)$, and the latter quantity is known to be of the order $\log n$ (see Proposition 2.2 and Schmidt's theorem 6.2). But obtaining tight bounds for Tusnády's problem seems quite hard, and the subsequent theorem gives nearly the best known upper bound.

[2] Here is a few-line proof of (4.7): By the binomial theorem, we have $(1+ax)^n \ge \sum_{0 \le i \le z} \binom{n}{i} a^i x^i$, so for $0 < x \le 1$ we get $\sum_{0 \le i \le z} \binom{n}{i} a^i \le \sum_{0 \le i \le z} \binom{n}{i} a^i x^{i-z} \le (1+ax)^n / x^z \le e^{axn}/x^z$ (since $1+y \le e^y$ for all real y). The estimate follows by substituting $x = z/an$.

4.14 Theorem. *The combinatorial discrepancy for axis-parallel rectangles satisfies*

$$\operatorname{disc}(n, \mathcal{R}_2) = O(\log^{5/2} n \sqrt{\log\log n}\,).$$

The $\sqrt{\log\log n}$ factor can be removed from the bound by a more sophisticated method (Exercise 5.5.2) but currently it is not known how to improve the exponent of $\log n$, let alone what the correct bound is.

Proof. First we construct a partial coloring. Let $P \subset \mathbf{R}^2$ be an n-point set, and let $p_1, p_2, \ldots, p_n$ be its points listed in the order of increasing x-coordinates (without loss of generality, we may assume that all the x-coordinates and all the y-coordinates of the points of P are pairwise distinct). Define a *canonical interval of P in the x-direction* as a subset of P of the form $\{p_{k2^q+1}, p_{k2^q+2}, \ldots, p_{(k+1)2^q}\}$. Here is a schematic illustration:

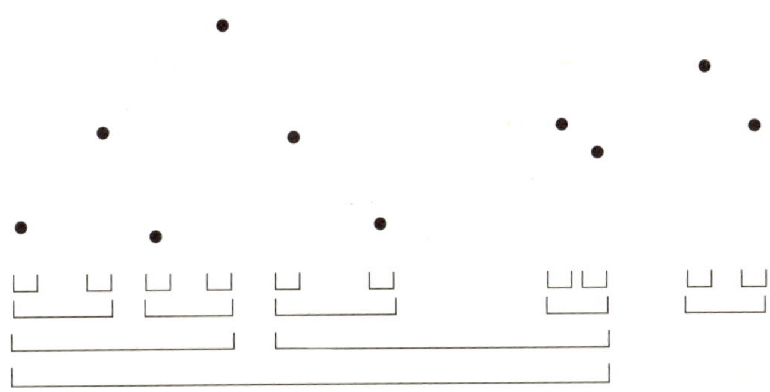

Let $\mathcal{C}$ be the collection of all canonical intervals of P in the x-direction, of all possible lengths 2^q with $1 \leq 2^q \leq n$. By considerations analogous to the ones in the proof of Proposition 2.2 (Claim II), we see that any interval in P, of the form $\{p_i, p_{i+1}, \ldots, p_{i+j}\}$, can be expressed as a disjoint union of at most $2\lfloor \log_2 n + 1 \rfloor \leq 2 \log 2n$ sets of $\mathcal{C}$.

For each canonical interval $C \in \mathcal{C}$, consider the collection of all canonical intervals of C in the y-direction (defined analogously to the canonical intervals in x-direction). Discard those of size smaller than t, where t is a threshold parameter (to be determined later). Call the collection of the remaining canonical intervals in the y-direction $\mathcal{F}_C$, and put $\mathcal{F} = \bigcup_{C \in \mathcal{C}} \mathcal{F}_C$.

Let $\mathcal{R}_2|_P$ be the set system defined on P by axis-parallel rectangles. We claim that for any rectangle $R \in \mathcal{R}_2$, the set $P \cap R$ can be written as a disjoint union of some sets from $\mathcal{F}$ plus a set of at most $s = 4t \log_2 2n$ extra points. To see this, we extend the rectangle R to an infinite vertical strip V. The set $P \cap V$ can be decomposed into at most $2 \log_2 2n$ disjoint sets from $\mathcal{C}$. For any C in this decomposition, $C \cap R$ is in fact an intersection of C with an infinite horizontal strip, and can thus be decomposed into disjoint sets from $\mathcal{F}_C$ plus an extra set consisting of at most $4t$ points. Such a decomposition is schematically depicted below:

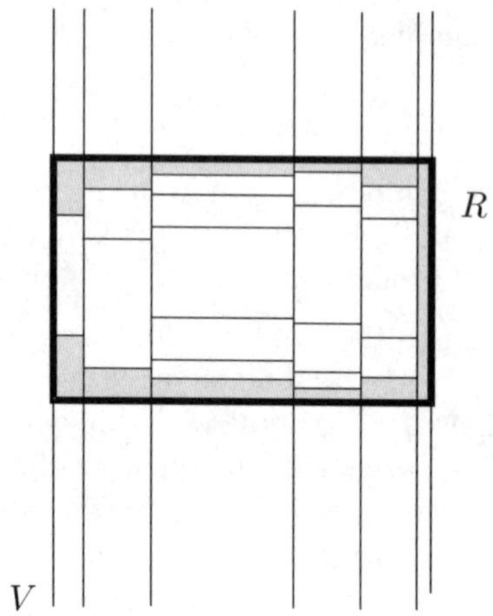

From this, the claim follows.

For the considered rectangle R, let M_R denote the set of the at most $s = O(t \log n)$ extra points, i.e. the points of $R \cap P$ that are not covered by the sets from $\mathcal{F}$ in the decomposition (these are the points of P in the gray region in the above schematic picture). Define $\mathcal{M} = \{M_R\colon R \in \mathcal{R}_2\}$. We have $|\mathcal{M}| \le |\mathcal{R}_2|_P| = O(n^4)$. We plan to apply the Partial coloring lemma for the set systems $\mathcal{F}$ and $\mathcal{M}$, so we need to choose the parameter t in such a way that $\prod_{\mathcal{F}}(|F|+1) \le 2^{(n-1)/5}$. If $C \in \mathcal{C}$ has 2^q points, then $\mathcal{F}_C$ contains 2^{q-i} sets of size 2^i, $\lceil \log_2 t \rceil \le i \le q$. Thus,

$$\log_2\Biggl(\prod_{F\in\mathcal{F}_C}(|F|+1)\Biggr) \le \sum_{i=\lceil \log_2 t\rceil}^{q} 2^{q-i}\log_2(2^i+1) = O\biggl(\frac{2^q \log t}{t}\biggr).$$

The system $\mathcal{C}$ contains $\lfloor n/2^q \rfloor$ sets C of size 2^q, so we have

$$\log_2\Biggl(\prod_{F\in\mathcal{F}}(|F|+1)\Biggr) \le \sum_{q=\lceil \log_2 t\rceil}^{\lfloor \log_2 n\rfloor} \frac{n}{2^q}\, O\biggl(\frac{2^q \log t}{t}\biggr) = O\biggl(\frac{n \log n \log t}{t}\biggr).$$

We see that in order to satisfy the assumption $\prod_{\mathcal{F}}(|F|+1) \le 2^{(n-1)/5}$ of the Partial coloring lemma, t should be chosen as $K \log n \log\log n$, for a sufficiently large constant K. Then the size of sets in $\mathcal{M}$ is bounded by $s = O(t \log n) = O(\log^2 n \log\log n)$.

From the Partial coloring lemma, we obtain a no-nonsense partial coloring χ satisfying $\mathrm{disc}(\chi, \mathcal{M}) = O(\sqrt{s \log n}) = O(\log^{3/2} n \sqrt{\log\log n}\,)$ and $\mathrm{disc}(\chi, \mathcal{F}) = 0$. For any rectangle $R \in \mathcal{R}_2$, we thus have $|\chi(P \cap R)| = |\chi(M_R)| = O(\log^{3/2} n \sqrt{\log\log n}\,)$.

To prove Theorem 4.14, we apply the construction described above iteratively, as the following drawing indicates:

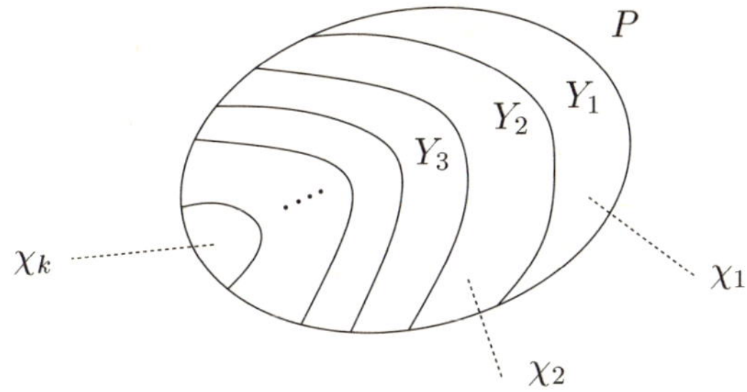

We set $P_1 = P$, and we construct a partial coloring χ_1 as above. Let Y_1 be the set of points colored by χ_1 and let $P_2 = P \setminus P_1$ be the uncolored points. We produce a partial coloring χ_2 of P_2 by applying the above construction to the set system $\mathcal{R}_2|_{P_2}$, and so on. We repeat this construction until the size of the set P_k becomes trivially small, say smaller than a suitable constant—this means $k = O(\log n)$. Then we define $Y_k = P_k$ and we let χ_k be the constant mapping with value 1 on Y_k. Finally we put $\chi(p) = \chi_i(p)$ for $p \in Y_i$.

Let $R \in \mathcal{R}_2$ be a rectangle. We have

$$|\chi(P \cap R)| \leq \sum_{i=1}^{k} |\chi_i(Y_i \cap R)| \leq \sum_{i=1}^{O(\log n)} O(\log^{3/2} n \sqrt{\log\log n}\,) = O(\log^{5/2} n \sqrt{\log\log n}\,).$$

□

Remark on Algorithms. The method of partial colorings is not algorithmic; the problem stems from the use of the pigeonhole principle in the proof of the Partial coloring lemma. (In fact, the problem mentioned earlier, with making Spencer's upper bound 4.2 effective, comes from the same source.) In some of the applications, the use of partial colorings can be replaced by the Beck–Fiala theorem 4.3. While one usually loses a few logarithmic factors, one obtains a polynomial-time algorithm—see Exercises 4 and 6.

Bibliography and Remarks. The partial coloring method was invented by Beck [Bec81b]; this is the paper with the first near-optimal upper bound for the discrepancy of arithmetic progressions (see the remarks to Section 4.2). The method was further elaborated by Beck in [Bec88a]. For other refinements see Section 4.6.

In 1980, Tusnády raised the question whether, in our terminology, the combinatorial discrepancy for axis-parallel rectangles is bounded by a constant (the question originated in an attempt to generalize results of Komlós et al. [KMT75] to higher dimensions). This was answered negatively by Beck [Bec81a], who also proved the upper bound of $O(\log^4 n)$. (The order of magnitude of the discrepancy is not interesting from the point of view of Tusnády's application but it is an intriguing problem in its own right.) This was improved to

$O((\log n)^{3.5+\varepsilon})$ by Beck [Bec89a] and to $O(\log^3 n)$ by Bohus [Boh90] via a bound for the "k-permutation problem" (Exercise 5). The possibility of a further slight improvement, to $O(\log^{5/2} n\sqrt{\log\log n}\,)$, was noted by the author of this book in a draft of this chapter. Independently, an $O(\log^{5/2} n)$ bound was recently proved by Srinivasan [Sri97] by a related but different method (see also Exercise 5.5.2). However, a generalization of the proof method shown above for Tusnády's problem gives an $O(\log^{d+1/2} n\sqrt{\log\log n}\,)$ bound in dimension d (see Exercise 1 or [Mat]), while the method of Srinivasan and the one indicated in Exercise 5.5.2 lead to worse bounds for $d > 2$. Another challenging problem is to determine the combinatorial L_2-discrepancy for axis-parallel boxes: while in the continuous setting, the L_2-discrepancy bounds are considerably better than the bounds for the worst-case discrepancy, no such improvement is known in the combinatorial setting.

Beck [Bec88a] investigated, as a part of more general questions, the Lebesgue-measure discrepancy for the family of translated and scaled copies of a fixed convex polygon P_0 in the plane and he proved an $O(\log^{4+\varepsilon} n)$ upper bound, with the constant of proportionality depending on ε and on P_0 (also see Beck and Chen [BC89]). His result in fact applies to a somewhat larger family. For a finite set H of hyperplanes in $\mathbf{R}^d$, let $\mathrm{POL}(H)$ denote the set of all polytopes $\bigcap_{i=1}^{\ell} \gamma_i$, where each γ_i is a halfspace with boundary parallel to some $h \in H$ (obviously, for each $h \in H$, it suffices to consider at most two γ_i parallel to h in the intersection). Beck's upper bound is valid for any family $\mathrm{POL}(H)$ with H a finite set of lines in the plane. Károlyi [Kár95a] studied a d-dimensional analogue of the problem and proved the upper bound $D(n, \mathrm{POL}(H)) = O((\log n)^{\max(3d/2+1+\varepsilon, 2d-1)})$ for any fixed H and an arbitrarily small $\varepsilon > 0$, with the constant of proportionality depending on H and on ε. He uses the partial coloring method plus a sophisticated way of decomposing the polytopes in $\mathrm{POL}(H)$ into "canonical" ones. Exercises 2 and 3 below indicate proofs of similar but quantitatively somewhat better bounds for the combinatorial discrepancy of $\mathrm{POL}(H)$ (at least for dimensions 2 and 3). A detailed discussion of these bounds is in [Mat]. The best estimates for the Lebesgue-measure discrepancy for $\mathrm{POL}(H)$ have recently been obtained by Skriganov [Skr98], whose results imply an $O(\log^{d-1} n(\log\log n)^{1+\varepsilon})$ upper bound, for any fixed finite H in $\mathbf{R}^d$ (this paper is discussed in the remarks to Section 2.5).

The 3-permutation problem discussed in Exercise 5 and Exercise 5.5.3 remains one of the most tantalizing questions in combinatorial discrepancy.

Exercises

1.* Consider a d-dimensional version of Tusnády's problem; generalize the method shown for the planar case to prove the upper bound $\mathrm{disc}(n,\mathcal{R}_d) = O(\log^{d+1/2} n\sqrt{\log\log n}\,)$ for any fixed d.

2. (Discrepancy for translates I)
(a)* Let T_0 be a triangle in the plane, and let $\mathcal{T}$ denote the family of all translated and scaled copies of $\mathcal{T}$ (no rotation allowed). Show that there is a plane $\rho \subset \mathbf{R}^3$ such that if $\mathbf{R}^2$ is identified with ρ then any triangle $T \in \mathcal{T}$ can be written as $T = \rho \cap R$ for some axis-parallel box $R \in \mathcal{R}_3$. By the result of Exercise 1, this implies that $\mathrm{disc}(n,\mathcal{T}) = O(\log^{3.5} n\sqrt{\log\log n})$.
(b) More generally, let H be a finite set of hyperplanes in $\mathbf{R}^d$, and define $\mathrm{POL}(H)$ as in the remarks above, i.e. as the set of all polytopes $\bigcap_{i=1}^{\ell} \gamma_i$, where each γ_i is a halfspace with boundary parallel to some $h \in H$. Using a suitable embedding of $\mathbf{R}^d$ into $\mathbf{R}^{|H|}$ and Exercise 1, derive that $\mathrm{disc}(n,\mathrm{POL}(H)) = O((\log n)^{|H|+1/2}\sqrt{\log\log n})$.

3. (Discrepancy for translates II)
(a) Modify the proof of Theorem 4.14 to show that if $H_1, H_2, \ldots, H_k$ are families consisting of two lines each, where k is considered as a constant, then
$$\mathrm{disc}(n,\mathrm{POL}(H_1)\cup\mathrm{POL}(H_2)\cup\cdots\cup\mathrm{POL}(H_k)) = O(\log^{5/2} n\sqrt{\log\log n})$$
(the same bound as for axis-parallel rectangles), with the constant of proportionality depending on k. The notation $\mathrm{POL}(H)$ is as in Exercise 2(a).
(b)* Using (a), improve the result of Exercise 2(a) to $\mathrm{disc}(n,\mathcal{T}) = O(\log^{5/2} n\sqrt{\log\log n})$.
(c)* More generally, if H is a set of k lines in the plane, with k a constant, prove $\mathrm{disc}(n,\mathrm{POL}(H)) = O(\log^{5/2} n\sqrt{\log\log n})$, with the constant of proportionality depending on k.
(d)** Generalize part (c) to dimension 3 (or even higher). That is, if H is a family of k planes in $\mathbf{R}^3$ with k fixed, then $\mathrm{disc}(n,\mathrm{POL}(H)) = O(\log^{3.5} n\sqrt{\log\log n})$. (Details of this can be found in [Mat].)

4.* Prove an upper bound $\mathrm{disc}(n,\mathcal{R}_2) = O(\log^4 n)$ by using the Beck–Fiala theorem 4.3 instead of the Partial coloring lemma.

5. (The k-permutation problem) Let $X = \{1,2,\ldots,n\}$, and let $\pi_1,\ldots,\pi_k$ be arbitrary permutations of X (bijective mappings $X \to X$). Define a set system $\mathcal{P}_k = \mathcal{P}(\pi_1)\cup\mathcal{P}(\pi_2)\cup\ldots\cup\mathcal{P}(\pi_k)$, where $\mathcal{P}(\pi)$ denotes the family of all initial segments along π; that is, $\mathcal{P}(\pi) = \{\{\pi(1),\pi(2),\pi(3),\ldots,\pi(q)\}\colon 1 \le q \le n\}$.
(a)* Show that for $k=2$, $\mathrm{disc}(\mathcal{P}_2) \le 1$ (for all choices of π_1, π_2).
(b)* Use the Partial coloring lemma to prove $\mathrm{disc}(\mathcal{P}_k) = O(\log n)$ for any fixed k. What is the dependence of the constant of proportionality on k in the resulting bound? (Also see Exercise 5.5.3.)

(c) Prove that $\operatorname{disc}(\mathcal{P}_k)$ is not bounded by a constant independent of k. Let us remark that the question whether $\operatorname{disc}(\mathcal{P}_3)$ is bounded by some constant is well-known and probably difficult (the *three-permutation problem*).

6.* Prove an upper bound of $O(\log^2 n)$ for the discrepancy of a set system defined by 3 permutations as in Exercise 5 using the Beck–Fiala theorem 4.3 instead of the Partial coloring lemma.

7. Consider the set system $\mathcal{A}_n$ consisting of all arithmetic progressions in $\{1, 2, \ldots, n\}$, that is,

$$\mathcal{A}_n = \{\{a_0, a_0 + d, a_0 + 2d, \ldots\} \cap \{1, 2, \ldots, n\} \colon a_0, d \in \mathbf{N}\}.$$

(a)* Prove that $\mathcal{A}_n$ has a no-nonsense partial coloring with discrepancy $O(n^{1/4} \log^{3/4} n)$ (if you can't get this try to get at least a bigger power of $\log n$).

(b) Explain why (a) cannot be used iteratively in a straightforward manner to conclude that $\operatorname{disc}(\mathcal{A}_n) = O(n^{1/4} \log^{7/4})$.

(c)* Let $X \subseteq \{1, 2, \ldots, n\}$ be an m-element set. Show that the restriction of $\mathcal{A}_n$ on X also has a no-nonsense partial coloring with discrepancy $O(n^{1/4} \log^{3/4} n)$. This already implies $\operatorname{disc}(\mathcal{A}_n) = O(n^{1/4} \log^{7/4} n)$.

(d) Why doesn't the Beck–Fiala theorem seem to be directly applicable for getting a bound close to $n^{1/4}$ in this problem?

Remark. A slightly better upper bound will be proved in Exercise 5.5.4.

4.6 The Entropy Method

We are going to discuss a refinement of the Partial coloring lemma 4.13 which can often save logarithmic factors in discrepancy bounds. For instance, suppose that we have a set system $\mathcal{S}$ and two auxiliary set systems $\mathcal{F}$ and $\mathcal{M}$ as in the Partial coloring lemma, such that any set of $\mathcal{S}$ can be expressed as a disjoint union of a set from $\mathcal{F}$ and a set from $\mathcal{M}$. If the assumptions of the Partial coloring lemma are met (in particular, this means that $\mathcal{F}$ has somewhat fewer than n sets) then the lemma gives us a partial coloring where the sets of $\mathcal{F}$ have zero discrepancy, while the sets of $\mathcal{M}$ have discrepancy roughly as if colored randomly. The exactly zero discrepancy of the sets of $\mathcal{F}$ is somewhat wasteful, however, since it would be quite sufficient to make their discrepancy of the same order as the discrepancy of the sets of $\mathcal{M}$. With this idea in mind, let us look at the proof of the Partial coloring lemma again.

In that proof, we have exhibited two (full) colorings χ_1 and χ_2 differning on many elements and satisfying $\chi_1(S) = \chi_2(S)$ for all sets $S \in \mathcal{F}$. We now want to relax the latter condition, and only require that $|\chi_1(S) - \chi_2(S)| < 2\Delta_S$, where Δ_S is the required bound for the discrepancy of S. If this condition is satisfied, then the "difference coloring" $\chi = \frac{1}{2}(\chi_1 - \chi_2)$ has $|\chi(S)| < \Delta_S$.

If we fix some discrepancy bound Δ_S for each of the considered sets S, we need not distinguish between sets of two types anymore as we did in the Partial coloring lemma. The sets $F \in \mathcal{F}$ in that lemma would simply have $\Delta_F = 1$, while the sets $M \in \mathcal{M}$ would have $\Delta_M = \sqrt{2s\ln(4|\mathcal{M}|)}$. So we work with a single set system $\mathcal{S}$, but we will typically need some knowledge about the distribution of the sizes of sets.

For an application of the pigeonhole principle as in the proof of the Partial coloring lemma, we need to replace the inequality $|\chi_1(S) - \chi_2(S)| < 2\Delta_S$ by an equality condition (so that we can assign a pigeonhole to every coloring). A suitable replacement for this inequality is

$$\mathrm{round}\left(\frac{\chi_1(S)}{2\Delta_S}\right) = \mathrm{round}\left(\frac{\chi_2(S)}{2\Delta_S}\right),$$

where $\mathrm{round}(x) = \lfloor x + \frac{1}{2} \rfloor$ denotes rounding to the nearest integer.

For a coloring $\chi: X \to \{-1, +1\}$ and a set $S \in \mathcal{S}$, let us put

$$b_S = b_S(\chi) = \mathrm{round}\left(\frac{\chi(S)}{2\Delta_S}\right),$$

and let $b = b(\chi)$ be the vector $(b_S\colon S \in \mathcal{S})$ (the sets of $\mathcal{S}$ are taken in some order fixed once and for all). The value of $b(\chi)$ is the pigeonhole where the pigeon χ is supposed to live.

The set $\mathcal{C}$ of all possible colorings $\chi: X \to \{-1, +1\}$ is partitioned into classes according to the value of the vector $b(\chi)$. We want to show that there exists a big class, since in a big class we have two colorings χ_1, χ_2 differing in sufficiently many points. Their difference coloring $\frac{1}{2}(\chi_1 - \chi_2)$ will be the partial coloring giving discrepancy below Δ_S to each set $S \in \mathcal{S}$. Up to the definition of the classes, this is the same argument as in the proof of the Partial coloring lemma, and we have already calculated how big a big class should be: a class containing at least $2^{4n/5}$ colorings has some two colorings χ_1, χ_2 differing in at least $\frac{n}{10}$ components, and hence provides a no-nonsense partial coloring, i.e. one that colors at least $\frac{n}{10}$ points.

Thus, it remains to show the existence of a big class. To this end, it is very convenient to use entropy.

Entropy. Let Z be a discrete random variable attaining values in a finite set V. (Our main example of such a variable will be the $b(\chi)$ defined above, which is a function of the random coloring χ.) For $v \in V$, let p_v denote the probability of the event "$Z = v$." The *entropy* of Z, denoted by $H(Z)$, is defined by

$$H(Z) = -\sum_{v \in V} p_v \log_2 p_v.$$

One can think of entropy as the average number of bits of information we gain by learning the value of Z. For instance, if there are m equally likely values, then by learning which one was actually attained we gain $\log_2 m$ bits of information. If there are only two possible values, rain tomorrow and no rain tomorrow, and we're in the middle of a rainy season, then rain tomorrow brings almost no information, and a sunny day is a big surprise but extremely unlikely, so the total entropy is small (here entropy measures the "expected surprise," so to speak).

We need three basic properties of entropy.

1. (Good chance) *If $H(Z) \leq K$ then some value $v \in V$ is attained by Z with probability at least 2^{-K}.*
2. (Uniformity optimal) *If Z attains at most k distinct values, then $H(Z) \leq \log_2 k$ (with equality when Z is uniformly distributed on k values).*
3. (Subadditivity) *Let $Z_1, Z_2, \ldots, Z_m$ be arbitrary discrete random variables, and let $Z = (Z_1, Z_2, \ldots, Z_m)$ be the random vector with components $Z_1, Z_2, \ldots, Z_m$. Then we have $H(Z) \leq H(Z_1) + H(Z_2) + \cdots + H(Z_m)$.*

The first property is immediate from the definition of entropy. The other two need some work to prove but are not difficult either. The subadditivity is intuitively obvious from the "average information" interpretation of entropy. The inequality may be strict, for instance if the vector consists of several copies of the same random variable.

Partial Coloring from Entropy. Let a set system $\mathcal{S}$ and numbers Δ_S be given, and let the vector $b = b(\chi)$ be defined as above. If χ is a random coloring, then $b(\chi)$ is a random variable. If we could prove that its entropy $H(b)$ is at most $\frac{n}{5}$, then by the first property of entropy (good chance), some value $\bar{b}$ is attained by $b(\chi)$ with probability at least $2^{-n/5}$. This means that the class of colorings with $b(\chi) = \bar{b}$ has at least $2^{4n/5}$ members. Together with the previous considerations, we obtain

4.15 Lemma. *If $H(b) \leq \frac{n}{5}$ then there exists a no-nonsense partial coloring χ such that $|\chi(S)| < \Delta_S$ holds for all $S \in \mathcal{S}$.* □

To apply the method in specific examples, we need to estimate $H(b)$. As a first step, we use subadditivity: $H(b) \leq \sum_{S \in \mathcal{S}} H(b_S)$. It remains to estimate the entropies $H(b_S)$ for the individual sets and sum up. A large part of this calculation can be done once and for all. The distribution of b_S, and thus also its entropy, only depends on the size of S and on Δ_S. Our aim is thus to estimate the function $h(s, \Delta)$ defined as the entropy of b_S with $|S| = s$ and $\Delta_S = \Delta$.

As we know, for large s and for χ random, $\chi(S)$ behaves roughly as a normal random variable with standard deviation $\sqrt{s}$. By passing from $\chi(S)$ to b_S, we "shrink" all values of $\chi(S)$ from an interval $[(2i-1)\Delta, (2i+1)\Delta)$ to the single value i, thus forgetting some information about $\chi(S)$ (and thereby lowering the entropy).

Let us put $\lambda = \Delta/\sqrt{s}$. For estimating $h(s, \Delta)$, we distinguish two cases. If $\lambda \geq 2$, then $\chi(S)$ almost always lies in the interval $[-\Delta, \Delta)$ where b_S is 0,

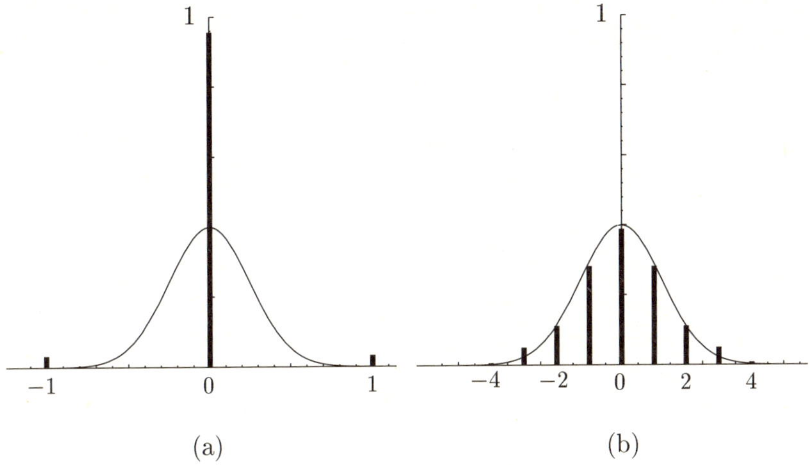

Fig. 4.1. Probability distributions of $\chi(S)/2\Delta$ and of b_S for $\lambda = 2$ (a), and for $\lambda = 0.4$ (b).

so the entropy will be small. Fig. 4.1(a) shows the probability distribution of $\chi(S)/2\Delta_S$ and the distribution of b_S for $\lambda = 2$ (with s large). On the other hand, for $\lambda < 2$, b_S has reasonable chance of attaining nonzero values, so the entropy will be larger (Fig. 4.1(b)). In other words, we do not violate the natural order of things so much by insisting that $|\chi(S)| < 10\sqrt{s}$ holds for the partial coloring, say, since a random coloring typically has this property anyway, and so we do not pay much entropy. On the other hand, requiring that $|\chi(S)| < \frac{\sqrt{s}}{10}$ already imposes quite a strong condition, so we need more entropy to compensate.

Having indicated what to expect, we do the actual calculation now.

The Case $\lambda \geq 2$. Let p_i denote the probability of $b_S = i$. We want to show that p_0 is very close to 1 and that the other p_i are small (Fig. 4.1(a)). For $i \geq 1$, we have

$$p_i \leq \Pr\left[\chi(S) \geq (2i-1)\Delta_S\right] = \Pr\left[\chi(S) \geq (2i-1)\lambda\sqrt{s}\right] \leq e^{-(2i-1)^2\lambda^2/2}$$

by Chernoff's inequality. By elementary calculus, the function $x \mapsto -x\log_2 x$ is nondecreasing on $(0, \frac{1}{e})$, and hence

$$-\sum_{i=1}^{\infty} p_i \log_2 p_i \leq \sum_{i=1}^{\infty} \frac{(2i-1)^2\lambda^2}{2\ln 2}\, e^{-(2i-1)^2\lambda^2/2}.$$

It is easy to check that the ratio of two successive terms in this series is smaller than $\frac{1}{4}$, and so by comparing the series with a geometric series we get that the sum is no larger than $\lambda^2 e^{-\lambda^2/2}$. By symmetry, the same bound applies for the contribution of the terms with $i \leq -1$.

For p_0 we derive

$$p_0 \geq 1 - \Pr\left[|\chi(S)| \geq \Delta_S\right] \geq 1 - 2e^{-\lambda^2/2}.$$

To estimate $\log_2 p_0$, we calculate that $2e^{-\lambda^2/2} < \frac{1}{3}$ for $\lambda \geq 2$, and we check that $\log_2(1-x) \geq -2x$ holds for $0 < x < \frac{1}{3}$ (more calculus). Therefore we have

$$-p_0 \log_2 p_0 \leq -\log_2 p_0 \leq -\log_2\left(1 - 2e^{-\lambda^2/2}\right) \leq 4e^{-\lambda^2/2}.$$

Altogether we obtain the estimate $h(s,\Delta) = H(b_S) \leq 6\lambda^2 e^{-\lambda^2/2}$. As one can check (by a computer algebra system, say), $6\lambda^2 e^{-\lambda^2/2}$ is bounded above by the simpler function $10e^{-\lambda^2/4}$ for all $\lambda \geq 2$.

The Case $\lambda < 2$. Here we can make use of the calculation done in the previous case, by the following trick. Let us decompose b_S into two parts $b_S = b'_S + b''_S$. The first addend b'_S is b_S rounded to the nearest integer multiple of $L = \lceil \frac{2}{\lambda} \rceil$; that is,

$$b'_S = L \operatorname{round}\left(\frac{b_S}{L}\right) = L \operatorname{round}\left(\frac{\chi(S)}{2L\Delta}\right).$$

Hence by the result of the $\lambda \geq 2$ case, we have $H(b'_S) = h(s, L\Delta) \leq 10e^{-(L\lambda)^2/4} \leq 4$, as $L\lambda \geq 2$. The second component, $b''_S = b_S - b'_S$, can only attain L different values, and thus its entropy is at most $\log_2 L$. Finally we obtain, by subadditivity,

$$\begin{aligned} h(s,\Delta) &= H(b_S) \leq H(b'_S) + H(b''_S) \leq 4 + \log_2 L \\ &\leq 4 + \log_2\left(\frac{2}{\lambda} + 1\right) \leq \log_2\left(16 + \frac{32}{\lambda}\right). \end{aligned}$$

The estimates in both cases can be combined into a single formula, as the reader is invited to check: $h(s,\Delta) \leq Ke^{-\lambda^2/4}\log_2(2 + 1/\lambda)$ for an absolute constant K. (This formula is a bit artificial, but it saves us from having to distinguish between two cases explicitly; it is a matter of taste whether it is better to write out the cases or not.) Plugging this into Lemma 4.15, we arrive at the following convenient device for applying the entropy method:

4.16 Proposition (Entropy method—quantitative version). *Let $\mathcal{S}$ be a set system on an n-point set X, and let a number $\Delta_S > 0$ be given for each $S \in \mathcal{S}$. Suppose that $\sum_{S \in \mathcal{S}} h(|S|, \Delta_S) \leq \frac{n}{5}$ holds, where the function $h(s,\Delta)$ can be estimated by*

$$h(s,\Delta) \leq Ke^{-\Delta^2/4s}\log_2\left(2 + \frac{\sqrt{s}}{\Delta}\right)$$

with an absolute constant K. Then there exists a no-nonsense partial coloring $\chi\colon X \to \{+1,-1,0\}$ such that $|\chi(S)| < \Delta_S$ for all $S \in \mathcal{S}$.

Often one only has upper bounds on the sizes of the sets in $\mathcal{S}$. In such a case, it is useful to know that the entropy contribution does not increase by decreasing the set size (while keeping Δ fixed). It suffices to check by elementary calculus that the function $s \mapsto e^{-\Delta^2/4s}\log_2(2+\sqrt{s}/\Delta)$ is nondecreasing in s.

Proof of Spencer's Upper Bound 4.2. We have a set system $\mathcal{S}$ on n points with m sets, $m \geq n$. We want to prove $\operatorname{disc}(\mathcal{S}) = O(\sqrt{n\ln(2m/n)})$. The desired coloring is obtained by iterating a partial coloring step based on Proposition 4.16.

To get the first partial coloring, we set $\Delta_S = \Delta = C\sqrt{n\ln(2m/n)}$ for all $S \in \mathcal{S}$, with a suitable (yet undetermined but sufficiently large) constant C. For the entropy estimate, we are in the region $\lambda \geq 2$, and so we have

$$\sum_{S\in\mathcal{S}} h(|S|,\Delta) \leq m\cdot h(n,\Delta) \leq m\cdot 10e^{-\Delta^2/4n} = m\cdot 10\left(\frac{n}{2m}\right)^{C^2} < \frac{n}{5}$$

for a sufficiently large C. Therefore, an arbitrary set system on n points with $m \geq n$ sets has a no-nonsense partial coloring with discrepancy at most $C\sqrt{n\ln(2m/n)}$.

Having obtained the first partial coloring χ_1 for the given set system $(\mathcal{S}, X_1)$, we consider the set system $\mathcal{S}_2$ induced on the set X_2 of points uncolored by χ_1, we get a partial coloring χ_2, and so on. The number of sets in $\mathcal{S}_i$ is at most m and the size of X_i is at most $\left(\frac{9}{10}\right)^i n$. We can stop the iteration at some step k when the number of remaining points drops below a suitable constant. The total discrepancy of the combined coloring is bounded by

$$\sum_{i=1}^{k} C\sqrt{\left(\tfrac{9}{10}\right)^i n\ln\left(\left(\tfrac{10}{9}\right)^i 2m/n\right)}.$$

After the first few terms, the series decreases geometrically, and thus the total discrepancy is $O\left(\sqrt{n\ln(2n/m)}\right)$ as claimed. □

Bibliography and Remarks. A refinement of Beck's partial coloring method similar to the one shown in this section was developed by Spencer [Spe85] for proving that the discrepancy of n sets on n points is $O(\sqrt{n})$. His method uses direct calculations of probability; the application of entropy, as suggested by Boppana for the same problem (see [AS92]), is a considerable technical simplification. As was remarked in Section 4.1, alternative geometric approaches to Spencer's result have been developed by Gluskin [Glu89] and by Giannopoulos [Gia97]; the latter paper can be particularly recommended for reading.

The possibility of taking set sizes into account and thus unifying the method, in a sense, with Beck's sophisticated applications of the partial coloring methods was noted in [Mat95]. Matoušek and Spencer

[MS96] use the method is a similar way, with an additional trick needed for a successful iteration of the partial coloring step, for proving a tight upper bound on the discrepancy of arithmetic progressions (see the remarks to Section 4.2). More applications of the entropy method can be found in [Sri97], [Mat96b], and [Mat98a].

Spencer's founding paper [Spe85] has the title "Six standard deviations suffice," indicating that the constant of proportionality can be made quite small: for instance, the discrepancy for n sets on n points is below $6\sqrt{n}$. The calculation shown above yields a considerably worse result. We were really wasteful only in the proof of the Partial coloring lemma, at the moment when we had a class of at least $2^{4n/5}$ colorings and concluded that it must contain two colorings at least $\frac{n}{10}$ apart. In reality, for instance, a class of this size contains colorings at least $0.48n$ apart. This follows from an isoperimetric inequality for the Hamming cube due to Kleitman [Kle66], which gives a tight bound on the number of points in a subset of the Hamming cube of a given diameter. Namely, any $\mathcal{C} \subseteq \{-1,+1\}^n$ of size bigger than $\sum_{j=0}^{\ell} \binom{n}{j}$ with $\ell \leq \frac{n}{2}$ contains two points differing in at least 2ℓ coordinates. This sum of binomial coefficients is conveniently bounded above by $2^{H(\ell/n)n}$, where $H(x) = -x\log_2 x - (1-x)\log_2(1-x)$ (here the customary notation $H(x)$ stands for a real function, not for the entropy of a random variable!). Using this result, it is sometimes possible to produce partial colorings with almost all points colored, say with at most $n^{0.99}$ uncolored points. Then much fewer than $\log n$ iterations of the partial coloring step are needed. An application, noted by Spencer, is given in Exercise 4.

Exercises

1. Prove the subadditivity property of entropy.
2. Prove that if a random variable Z attains at most k distinct values then $H(Z) \leq \log_2 k$.
3. (a) Let $\mathcal{S}$ be a system of n sets on an n-point set, and suppose that each set of $\mathcal{S}$ has size at most s. Check that the entropy method provides a no-nonsense partial coloring where each set of $\mathcal{S}$ has discrepancy at most $O(\sqrt{s})$.
   (b) Why can't we in general conclude that $\mathrm{disc}(\mathcal{S}) = O(\sqrt{s})$ for an $\mathcal{S}$ as in (a)? Show that this estimate is false in general.
4. (The discrepancy for m sets of size s) The goal is to show that for any m sets of size at most s, where $s \leq m$, the discrepancy is $O(\sqrt{s\log(2m/s)}\,)$. The important case, dealt with in (c), is an unpublished result of Spencer (private communication from September 1998).
   (a) Show that this bound, if valid, is asymptotically the best possible (at least for m bounded by a polynomial function of s, say).

(b) Why can we assume that n, the size of the ground set, equals m, and that $Cs \leq m \leq s^{1+\varepsilon}$ for arbitrary constants C and $\varepsilon > 0$?
(c)* With the assumptions as in (b), use Kleitman's isoperimetric inequality mentioned at the end of the remarks to this section to show that there is a partial coloring with at most s uncolored points for which each set has discrepancy at most $O(\sqrt{s \log(2m/s)})$.
(d) Using (a)–(c), prove the bound claimed at the beginning of this exercise.

5. VC-Dimension and Discrepancy

In this chapter, we introduce combinatorial parameters measuring the complexity of a set system: the shatter functions and the Vapnik–Chervonenkis dimension. These concepts have recently become quite important in several branches of pure and applied mathematics and of theoretical computer science, such as statistics, computational learning theory, combinatorial geometry, and computational geometry.

We will mainly consider them in connection with discrepancy. Recall that a general system of, say, n^2 sets on n points may have discrepancy as high as $c\sqrt{n \log n}$, i.e. the random coloring is essentially optimal. But if, moreover, the shatter functions of the considered system are polynomially bounded, which is the case in most geometric settings, we obtain a discrepancy upper bound of $O(n^{1/2-\delta})$ for some fixed $\delta > 0$. It even turns out that many of the current best upper bounds on the discrepancy for various families of geometric shapes, and sometimes provably tight bounds, can be derived solely from two general results using the shatter functions of the considered geometric families.

In Section 5.1, we define the shatter functions, we state the two upper bounds for discrepancy, and we review theorems that can be helpful for bounding the shatter functions for specific geometric families.

In Section 5.2, we introduce the Vapnik–Chervonenkis dimension, relate it to the shatter functions, and prove basic results about these concepts. Section 5.3 presents another auxiliary result with a more complicated proof. Then we finally get back to discrepancy and, in Sections 5.4 and 5.5, we prove the two upper bounds stated in Section 5.1.

5.1 Discrepancy and Shatter Functions

We have already derived the upper bound $D(n, \mathcal{B}_2) = O(n^{1/4}\sqrt{\log n})$ for the discrepancy for discs in the plane in Section 3.1. Such a bound holds for the combinatorial discrepancy for discs as well. It turns out that this bound is implied by a simple combinatorial property of the set systems induced by discs on finite point sets, and that upper bounds of this type can be applied for quite general geometric situations. In order to describe these results, we begin with the necessary definitions.

Let $(X, \mathcal{S})$ be a set system, where the ground set X may be finite or infinite.

5.1 Definition (Primal shatter function). *The* primal shatter function *of $(X, \mathcal{S})$ is a function, denoted by $\pi_{\mathcal{S}}$, whose value at m ($m = 0, 1, 2, \ldots$) is defined by*

$$\pi_{\mathcal{S}}(m) = \max_{Y \subseteq X,\ |Y|=m} \left|\mathcal{S}|_Y\right|.$$

In words, $\pi_{\mathcal{S}}(m)$ is the maximum possible number of distinct intersections of the sets of $\mathcal{S}$ with an m-point subset of X. If X is finite, then the domain of $\pi_{\mathcal{S}}$ is $\{0, 1, 2, \ldots, |X|\}$, and for X infinite, the domain is $\mathbf{N}$.

Another way to understand the primal shatter function is via the incidence matrix A of the set system $(X, \mathcal{S})$, with rows indexed by sets of $\mathcal{S}$ and columns indexed by points of X. The value $\pi_{\mathcal{S}}(m)$ is the maximum number of distinct rows appearing in any m-column submatrix of A.

For example, in Lemma 3.3 we have we have shown that at most $O(m^3)$ subsets can be defined by circular discs on an m-point set in the plane. In terms of the primal shatter function, this says that $\pi_{\mathcal{B}_2}(m) = O(m^3)$, where $\mathcal{B}_2$ denotes the family of all closed discs in the plane. A similar (but simpler) argument can be used for showing that for the system $\mathcal{H}_2$ of all closed halfplanes in the plane, we have $\pi_{\mathcal{H}_2}(m) = O(m^2)$. That is, for any m-point set Y in the plane, at most $O(m^2)$ subsets of Y can be "cut off" by a halfplane. Below we describe general tools for bounding shatter functions of various geometric families, and the bounds for halfplanes and for discs are simple consequences.

An example of a different nature is the family of all convex sets in the plane. Here the primal shatter function is 2^m (see Exercise 5.2.2).

Next, we define the dual shatter function. This is just the primal shatter function of the set system dual to $\mathcal{S}$, whose incidence matrix arises by transposing the incidence matrix of $\mathcal{S}$ (and possibly deleting repeated rows). Formulating this explicitly, we get

5.2 Definition (Dual shatter function). *The* dual shatter function *of a set system $(X, \mathcal{S})$ is a function, denoted by $\pi^*_{\mathcal{S}}$, whose value at m is defined as the maximum number of equivalence classes on X defined by an m-element subfamily $\mathcal{Y} \subseteq \mathcal{S}$, where two points $x, y \in X$ are* equivalent with respect to $\mathcal{Y}$ *if x belongs to the same sets of $\mathcal{Y}$ as y does. In other words, $\pi^*_{\mathcal{S}}(m)$ is the maximum number of nonempty cells in the Venn diagram of m sets of $\mathcal{S}$. If $\mathcal{S}$ is finite, the domain of $\pi^*_{\mathcal{S}}$ is $\{0, 1, 2, \ldots, |\mathcal{S}|\}$, and for an infinite $\mathcal{S}$ the domain is $\mathbf{N}$.*

The dual shatter function is perhaps more intuitive in geometric setting than the primal shatter function. For example, for determining $\pi^*_{\mathcal{H}_2}(m)$, the dual shatter function for (closed) halfplanes, we are simply asking what is the maximum number of regions into which m halfplanes partition the plane.

Here two points belong to the same region if they lie in the same subset of the given m halfplanes. Such regions for 4 halfplanes are illustrated below:

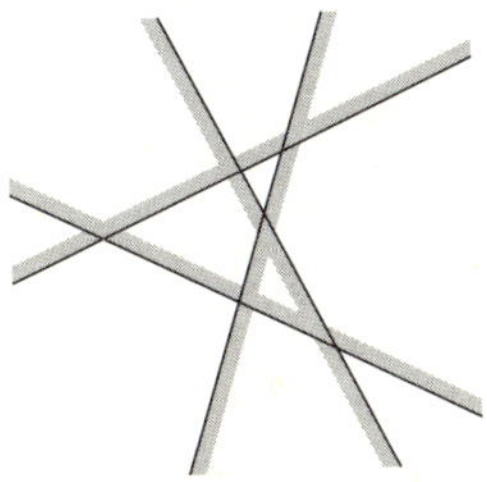

Since adding an mth halfplane to any given $m-1$ halfplanes divides at most m of the existing regions into two pieces, induction gives $\pi^*_{\mathcal{H}_2}(m) = O(m^2)$. Similarly, for discs one finds that $\pi^*_{\mathcal{B}_2}(m) = O(m^2)$.

Discrepancy Bounds. The following two theorems bound the discrepancy of a set system on an n-point set in terms of its shatter functions. Perhaps surprisingly, these bounds are often tight or near-tight for many geometrically defined set systems.

5.3 Theorem (Primal shatter function bound). *Let $d > 1$ and C be constants, and let $\mathcal{S}$ be a set system on an n-point set X with primal shatter function satisfying $\pi_{\mathcal{S}}(m) \le Cm^d$ for all $m \le n$. Then*

$$\mathrm{disc}(\mathcal{S}) \le C' n^{1/2-1/2d},$$

where the constant C' depends on C and d.

Since $\pi_{\mathcal{H}_2}(m) = O(m^2)$, this theorem implies an $O(n^{1/4})$ upper bound for the combinatorial discrepancy of halfplanes. In Section 6.4, we show that this bound is the best possible up to the multiplicative constant. Similarly, for halfspaces in $\mathbf{R}^d$, the primal shatter function is $O(m^d)$, and so we get an $O(n^{1/2-1/2d})$ bound for their discrepancy, which is also asymptotically tight. Therefore, the bound in Theorem 5.3 is the best possible in general for all integers $d \ge 2$. In Exercise 5 below, we indicate how to prove this last fact in a more direct way, using a purely combinatorial argument instead of a geometric one.

On the other hand, discs in the plane have the primal shatter function of the order m^3, and so Theorem 5.3 only gives an $O(n^{1/3})$ discrepancy upper bound, which is way off the mark—we already know that an $O(n^{1/4}\sqrt{\log n})$ bound holds true, at least for the Lebesgue-measure discrepancy. The same bound in the more general combinatorial setting follows from the next theorem.

5.4 Theorem (Dual shatter function bound). *Let $d > 1$ and C be constants and let $\mathcal{S}$ be a set system on an n-point set X with dual shatter function satisfying $\pi^*_{\mathcal{S}}(m) \le Cm^d$ for all $m \le |\mathcal{S}|$. Then*

$$\mathrm{disc}(\mathcal{S}) \leq C' n^{1/2-1/2d}\sqrt{\log n},$$

where the constant C' depends on C and d.

For halfplanes in the plane, the dual shatter function is $O(m^2)$, and so this theorem gives a slightly worse bound than Theorem 5.3 above, only $O(n^{1/4}\sqrt{\log n})$. On the other hand, the dual shatter function for discs is $O(m^2)$ as well, and so we get the bound of $O(n^{1/4}\sqrt{\log n})$ for the discrepancy for discs from Theorem 5.4, and this is currently the best result. It is a tantalizing open problem whether it can be improved, perhaps by removing the $\sqrt{\log n}$ factor. The best known lower bound is $\Omega(n^{1/4})$ only, the same as that for halfplanes.

Concerning Theorem 5.4 itself, it is known that it is asymptotically tight for all integers $d \geq 2$, including the $\sqrt{\log n}$ factor. A proof for $d = 2$ is indicated in Exercise 6 below. The set systems used in the proof are not of a geometric origin, and so they shed no light on the question of the discrepancy for discs in the plane.

Bounding the Dual Shatter Function for Geometric Families. A halfplane can be defined by a linear inequality. A circular disc in the plane can be written as $\{(x,y) \in \mathbf{R}^2\colon (x-a)^2 + (y-b)^2 \leq r^2\}$ for three real parameters a, b, r. Many other simple geometric shapes can be defined by a polynomial inequality with several real parameters, or sometimes by a Boolean combination of several such polynomial inequalities (for example, a square in the plane is defined by 4 linear inequalities).

Let us first consider bounding the dual shatter function of a set system $(\mathcal{S}, \mathbf{R}^d)$, where each set $S \in \mathcal{S}$ is defined by a single polynomial inequality of degree at most D. That is, we have $S = \{x \in \mathbf{R}^d\colon f_S(x) \geq 0\}$ where $f_S \in \mathbf{R}[x_1, \ldots, x_d]$ is a d-variate real polynomial of degree at most D. For example, if S were the disc of radius 2 centered at $(4,5)$ then $f_S(x) = 2^2 - (x_1-4)^2 - (x_2-5)^2$.

For determining the dual shatter function, we ask, what is the largest number of regions that can be induced in $\mathbf{R}^d$ by m sets $S_1, S_2, \ldots, S_m \in \mathcal{S}$? Two points $x, y \in \mathbf{R}^d$ lie in the same region if and only if $f_{S_i}(x)$ and $f_{S_i}(y)$ are both nonnegative or both negative, for all $i = 1, 2, \ldots, m$. This is closely related to the notion of *sign patterns* for a collection of polynomials.

Consider m real polynomials $f_1(x_1, x_2, \ldots, x_d), \ldots, f_m(x_1, x_2, \ldots, x_d)$ in d variables, each f_i of degree at most D. Let us call a vector $\sigma \in \{-1, 0, +1\}$ a *sign pattern* of $f_1, f_2, \ldots, f_m$ if there exists an $x \in \mathbf{R}^d$ such that the sign of $f_i(x)$ is σ_i, for all $i = 1, 2, \ldots, m$. Trivially, the number of sign patterns for any m polynomials is at most 3^m. For $d = 1$, it is easy to see that the actual number of sign patterns is much smaller than 3^m for m large. Namely, m univariate polynomials of degree D have altogether at most mD real roots which partition the real axis into at most $mD + 1$ intervals. In each of these intervals, the sign pattern is fixed. Hence there are $O(m)$ sign patterns for $d = 1$ and D fixed. The following very important theorem of real-algebraic geometry shows that in any fixed dimension d, there are at most $O(m^d)$ sign patterns:

5.5 Theorem. *The maximum number of sign patterns for a collection* $f_1, f_2, \ldots, f_m$ *of d-variate polynomials of degree at most D is bounded by*

$$\left(\frac{CDm}{d}\right)^d,$$

where C is an absolute constant.

This is not an easy result, and we omit the proof. We stated the bound more precisely than we need, as it is interesting in various other applications. We only need the fact that the number of sign patterns is $O(m^d)$ for D and d fixed. This also follows from older, somewhat easier, and less precise estimates.

To have a geometric picture of the situation in the theorem, consider the zero sets of the f_i: $Z_i = \{x \in \mathbf{R}^d\colon f_i(x) = 0\}$. We can think of them as hypersurfaces in $\mathbf{R}^d$ (although in some cases they can be disconnected, have singularities, or have dimension smaller than $d-1$). These sets partition $\mathbf{R}^d$ into cells of various dimensions; the following drawing illustrates a possible situation with 3 polynomials in $\mathbf{R}^2$:

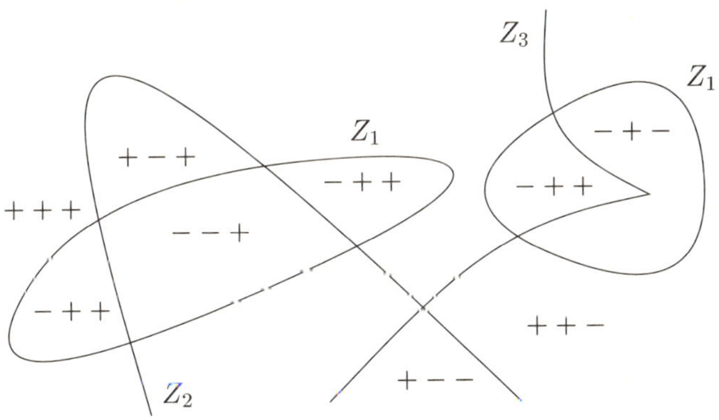

The sign patterns remain constant within each cell (they are marked in the picture for the 2-dimensional cells), and so the number of sign patterns is no bigger than the number of cells. On the other hand, several cells may have the same sign pattern. It is known that the maximum number of cells can be bounded by the same expression as in the theorem.

Important special cases of the theorem can be proved by elementary geometric arguments. For example, if all the f_i are linear polynomials, it suffices to show that m hyperplanes partition $\mathbf{R}^d$ into $O(m^d)$ cells. This is a well-known fact about the so-called *arrangements of hyperplanes*, and it can be proved easily by double induction on d and on m.

Returning to the dual shatter function, Theorem 5.5 and the discussion above it give the following corollary.

5.6 Corollary. *If $(\mathcal{S}, \mathbf{R}^d)$ is a set system in which each set is defined as a Boolean combination of at most k polynomial inequalities of degree at most D, where k and D are considered as constants, then $\pi^*_{\mathcal{S}}(m) = O(m^d)$. In particular, the dual shatter function for halfspaces and balls in $\mathbf{R}^d$ is $O(m^d)$.*

(Strictly speaking, we have only discussed the relation of sign patterns to the dual shatter function in the case of sets defined by a single polynomial inequality, but the extension to sets defined by Boolean combinations of several polynomial inequalities is immediate and we leave it to the reader.) Roughly, this theorem says that for "reasonable" geometric families in $\mathbf{R}^d$, the dual shatter function is essentially determined by the space dimension d.

Bounding the Primal Shatter Function. The primal shatter function depends mainly on the number of real parameters determining a set in a given family. For example, a disc in the plane has 3 parameters, since it can be written as $\{(x,y) \in \mathbf{R}^2\colon (x-a)^2 + (y-b)^2 \leq r^2\}$ for some $a, b, r \in \mathbf{R}$.

Let us again consider a family of geometric shapes determined by a single polynomial inequality. This time we let $f(x_1, x_2, \ldots, x_d, a_1, a_2, \ldots, a_p)$ be a fixed real polynomial in $d+p$ variables, and for a parameter vector $a = (a_1, a_2, \ldots, a_p) \in \mathbf{R}^p$ we put

$$S_f(a) = \{x = (x_1, x_2, \ldots, x_d) \in \mathbf{R}^d\colon f(x_1, \ldots, x_d, a_1, \ldots, a_p) \geq 0\}.$$

We let $\mathcal{S}_f$ be the set system $\{S_f(a)\colon a \in \mathbf{R}^p\}$. For example, the system of all discs in the plane can be written as $\mathcal{S}_f$ with

$$f(x_1, x_2, a_1, a_2, a_3) = a_3^2 - (x_1 - a_1)^2 - (x_2 - a_2)^2.$$

How many distinct subsets can be defined by $\mathcal{S}_f$ on an m-point set $\{x^{(1)}, \ldots, x^{(m)}\} \subset \mathbf{R}^d$? This number is bounded by the number of sign patterns of the p-variate polynomials $f_1, f_2, \ldots, f_m \in \mathbf{R}[a_1, a_2, \ldots, a_p]$, where $f_i(a_1, \ldots, a_p) = f(x_1^{(i)}, \ldots, x_d^{(i)}, a_1, \ldots, a_p)$. According to Theorem 5.5, we thus obtain

5.7 Corollary. *Let $f(x_1, \ldots, x_d, a_1, \ldots, a_p)$ be a $(d+p)$-variate polynomial of degree bounded by a constant. Then we have $\pi_{\mathcal{S}_f}(m) = O(m^p)$, where the set system $\mathcal{S}_f$ is defined as above. In particular, the primal shatter function for halfspaces in $\mathbf{R}^d$ is $O(m^d)$, and for balls in $\mathbf{R}^d$ it is $O(m^{d+1})$.*

The result for halfspaces can again be derived by an elementary geometric argument; in this case, the primal and dual shatter functions are exactly equal (this follows from geometric duality). But the bound for halfspaces, too, follows from the general result stated in the corollary. Although the family $\mathcal{H}_d$ of all halfspaces cannot be written as $\mathcal{S}_f$ for a $(d+d)$-variate polynomial f, the family of all upper halfspaces, say, can be so expressed, and for d fixed, $\mathcal{H}_d$ is a union of a constant-bounded number of such families.

The bound follows since, clearly, $\pi_{\mathcal{S}_1 \cup \mathcal{S}_2}(m) \le \pi_{\mathcal{S}_1}(m) + \pi_{\mathcal{S}_2}(m)$ for any set systems $(X, \mathcal{S}_1)$ and $(X, \mathcal{S}_2)$.

Also, we should remark that the result in Corollary 5.7 does not need the assumption that $f(x_1, \ldots, x_d, a_1, \ldots, a_p)$ depend polynomially on $x_1, \ldots, x_d$. It suffices to assume that it is a polynomial of a constant-bounded degree in $a_1, \ldots, a_p$ for each fixed value of $x = (x_1, \ldots, x_d) \in \mathbf{R}^d$.

Bibliography and Remarks. The history of the concept of the shatter functions will be briefly considered in the remarks to the next section.

The relation of the shatter functions to discrepancy was observed by Matoušek, Wernisch, and Welzl [MWW93] who proved the stated bound in terms of the dual shatter function (Theorem 5.4) and a weaker version of the primal bound (Theorem 5.3). Some of the ideas used in the proof of the dual bound were noted independently by Beck [Bec91b] in a particular geometric situation. The tight bound in Theorem 5.3 was obtained in Matoušek [Mat95]. The tightness of the dual bound for $d = 2, 3$ was proved in [Mat97], and results of Alon et al. [ARS99] (based on previous work by Kollár et al. [KRS96]) supplied a missing combinatorial ingredient for extending the proof to an arbitrary integer $d \ge 2$.

Theorem 5.5 is from Pollack and Roy [PR93]. Basu et al. [BPR96] showed that the number of sign patterns defined by m polynomials of degree at most D on a k-dimensional algebraic variety $V \subseteq \mathbf{R}^d$, where V can be defined by polynomials of degree at most D, is at most $\binom{m}{k} O(D)^d$.

The original bounds on the number of sign patterns, less precise than Theorem 5.5 but still implying Corollaries 5.6 and 5.7, were given independently by Oleinik and Petrovskiĭ [OP49] (also see [Ole51]), Milnor [Mil64], and Thom [Tho65]. Quantitatively somewhat weaker versions of Corollaries 5.6 and 5.7 can be derived without the real-algebraic machinery, by embedding the set system $\mathcal{S}_f$ into a set system defined by halfspaces in a suitable higher-dimensional space.

A far-reaching generalization of the results on the number of sign-patterns for polynomials has recently been achieved by Wilkie [Wil96], culminating a long development in model theory. These results consider geometric shapes defined by functions from more general classes than polynomials. Perhaps the most important of such classes are the *Pfaffian functions*; roughly speaking, these are functions definable by first-order differential equations, and a prominent example is the exponential function e^x. Wilkie's results imply that the shatter functions for geometric shapes defined by bounded-size formulas from Pfaffian functions are polynomially bounded. More special but quantitatively sharper results in this spirit were obtained by Karpinski and Macintyre ([KM97b]; also see [KM97a] for algorithmic applications). The subtlety of these results can be illustrated by remarking that the function $\sin x$, which is not a Pfaffian function but seems to be of nature similar to e^x (at least in the complex domain), leads to families with non-polynomial shatter functions (see Exercise 5.2.3 in the next section).

Exercises

1. Show that the primal shatter function $\pi_{\mathcal{B}_2}(m)$ for discs is at least $\Omega(m^3)$.
2. For each of the following classes of shapes, determine the discrepancy upper bounds provided by Theorems 5.3 and 5.4 (if you can't bound the shatter functions exactly give at least some estimates):
   (a) all axis-parallel rectangles in the plane;
   (b) all rectangles (arbitrarily rotated) in the plane.
3. Define the discrepancy of a set D of n circular discs in the plane as the minimum, over all red-blue colorings of the discs in D, of the maximum possible difference of the number of red discs containing a point x and the number of blue discs containing that x. (Note that here, exceptionally, we color discs rather than points.) What upper bounds can be obtained for this discrepancy from Theorems 5.3 and 5.4?
4. Let $\Phi(X_1, X_2, \ldots, X_t)$ be a fixed set-theoretic expression (using the operations of union, intersection, and difference) with variables $X_1, \ldots X_t$ standing for sets, for instance,
$$\Phi(X_1, X_2, X_3) = (X_1 \cup X_2 \cup X_3) \setminus (X_1 \cap X_2 \cap X_3).$$
   Let $\mathcal{S}$ be a set system on a set X. Let $\mathcal{T}$ consist of all sets $\Phi(S_1, \ldots, S_t)$, for all possible choices of $S_1, \ldots, S_t \in \mathcal{S}$.
   (a) Suppose that $\pi^*_{\mathcal{S}}(m) \leq Cm^d$ for all m. Prove $\pi^*_{\mathcal{T}}(m) = O(m^d)$, with the constant of proportionality depending on C, d, and Φ.
   (b) Suppose that $\pi_{\mathcal{S}}(m) \leq Cm^d$ for all m. Prove $\pi_{\mathcal{T}}(m) = O(m^{td})$.
   (c) Show that the bound in (b) is asymptotically tight in the worst case.
   This exercise consists of variations on ideas appearing in Dudley [Dud78].
5. Let $(X, \mathcal{P})$ be a finite projective plane of order q (i.e. a system of $n = q^2 + q + 1$ sets of size $q + 1$ on n points such that any two sets intersect at exactly one point and for every two points, there is exactly one set containing both).
   (a) Determine the order of magnitude of the shatter functions $\pi_{\mathcal{P}}$ and $\pi^*_{\mathcal{P}}$.
   (b) Show that $\mathrm{disc}(\mathcal{P}) = \Omega(\sqrt{q})$ (follow the proof method of Proposition 4.4). Deduce that for $d = 2$, the bound in Theorem 5.3 is asymptotically the best possible in general, and that the bound in Theorem 5.3 is tight up to the $\sqrt{\log n}$ factor.
   (c) Show that the L_1-discrepancy of $\mathcal{P}$ is upper-bounded by a constant independent of q.

(d)* Generalize (a) and (b) for the system of all hyperplanes in a finite projective d-space of order q.

6. Let F be a q-element finite field. Let $\mathcal{P}_2$ be the set of all univariate quadratic polynomials over F. For a polynomial $p \in \mathcal{P}_2$, define a set $S_p \subseteq X = F \times F$ as the graph of p, i.e. $S_p = \{(x, p(x)) \colon x \in F\}$. Let $\mathcal{S} = \{S_p \colon p \in \mathcal{P}_2\}$.
(a) Show that $|S_p \cap S_{p'}| \leq 2$ for any two distinct $p, p' \in \mathcal{P}_2$. Infer that $\pi^*_{\mathcal{S}}(m) = O(m^2)$.
(b) For each $p \in \mathcal{P}_2$, define R_p as a random subset of S_p, where all subsets are taken with equal probability and the choices are mutually independent for distinct p. Put $\mathcal{R} = \{R_p \colon p \in \mathcal{P}_2\}$. Show that $\pi^*_{\mathcal{R}}(m) = O(m^2)$.
(c)* Use the method and results of Exercise 4.1.1 to prove that $\operatorname{disc}(\mathcal{R}) = \Omega(\sqrt{q \log q})$ holds with a positive probability. This means that Theorem 5.4 is tight for $d = 2$.

5.2 Set Systems of Bounded VC-Dimension

In this section we introduce the concept of Vapnik–Chervonenkis dimension, which is closely related to the shatter functions and provides new insight into their behavior. Then we prove some important results concerning these notions, which will be used later on in the proofs of the discrepancy bounds.

5.8 Definition (VC-dimension). *Let $\mathcal{S}$ be a set system on a set X. Let us say that a subset $A \subseteq X$ is* shattered *by $\mathcal{S}$ if each of the subsets of A can be obtained as the intersection of some $S \in \mathcal{S}$ with A, i.e. if $\mathcal{S}|_A = 2^A$. We define the* Vapnik–Chervonenkis dimension *(or VC-dimension for short) of $\mathcal{S}$, denoted by $\dim(\mathcal{S})$, as the supremum of the sizes of all finite shattered subsets of X. If arbitrarily large subsets can be shattered, the VC-dimension is ∞.*

An immediate reformulation of the definition is

$$\dim(\mathcal{S}) = \sup\{m \colon \pi_{\mathcal{S}}(m) = 2^m\}.$$

Somewhat surprisingly, the order of growth of shatter functions cannot be quite arbitrary. It turns out that they can be either polynomially bounded or exponential, but nothing in between. This is a consequence of a key lemma below relating the VC-dimension and the primal shatter function.

5.9 Lemma (Shatter function lemma). *For any set system $\mathcal{S}$ of VC-dimension at most d, we have*

$$\pi_{\mathcal{S}}(m) \leq \Phi_d(m)$$

for all m, where

$$\Phi_d(m) = \binom{m}{0} + \binom{m}{1} + \cdots + \binom{m}{d}.$$

Thus, the primal shatter function for any set system is either 2^m (the case of infinite VC-dimension) or it is bounded by a fixed polynomial.

Proof. Since VC-dimension does not decrease by passing to a subsystem, it suffices to show that any set system of VC-dimension $\leq d$ on an n-point set has no more than $\Phi_d(n)$ sets. We proceed by induction on d, and for a fixed d we use induction on n.

Consider a set system $(X, \mathcal{S})$ of VC-dimension d, with $|X| = n$, and fix some $x \in X$. In the induction step, we would like to remove x and pass to the set system $\mathcal{S}_1 = \mathcal{S}|_{X \setminus \{x\}}$ on $n-1$ points. This $\mathcal{S}_1$ has VC-dimension at most d, and hence $|\mathcal{S}_1| \leq \Phi_d(n-1)$ by the inductive hypothesis. How many more sets can $\mathcal{S}$ have compared to $\mathcal{S}_1$? The only way the number of sets decreases by removing the element x is when two sets $S, S' \in \mathcal{S}$ give rise to the same set in $\mathcal{S}_1$, which means that $S' = S \dot\cup \{x\}$ (or the other way round). This suggests that we define an auxiliary set system $\mathcal{S}_2$ consisting of all sets in $\mathcal{S}_1$ that correspond to such pairs $S, S' \in \mathcal{S}$, that is, we set $\mathcal{S}_2 = \{S \in \mathcal{S}\colon x \notin S, S \cup \{x\} \in \mathcal{S}\}$.

By the above discussion, we have $|\mathcal{S}| = |\mathcal{S}_1| + |\mathcal{S}_2|$. Crucially, we observe that $\dim(\mathcal{S}_2) \leq d-1$, since if $A \subseteq X \setminus \{x\}$ is shattered by $\mathcal{S}_2$ then $A \cup \{x\}$ is shattered by $\mathcal{S}$. Therefore $|\mathcal{S}_2| \leq \Phi_{d-1}(n-1)$. This gives a recurrence from which the asserted formula is verified by an easy manipulation of binomial coefficients. □

Another Proof. Lemma 5.9 can also be proved using linear algebra. It would be a pity to omit this pretty proof.

Suppose that $X = \{1, 2, \ldots, n\}$. For each set $S \in \mathcal{S}$, define a polynomial p_S in variables $x_1, x_2, \ldots, x_n$:

$$p_S(x_1, x_2, \ldots, x_n) = \Bigl(\prod_{i \in S} x_i\Bigr)\Bigl(\prod_{i \notin S}(1 - x_i)\Bigr).$$

For a set $S \in \mathcal{S}$, let $\mathbf{v}_S \in \{0,1\}^n$ denote the characteristic vector of S, with $(\mathbf{v}_S)_i = 1$ if $i \in S$ and $(\mathbf{v}_S)_i = 0$ otherwise, and let $V_{\mathcal{S}} = \{\mathbf{v}_S\colon S \in \mathcal{S}\}$. Each polynomial p_S can be regarded as a real function defined on the set $V_{\mathcal{S}}$; the value at a vector $\mathbf{v}_T \in V_{\mathcal{S}}$ is naturally $p_S((\mathbf{v}_T)_1, (\mathbf{v}_T)_2, \ldots, (\mathbf{v}_T)_n)$. All real functions with domain $V_{\mathcal{S}}$ constitute a vector space (with pointwise addition and multiplication by a real number), which we denote by F, and we let L be the linear span in F of the set $\{p_S\colon S \in \mathcal{S}\}$.

First we note that the p_S's are all linearly independent in F. This is because p_S has value 1 at $\mathbf{v}_S$ and value 0 at the characteristic vectors of all other $S' \in \mathcal{S}$; hence no p_S can be a linear combination of others. Therefore $\dim(L) = |\mathcal{S}|$.

To prove the lemma, we now show that $\dim(L) \leq \Phi_d(n)$. We claim that L is contained in the linear span of all monomials of the form $x_{i_1} x_{i_2} \cdots x_{i_k}$ with

$1 \le i_1 < i_2 < \cdots < i_k \le n$ and $k \le d$ (call them *multilinear monomials of degree at most d*). Note that once we show this we are done since the number of such monomials is just $\Phi_d(n)$.

By multiplying out the parentheses in the definition of p_S we see that each p_S is a linear combination of some multilinear monomials. It suffices to show that any multilinear monomial of degree $d+1$, regarded as a function on V_S, is a linear combination of multilinear monomials of degree $\le d$. So let $x_{i_1}x_{i_1}\cdots x_{i_{d+1}}$ be a multilinear monomial of degree $d+1$. At this moment we use (finally) the definition of VC-dimension: there exists a set $B \subseteq \{i_1, i_2, \ldots, i_{d+1}\} = A$ that is not of the form $S \cap A$ for any $S \in \mathcal{S}$, for otherwise A would be shattered. Define a polynomial $q(x_1, x_2, \ldots, x_n) = \left(\prod_{j\in B} x_j\right)\left(\prod_{j\in A\setminus B}(x_j - 1)\right)$. A little thought reveals that the value of q is 0 for all characteristic vectors of sets from $\mathcal{S}$; so q regarded as an element of F is the zero function. At the same time, q can be written as $q(x_1, x_2, \ldots, x_n) = x_{i_1}x_{i_2}\cdots x_{i_{d+1}} + r(x_1, x_2, \ldots, x_n)$, where r is a linear combination of multilinear monomials of degree at most d. Hence our monomial $x_{i_1}x_{i_1}\cdots x_{i_{d+1}}$ is a linear combination of multilinear monomials of degree at most d as claimed. This proves the Shatter function lemma. □

Taking all subsets of X of size at most d for $\mathcal{S}$ shows that the bound in the Shatter function lemma 5.9 is tight in the worst case. However, the primal shatter function is often considerably smaller than the upper bound derived from the VC-dimension. For instance, the set system $\mathcal{H}_2$ of all halfplanes has VC-dimension 3, as is easily checked, but we know that $\pi_{\mathcal{H}_2}$ is only quadratic.

Another interesting result obtained via VC-dimension is the following:

5.10 Lemma (Dual set system lemma). *Let $(X, \mathcal{S})$ be a set system, and let $(\mathcal{S}, \mathcal{S}^*)$ denote the* dual set system*: $\mathcal{S}^* = \{\{S \in \mathcal{S}: x \in S\}: x \in X\}$ (the incidence matrix of $\mathcal{S}^*$ is a transpose of the incidence matrix of $\mathcal{S}$, with repeated rows deleted). Then*

$$\dim(\mathcal{S}^*) < 2^{\dim(\mathcal{S})+1}.$$

Thus, if $\pi_{\mathcal{S}}(m) \le Cm^d$ for all m and for some constants C, d, then $\pi^*_{\mathcal{S}}(m) = \pi_{\mathcal{S}^*}(m) \le C'm^{d'}$, where C', d' are constants depending on C, d only. The primal and dual shatter function are either both polynomially bounded or both equal to 2^m.

We leave a proof as Exercise 4.

Epsilon-Nets. For a set system $(X, \mathcal{S})$, we often need a set intersecting all sets in $\mathcal{S}$, the so-called *transversal* of $\mathcal{S}$. Of course, the whole X is a transversal, but one of the key problems in combinatorics of set systems is the existence of a *small* transversal. The ε-nets we are going to consider now are transversals for all "large" sets in $\mathcal{S}$.

Let $(X, \mathcal{S})$ be a set system with X finite. A set $N \subseteq X$ (not necessarily one of the sets of $\mathcal{S}$) is called an *ε-net for $(X, \mathcal{S})$*, where $\varepsilon \in [0, 1]$ is a real number, if each set in $S \in \mathcal{S}$ with $|S| \geq \varepsilon|X|$ intersects N. Sometimes it will be convenient to write $\frac{1}{r}$ instead of ε, with $r > 1$ a real parameter.

More generally, an ε-net can be defined for a set system $(X, \mathcal{S})$ with a probability measure μ on X (i.e. $\mu(X) = 1$). A set $N \subseteq X$ is an ε-net for $(X, \mathcal{S})$ with respect to μ if it intersects all $S \in \mathcal{S}$ with $\mu(S) \geq \varepsilon$.

The ε-nets are a concept somewhat akin to the ε-approximations introduced in Section 1.3. Recall that $Y \subseteq X$ is an ε-approximation if

$$\left| \frac{|Y \cap S|}{|Y|} - \mu(S) \right| \leq \varepsilon$$

for all $S \in \mathcal{S}$. It is easy to see that an ε-approximation is also an ε-net, but the converse need not be true in general.

First we prove the following simple probabilistic bound for the size of ε-nets:

5.11 Lemma (Easy ε-net lemma). *Let X be a set, μ a probability measure on X, and $\mathcal{S}$ a finite system of μ-measurable subsets of X. Then, for every real number $r > 1$, there exists a $\frac{1}{r}$-net N for $(X, \mathcal{S})$ with respect to μ with $|N| \leq r \ln |\mathcal{S}|$.*

Proof. Let N be a random sample drawn from X by s independent random draws (thus, elements may be drawn several times); each of the s elements is sampled according to the distribution μ. Then for any given $S \in \mathcal{S}$ with $\mu(S) \geq \frac{1}{r}$, $\Pr[S \cap N = \emptyset] \leq (1 - \frac{1}{r})^s < e^{-s/r}$. The probability that any of the sets of $\mathcal{S}$ is missed by N is thus smaller than $|\mathcal{S}| \cdot e^{-s/r}$. For $s \geq r \ln(|\mathcal{S}|)$ this is at most 1, so N is a $\frac{1}{r}$-net with a positive probability. □

It turns out that for set systems of bounded VC-dimension, a significant improvement over the Easy ε-net lemma 5.11 is possible. Namely, if the VC-dimension is bounded, one can get ε-nets whose size depends on ε and on the VC-dimension but not on the size of X or $\mathcal{S}$. So, for example, there is an absolute constant C such that for any finite set X in the plane, there exists a C-point subset $N \subseteq X$ such that any triangle containing at least 1% of the points of X intersects N. Note that the Easy ε-net lemma would only give an $O(\log |X|)$ bound for the size of such an N.

5.12 Theorem (Epsilon-net theorem). *For every $d \geq 1$ there exists a constant $C(d)$ such that if X is a set with a probability measure μ, $\mathcal{S}$ is a system of μ-measurable subsets of X with $\dim(\mathcal{S}) \leq d$, and $r \geq 2$ is a parameter, then there exists a $\frac{1}{r}$-net for $(X, \mathcal{S})$ with respect to μ of size at most $C(d) r \ln r$.*

It is known that one can take $C(d) = d + o(d)$. More precisely, for any $d > 1$ there exists an $r_0 > 1$ such that for any $r > r_0$, each set system of VC-dimension d admits a $\frac{1}{r}$-net of size at most $dr \ln r$.

We will only prove an important special case of this result, where X is finite and μ is the counting measure on X. The following proof is conceptually simple and related to discrepancy, but it gives a somewhat worse value of $C(d)$ than the best known bounds.

First we establish an analogous result with ε-approximations instead of ε-nets. For simplicity, we will assume that the size of the ground set X is a power of 2 (but the proof can be modified to the general case without much work).

5.13 Lemma. *Let X be a set of $n = 2^k$ points, let $\mathcal{S}$ be a set system of VC-dimension at most d on X, and let $r \geq 2$. Then a $\frac{1}{r}$-approximation for $(X, \mathcal{S})$ exists of size at most $C(d)r^2 \log r$.*

We should remark that this lemma, even with a better bound for the size of the $\frac{1}{r}$-approximation, is an immediate consequence of Theorem 5.3, the Shatter function lemma 5.9, and Lemma 1.6(ii) (on the relation of combinatorial discrepancy and ε-approximations); see Exercise 7. But since the proof of Theorem 5.3 is not simple, we give another, self-contained proof.

Proof. The proof resembles the proof of the transference lemma (Proposition 1.8)—we proceed by repeated halving, using a random coloring at each halving step. Let us set $Y_0 = X$ and $\mathcal{S}_0 = \mathcal{S}$. Having constructed the set system $(Y_i, \mathcal{S}_i)$, with $n_i = |Y_i| = 2^{k-i}$, we apply the Random coloring lemma 4.1 to $(Y_i, \mathcal{S}_i)$, obtaining

$$\operatorname{disc}(\mathcal{S}_i) \leq \sqrt{2n_i \ln(4|\mathcal{S}_i|)} \leq \sqrt{2n_i \ln(4\pi_{\mathcal{S}}(n_i))} = O(\sqrt{n_i \ln n_i}\,)$$

by the Shatter function lemma 5.9 (the constant of proportionality depends on d). From Lemma 1.6, we know that a low-discrepancy coloring can be converted to an ε-approximation. In our case, there exists an ε_i-approximation Y_{i+1} for $(Y_i, \mathcal{S}_i)$ of size $|Y_{i+1}| = \frac{n_i}{2}$, where $\varepsilon_i \leq \operatorname{disc}(\mathcal{S}_i)/n_i = O(\sqrt{\ln n_i / n_i}\,)$.

We stop the construction as soon as n_{i+1}, the size of Y_{i+1}, drops below $Cr^2 \ln r$ for a sufficiently large constant C (depending on d). Suppose that we have stopped after the ℓth step. The resulting set $Y_{\ell+1}$ has size $n_{\ell+1} < Cr^2 \ln r$ and it is an ε-approximation for $(X, \mathcal{S})$ by Observation 1.7 (on iterated approximations), where

$$\begin{aligned}
\varepsilon &\leq \sum_{i=0}^{\ell} \varepsilon_i = O(1) \cdot \sum_{i=0}^{\ell} \sqrt{\frac{\ln n_i}{n_i}} = O\left(\sqrt{\frac{\ln n_\ell}{n_\ell}}\right) \\
&= O\left(\sqrt{\frac{\ln(2Cr^2 \ln r)}{Cr^2 \ln r}}\right) = O\left(\sqrt{\frac{\ln C}{C}} \cdot \frac{1}{r}\right) \leq \frac{1}{r}
\end{aligned}$$

if C was chosen sufficiently large. Lemma 5.13 is proved. □

Proof of Theorem 5.12 for the counting measure case. Let $(X, \mathcal{S})$ be the considered set system with $n = |X|$. Add at most $n-1$ more dummy points to X, lying in no set of $\mathcal{S}$, obtaining a set X' with $n' = 2^k$ points. Obviously, a $\frac{1}{2r}$-net for $(X', \mathcal{S})$ is also a $\frac{1}{r}$-net for X.

Using Lemma 5.13, we find a $\frac{1}{4r}$-approximation Y for the set system $(X', \mathcal{S})$ of size $O(r^2 \log r)$. Next, apply the Easy ε-net lemma 5.11 to the set system $(Y, \mathcal{S}|_Y)$, obtaining a $\frac{1}{4r}$-net N for $(Y, \mathcal{S}|_Y)$, of size $O(r \ln \pi_{\mathcal{S}}(|Y|)) = O(r \ln r)$. It is easy to check that N is also a $\frac{1}{2r}$-net for $(X', \mathcal{S})$ (this is analogous to Observation 1.7) and consequently a $\frac{1}{r}$-net for $(X, \mathcal{S})$. Theorem 5.12 is proved. □

Bibliography and Remarks. The notion now commonly called VC-dimension originated in statistics. It was introduced by Vapnik and Chervonenkis [VC71]. Under different names, it also appeared in other papers (I am aware of Sauer [Sau72] and Shelah [She72]), but the work [VC71] was probably the most influential for the subsequent developments. The name VC-dimension and some other, by now more or less standard terminology was introduced in a key paper of Haussler and Welzl [HW87].

The VC-dimension and related concepts have been applied and further developed in several areas of mathematics and computer science. For most of them, we only give a few pointers to the extensive literature. In statistics, the VC-dimension is used mainly the theory of so-called empirical processes ([Vap82], [Dud84], [Dud85], [Pol90]). In computational learning theory, VC-dimension is one of the main concepts ([BEHW89], [AB92], [Hau92]). In combinatorics of hypergraphs, set systems of VC-dimension d can be viewed as a class of hypergraphs with a certain forbidden subhypergraph (the complete hypergraph on $d+1$ points), which puts this topic into the broader context of extremal hypergraph theory (see for instance [Fra83], [WF94], [DSW94]).

The above-mentioned paper of Haussler and Welzl [HW87] belongs to computational geometry (this field was discussed a little in the remarks to Section 1.4). Chazelle and Friedman [CF90] is another significant paper applying combinatorial properties of geometric set systems (similar to polynomially bounded shatter functions) in computational geometry.

The Shatter function lemma 5.9 was independently found in the three already mentioned papers [VC71], [Sau72], [She72]. The linear-algebraic proof is due to Frankl and Pach [FP83]. A forthcoming monograph by Babai and Frankl [BF92] is an excellent source for applications of linear algebra in this spirit. The shatter functions were defined and applied by Welzl [Wel88]. I am not aware of any earlier explicit reference but implicitly these notions have been used much earlier, for instance by Dudley [Dud78]. In the literature, the shatter functions appear under various names, such as the *growth functions*, and a related dimension concept, where the dimension is defined as the degree of a polynomial bounding the primal shatter function, was re-discovered several times. The Dual set system lemma 5.10 was noted by Assouad [Ass83].

The notions of shattering and of VC-dimension can also be used for a family $\mathcal{F}: X \to \{0,1\}$ of two-valued functions on a set X, since such functions are in a one-to-one correspondence with subsets of X. There are several generalizations of these notions to k-valued functions. Two of them are mentioned in Exercise 10, and more information and references can be found, e.g., in Haussler and Long [HL95].

The ε-approximations were introduced by Vapnik and Chervonenkis [VC71]. They also proved a more general version of Lemma 5.13, analogous to Theorem 5.12: If X is a set with a probability measure μ then there exist $\frac{1}{r}$-approximations for $(X, \mathcal{S})$ with respect to μ of size at most $C(d)r^2 \log r$, where $d = \dim(\mathcal{S})$. The idea of their proof is indicated in Exercise 8, for the technically simpler case of ε-nets. The notion of ε-net and the Epsilon-net theorem 5.12 (for the case of the counting measure) are due to Haussler and Welzl [HW87]. They proved the theorem directly, imitating the proof of Vapnik and Chervonenkis for ε-approximations, instead deriving the ε-net result from the ε-approximation result as we did above. The dependence of the bound in Theorem 5.12 on d was subsequently improved by Blumer et al. [BEHW89] and then by Komlós et al. [KPW92], who also give a matching lower bound (Exercise 6). The proof of the Epsilon-net theorem 5.12 presented in the text is a simplification of an algorithmic method for computing ε-approximations from [Mat96a].

Let us remark that finite VC-dimension characterizes set systems $(X, \mathcal{S})$ where all subsystems induced by finite sets have a sublinear combinatorial discrepancy, i.e. such that $\mathrm{disc}(m, \mathcal{S}) = o(m)$ as $m \to \infty$. Indeed, if $\dim(\mathcal{S}) = d < \infty$, then $\mathrm{disc}(m, \mathcal{S}) = O(\sqrt{m \log m})$ by the Random coloring lemma 4.1 and by the Shatter function lemma 5.9 (or, we can even do slightly better using Theorem 5.3). On the other hand, if $\dim(\mathcal{S}) = \infty$ then there is an m-point shattered shattered subset for all m, and so $\mathrm{disc}(m, \mathcal{S}) = \lceil m/2 \rceil$ for all m.

An interesting example in which small ε-approximations exist, although the VC-dimension is not bounded, was discovered by Chazelle [Cha93]. Let L be a set of n lines in the plane in general position, and let $A \subseteq L$ be an ε-approximation for the set system induced on L by line segments (that is, for each segment s, the set $\{\ell \in L: \ell \cap s \neq \emptyset\}$ is included in the set system). Then A can be used to estimate the number of intersections of the lines of L within any convex set $C \subseteq \mathbf{R}^2$.

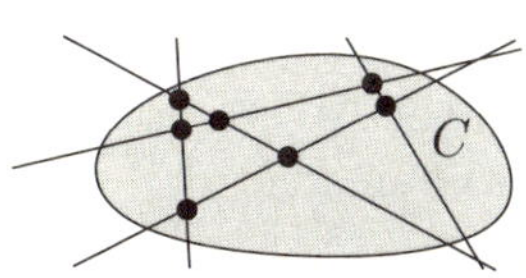

Namely, if $V(L)$ denotes the set of all intersections of the lines of L, and similarly for $V(A)$, we have

$$\left|\frac{|V(A) \cap C|}{|V(A)|} - \frac{|V(L) \cap C|}{|V(L)|}\right| \le 2\varepsilon.$$

Note that the set system induced by convex sets on $V(L)$ has an arbitrarily large VC-dimension. But, for set systems of small VC-dimension, small ε-approximations exist for subsystems induced by arbitrary subsets of the ground set, whereas in the above case, only some special subsets of $V(L)$ can be obtained as $V(L')$ for some $L' \subseteq L$. A similar result is valid for hyperplanes in $\mathbf{R}^d$. A more general version of Chazelle's result is treated in [BCM]. Given two set systems $(X, \mathcal{S})$ and $(Y, \mathcal{T})$, their *product set system* has $X \times Y$ as the ground set and it contains all sets $Q \subseteq X \times Y$ all of whose "slices" lie in $\mathcal{S}$ or in $\mathcal{T}$. Formally, if we put $xQ = \{y \in Y\colon (x, y) \in Q\}$ for $x \in X$, and $Qy = \{x \in X\colon (x, y) \in Q\}$ for $y \in Y$, then the product set system contains all the $Q \subseteq X \times Y$ such that $xQ \in \mathcal{T}$ for all $x \in X$ and $Qy \in \mathcal{S}$ for all $y \in Y$. Now if $A \subseteq X$ is an ε-approximation for $(X, \mathcal{S})$ and $B \subseteq Y$ is a δ-approximation for $(Y, \mathcal{T})$ then $A \times B$ is an $(\varepsilon + \delta)$-approximation for the product set system.

Exercises

1. Let $\mathcal{S}_1, \mathcal{S}_2$ be set systems on a set X and let $\mathcal{S} = \mathcal{S}_1 \cup \mathcal{S}_2$.
   (a) Bound $\pi_{\mathcal{S}}$ in terms of $\pi_{\mathcal{S}_1}$ and $\pi_{\mathcal{S}_2}$.
   (b) If $\dim(\mathcal{S}_1) = d_1$ and $\dim(\mathcal{S}_2) = d_2$, what is the maximum possible value of $\dim(\mathcal{S})$?
   (c) Bound $\pi^*_{\mathcal{S}}$ in terms of $\pi^*_{\mathcal{S}_1}$ and $\pi^*_{\mathcal{S}_2}$.
2. (a) Show that for the set system consisting of all convex sets in the plane, the VC-dimension is infinite.
   (b)* Show that for any integer d there exists a convex set C in the plane such that the family of all isometric copies of C has VC-dimension at least d.
   (c)** Show that for the family of all translated and scaled copies of a fixed convex set in the plane, the VC-dimension is bounded by a universal constant.
3.* Let $\mathcal{S}$ be the family of all subsets of the real line of the form $S_a = \{x \in \mathbf{R}\colon \sin(ax) \ge 0\}$, $a \in \mathbf{R}$. Prove that $\dim(\mathcal{S}) = \infty$.
4. (a) Prove the Dual set system lemma 5.10.
   (b) Prove that the bound $2^{\dim(\mathcal{S})+1}$ for the VC-dimension of the dual system cannot be improved in general.
5. (a) Prove that any set system $\mathcal{S}$ on n points with hereditary discrepancy at most d has no more than $\Phi_{2d}(n) = \sum_{i=0}^{2d} \binom{n}{i}$ distinct sets.

(b)** Show that the upper bound in (a) is the best possible.

The results in this exercise are from Lovász and Vesztergombi [LV89].

6.* (Lower bounds for ε-net size) Let X be an n-point set (n large) and let $\mathcal{R}$ be a set system obtained by drawing a random subset of size s from X m times (independently and with possible repetitions). Show that with n large, $m = n^{3/4}$, and $s = n^{0.3}$, the following holds with a positive probability.
(i) There exists no k-element set $N \subseteq X$ intersecting each set of $\mathcal{R}$ with $k \leq c\frac{n}{s} \log \frac{n}{s}$, $c > 0$ a constant.
(ii) No 3-point set $A \subseteq X$ is shattered by $\mathcal{R}$.
Thus, the bound for ε-net size in the Epsilon-net theorem 5.12 is tight for $d = 3$ up to the value of $C(3)$. A similar lower bound proof works for an arbitrary $d > 1$.

7. (Improved bounds for the size of ε-approximations) Use Theorems 5.3 and 5.4 to show the following improvements over the $O(r^2 \log r)$ bound for the size of a $\frac{1}{r}$-approximation (Lemma 5.13). If the primal shatter function of a set system $(X, \mathcal{S})$ is bounded by Cm^d for all m and for some constants $C, d > 1$, then $\frac{1}{r}$-approximations exist of size $O(r^{2-2/(d+1)})$, with the constant of proportionality depending on C and d. Similarly, if the dual shatter function is bounded by Cm^d, then $\frac{1}{r}$-approximations exist of size $O(r^{2-2/(d+1)}(\log r)^{1-1/(d+1)})$.

8. (An alternative proof of the Epsilon-net theorem) Complete the following outline of the Haussler-Welzl original proof of the Epsilon-net theorem 5.12.
Pick a random sample $N \subseteq X$ by s independent random draws (according to the probability distribution μ), where $s = C(d)r \log r$. The goal is to show that N is a $\frac{1}{r}$-net with high probability.
(a)* By s more independent random draws, pick another sample M. Regard both N and M as multisets. Show that the probability of N *not* being a $\frac{1}{r}$-net is at most

$$O\left(\Pr\left[\exists S \in \mathcal{S}\colon S \cap N = \emptyset \text{ and } |S \cap M| \geq \frac{s}{r}\right]\right).$$

(b)* Let N_0 be a fixed multiset of $2s$ elements of X. Put s randomly selected elements of N_0 into a multiset N and the remaining s elements into a multiset M. For any fixed subset $R \subseteq N_0$, show that the probability of $R \cap N = \emptyset$ and $|R \cap M| \geq \frac{s}{r}$ is $o(s^{-d})$.
(c)* Find how (a), (b), and the Shatter function lemma 5.9 imply that a multiset N obtained by s independent random draws from X according to μ is a $\frac{1}{r}$-net with a positive probability.

9. (A quantitative Ramsey-type result for bipartite graphs)
(a) Let A be a d-point set, and let G be the bipartite graph of incidence of the set system $(A, 2^A)$. That is, the vertices of one class are the points of A, the vertices of the other class are the subsets of A, and edges

correspond to membership. Let H be any bipartite graph with classes of size a and b, where $a + \log_2 b \leq d$. Prove that G contains an induced copy of H.

(b)* Let H be a fixed bipartite graph with classes of size a and b, and let $d = a + \lceil \log_2 b \rceil$. Let m be an integer and let $n > (m-1)\Phi_{d-1}(2m-1) = O(m^d)$. Prove that any bipartite graph G with $2m-1$ vertices in one class and n vertices in the other class and with no induced copy of H contains a homogeneous subgraph on $m+m$ vertices (that is, the complete bipartite graph $K_{m,m}$ or the discrete bipartite graph on $m+m$ vertices).

Remark. There is a general Ramsey-type result saying that a bipartite graph on $n+n$ vertices contains a homogeneous subgraph on $m+m$ vertices if n is sufficiently large in terms of m, but the quantitative bound is only $m \approx \log n$. The above result, due to Hajnal and Pach (oral communication), shows that if any small induced subgraph H as above is forbidden, the guaranteed size of a homogeneous subgraph is at least $\Omega(n^{1/d})$. Erdős and Hajnal [EH77] conjectured a similar result for non-bipartite graphs: Given any fixed graph H, there is a $\delta > 0$ such that any graph on n vertices with no induced copy of H contains a complete graph on $\Omega(n^\delta)$ vertices or its complement. Also see Erdős and Hajnal [EH89] for partial results.

10. (Generalization of VC-dimension to k-valued functions) Let X be an n-point set and let $\mathcal{F}$ be a family of functions $X \to \{1, 2, \ldots, k\}$, where $k \geq 2$ is an integer parameter.

(a) We say that a subset $A \subseteq X$ is *k-shattered* by $\mathcal{F}$ if for each function $f: A \to \{1, 2, \ldots, k\}$, a function $\bar{f} \in \mathcal{F}$ exists such that $\bar{f}|_A = f$. For $k = 2$, how does this correspond to the notion of shattering of a set by a set system on X?

(b)* The family $\mathcal{F}_0$ consisting of all functions $f: X \to \{1, 2, \ldots, k\}$ attaining the value k at most d times shows that for $k \geq 3$, there exist families of size exponential in n with no k-shattered subset of size $d+1$, even for $d = 0$. (This makes the notion of dimension based on the k-shattering for $k \geq 3$ much less useful than the VC-dimension.) Extend the linear algebra proof of the Shatter function lemma 5.9 to show that no family without a k-shattered subset of size $d+1$ has more functions than the family $\mathcal{F}_0$ defined above.

(c)* Let us say that a subset $A \subseteq X$ is *2-shattered* by a family $\mathcal{F}$ as above if for each $x \in A$, there exist a 2-element subset $V_x \subseteq \{1, 2, \ldots, k\}$ such that for each combination of choices $v_x \in V_x$, $x \in A$, the family $\mathcal{F}$ contains a function f with $f(x) = v_x$ for all $x \in A$. Generalizing the first proof of the Shatter function lemma 5.9, prove that for d fixed, the maximum size of a family $\mathcal{F}$ on an n-point set X with no 2-shattered subset of size $d+1$ is $O(k^{2d} n^d)$.

(d) In the situation as in (c), show that $\mathcal{F}$ can have size at least $\Omega(k^d n^d)$.

Remark. It is not known whether the bound in (c) is tight, even for $d = 1$ and $n = 3$. The construction in the hint to (d) can be improved by a factor somewhat smaller than k but nothing better is known.

5.3 Packing Lemma

As a motivating example, consider n lines in the plane in general position. If they are removed from the plane, we obtain a collection $\mathcal{R}$ of convex regions (a simple induction on the number of lines shows that $|\mathcal{R}| = \binom{n+1}{2} + 1$). One can naturally define a metric on these regions: the distance of two regions $R_1, R_2 \in \mathcal{R}$ is the number of lines crossed by a segment connecting a point in R_1 to a point in R_2.

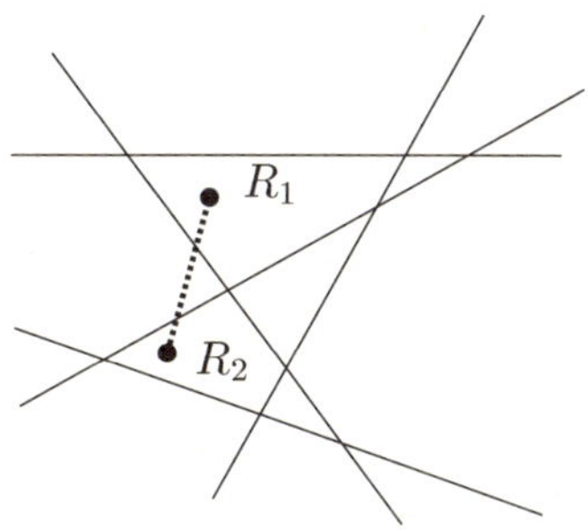

In the proofs of the upper bounds in Theorems 5.3 and 5.4, we are interested in properties of this metric space. In particular, we ask the following question. Suppose that for a given integer δ, we have a subset $\mathcal{P} \subseteq \mathcal{R}$ of regions such that any two regions in $\mathcal{P}$ have distance greater than δ (we call such a $\mathcal{P}$ *δ-separated*). What is the maximum possible cardinality of $\mathcal{P}$?

This is actually a question about packing of disjoint balls. For a region $R \in \mathcal{P}$, let $B(R, \rho)$ be the ball of radius ρ centered at R; that is, the set of all regions $R' \in \mathcal{R}$ at distance at most ρ from R. If $\mathcal{P}$ is a 2ρ-separated set, all the balls $B(R, \rho)$ for $R \in \mathcal{P}$ are disjoint, and conversely, a disjoint collection of ρ-balls defines a (2ρ)-separated set.

This ball-packing question can be answered concisely: as far as ball packing is concerned, the metric space of regions in a line arrangement behaves in a way similar to the square Q of side n with the usual Euclidean metric. In the Euclidean plane, a disc B of radius ρ has area $\pi\rho^2$, and if its center lies in the square Q, then the area of $B \cap Q$ is at least $\frac{\pi}{4}\rho^2 = \Omega(\rho^2)$. Consequently, one cannot place more than $O((n/\rho)^2)$ centers of disjoint discs of radius ρ into Q. Similarly, as a special case of the Packing lemma below, we will see that no more than $O((n/\rho)^2)$ disjoint ρ-balls can be placed into the considered metric space $\mathcal{R}$ of regions.

The Packing lemma is stated and proved for general set systems. Let $(X, \mathcal{S})$ be a set system on a finite set X. We define a metric on $\mathcal{S}$: the distance of two sets $S_1, S_2 \in \mathcal{S}$ is $|S_1 \triangle S_2|$, where $S_1 \triangle S_2 = (S_1 \cup S_2) \setminus (S_1 \cap S_2)$ denotes the symmetric difference. In other words, this is the L_1-distance, or Hamming distance, of the characteristic vectors of S_1 and S_2.

The Packing lemma says that a set system on an n-point set with the primal shatter function bounded by $O(m^d)$ behaves, as far as the ball packing problem is concerned, in a way similar to the cube with side n in the Euclidean space $\mathbf{R}^d$.

5.14 Lemma (Packing lemma). *Let $d > 1$ and C be constants, and let $(X, \mathcal{S})$ be a set system on an n-point set whose primal shatter function satisfies $\pi_{\mathcal{S}}(m) \le Cm^d$ for all $m = 1, 2, \ldots, n$. Let δ be an integer, $1 \le \delta \le n$, and let $\mathcal{P} \subseteq \mathcal{S}$ be δ-separated (any two distinct sets of $\mathcal{P}$ have distance greater than δ). Then $|\mathcal{P}| = O((n/\delta)^d)$.*

Let us remark that the original formulation of the Packing lemma, due to Haussler, assumes that the set system has VC-dimension d and yields the bound $|\mathcal{P}| \le (cn/(\delta + d))^d$, with c an absolute constant (independent of d).

Proof of a Weaker Bound. It is instructive to prove a weaker result first, namely

$$|\mathcal{P}| = O\left(\left(\frac{n}{\delta}\right)^d \log^d \frac{n}{\delta}\right),$$

using the Epsilon-net theorem 5.12. So let d be a constant, and let $\mathcal{P}$ satisfy $|S_1 \triangle S_2| > \delta$ for all $S_1 \neq S_2 \in \mathcal{P}$.

Consider the set system $\mathcal{D} = \{S_1 \triangle S_2 \colon S_1, S_2 \in \mathcal{S}\}$. This $\mathcal{D}$ has a bounded VC-dimension since its primal shatter function is polynomially bounded. Set $r = \frac{n}{\delta}$, and fix a $\frac{1}{r}$-net N of size $O(r \log r)$ for $\mathcal{D}$, according to the Epsilon-net theorem 5.12.

Whenever the symmetric difference of any two sets $S_1, S_2 \in \mathcal{S}$ has more than $\frac{n}{r} = \delta$ elements, it contains a point of N. In particular, we get $S_1 \cap N \neq S_2 \cap N$ for any two distinct sets $S_1, S_2 \in \mathcal{P}$. Therefore, the set system induced by $\mathcal{S}$ on N has at least $|\mathcal{P}|$ elements, and so we get $|\mathcal{P}| \le \pi_{\mathcal{S}}(|N|) = O((n/\delta)^d \log^d(n/\delta))$ as claimed. □

To get the better bound as in the Packing lemma, we first prove an auxiliary result. For a set system $(X, \mathcal{S})$, we define the *unit distance graph* $\mathrm{UD}(\mathcal{S})$. The vertex set of $\mathrm{UD}(\mathcal{S})$ is $\mathcal{S}$, and a pair $\{S, S'\}$ is an edge if S and S' have distance 1; that is, $|S \triangle S'| = 1$. Thus, each edge e can be written as $e = \{S, S \dot\cup \{y\}\}$.

5.15 Lemma. *If $\mathcal{S}$ is a set system of VC-dimension d_0 on a finite set X then the unit distance graph $UD(\mathcal{S})$ has at most $d_0|\mathcal{S}|$ edges.*

Proof. This is very similar to the proof of the Shatter function lemma 5.9. We proceed by induction on $n = |X|$ and d_0. The case $n = 1$ is trivial, and the case $d_0 = 1$ is easy to discuss. Hence we assume $n > 1$, $d_0 > 1$.

Fix an element $x \in X$, and define set systems $\mathcal{S}_1$ and $\mathcal{S}_2$ as in the proof of the Shatter function lemma 5.9:

$$\mathcal{S}_1 = \mathcal{S}|_{X \setminus \{x\}}, \quad \mathcal{S}_2 = \{S \in \mathcal{S}\colon x \notin S, S \cup \{x\} \in \mathcal{S}\}.$$

We know that $|\mathcal{S}| = |\mathcal{S}_1| + |\mathcal{S}_2|$. Let E be the edge set of $\mathrm{UD}(\mathcal{S})$, E_1 the edge set of $\mathrm{UD}(\mathcal{S}_1)$, and E_2 the edge set of $\mathrm{UD}(\mathcal{S}_2)$. By the inductive hypothesis, we have $|E_1| \le d_0|\mathcal{S}_1|$, and $|E_2| \le (d_0 - 1)|\mathcal{S}_2|$ since $\mathcal{S}_2$ has VC-dimension $\le d_0 - 1$.

Consider an edge $e = \{S, S'\}$ of $\mathrm{UD}(\mathcal{S})$, and let y_e be the element with $S \triangle S' = \{y_e\}$. There are at most $|\mathcal{S}_2|$ edges e with $y_e = x$. Next, we assume that $y_e \ne x$. We want to find an edge e' in E_1 or in E_2 to pay for e, in such a way that no edge in E_1 or in E_2 pays for more than one $e \in E$. If we succeed, we obtain $|E| \le |\mathcal{S}_2| + |E_1| + |E_2| \le |\mathcal{S}_2| + d_0|\mathcal{S}_1| + (d_0 - 1)|\mathcal{S}_2| \le d_0|\mathcal{S}|$, and the induction step will be finished.

A natural candidate for paying for the edge $e = \{S, S'\}$ is the edge $e' = \{S \setminus \{x\}, S' \setminus \{x\}\}$. We have $e' \in E_1$, but it may happen that two distinct edges $e_1, e_2 \in E$ lead to the same edge e' in this way. This happens if and only if $e_1 = \{S, S'\}$ and $e_2 = \{S \setminus \{x\}, S' \setminus \{x\}\}$. But this is exactly the case when e' is present in E_2 as well, and so it can pay for both e_1 (as a member of E_1) and for e_2 (as a member of E_2). Lemma 5.15 is proved. □

Proof of the Packing Lemma 5.14. The Packing lemma is proved by a probabilistic argument which looks like a magician's trick. Let $(X, \mathcal{S})$ be a set system and let $\mathcal{P} \subseteq \mathcal{S}$ be a δ-separated subsystem of $\mathcal{S}$. We choose a random s-element subset $A \subseteq X$, where the size s is chosen suitably; later on we will see that a good choice is $s = \lceil 4d_0 n/\delta \rceil$, where d_0 is the VC-dimension of $\mathcal{P}$. Set $\mathcal{Q} = \mathcal{P}|_A$, and for each set $Q \in \mathcal{Q}$ define its *weight* $w(Q)$ as the number of sets $S \in \mathcal{P}$ with $S \cap A = Q$. Note that $\sum_{Q \in \mathcal{Q}} w(Q) = |\mathcal{P}|$.

Let E be the edge set of the unit distance graph $\mathrm{UD}(\mathcal{Q})$, and define the *weight* of an edge $e = \{Q, Q'\}$ as $\min(w(Q), w(Q'))$. Put $W = \sum_{e \in E} w(e)$; note that W is a random variable depending on the random choice of A. The bound on $\mathcal{P}$ in the Packing lemma is obtained by estimating the expectation of W in two ways.

First, we claim that for any set $A \subseteq X$, we have

$$W \le 2d_0 \sum_{Q \in \mathcal{Q}} w(Q) = 2d_0|\mathcal{P}|. \tag{5.1}$$

This is because by Lemma 5.15, the unit distance graph $\mathrm{UD}(\mathcal{Q})$ has some vertex Q_0 of degree at most $2d_0$. By removing Q_0, the total vertex weight drops by $w(Q_0)$ and the total edge weight by at most $2d_0 w(Q_0)$. Repeating this until no vertices are left we see that the sum of edge weights is at most $2d_0$ times the sum of vertex weights as claimed.

Next, we bound the expectation $\mathbf{E}[W]$ from below. Imagine the following random experiment. First, we choose a random $(s-1)$-element set $A' \subset X$,

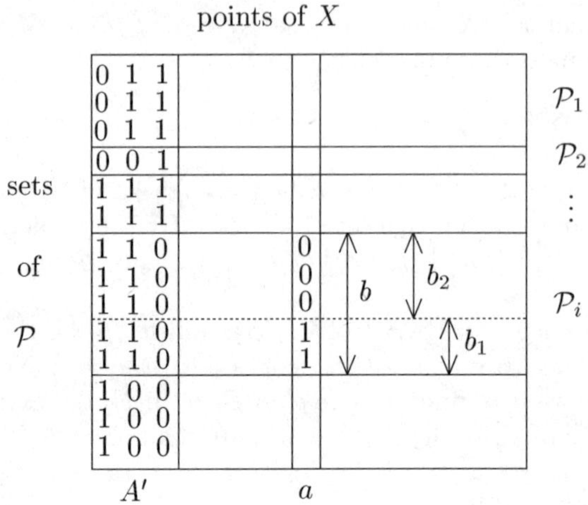

Fig. 5.1. Dividing the sets of $\mathcal{P}$ into classes according to A', and then refining by a random $a \in X \setminus A'$.

and then we choose a random element $a \in X \setminus A'$. The set $A = A' \cup \{a\}$ is a random s-element subset of X, and we consider the corresponding unit distance graph on $\mathcal{Q} = \mathcal{P}|_A$ as above. Each edge of this graph is a pair of sets of $\mathcal{Q}$ differing in exactly one element of A. We let $E_1 \subseteq E$ be the edges for which the difference element is a, and let W_1 be the sum of their weights. By symmetry, we have $\mathbf{E}[W] = s \cdot \mathbf{E}[W_1]$.

We are going to bound $\mathbf{E}[W_1]$ from below. Let $A' \subset X$ be an arbitrary but fixed $(s-1)$-element set. We estimate the conditional expectation $\mathbf{E}[W_1|A']$; that is, the expected value of W_1 when A' is fixed and a is random.

Divide the sets of $\mathcal{P}$ into equivalence classes $\mathcal{P}_1, \mathcal{P}_2, \ldots, \mathcal{P}_t$ according to their intersection with the set A'. We have $t \leq \pi_{\mathcal{S}}(s-1) \leq C_2(n/\delta)^d$ for a constant C_2.

Let $\mathcal{P}_i$ be one of the equivalence classes, and let $b = |\mathcal{P}_i|$. (The situation is perhaps best visualized using the incidence matrix of the set system $\mathcal{P}$; see Fig. 5.1.) Suppose that an element $a \in X \setminus A'$ has been chosen. If b_1 sets of $\mathcal{P}_i$ contain a and $b_2 = b - b_1$ sets do not contain a then the class $\mathcal{P}_i$ gives rise to an edge of E_1 of weight $\min(b_1, b_2)$. (Note that b_1 and b_2 depend on the choice of a while b does not.)

For any nonnegative real numbers b_1, b_2 with $b_1 + b_2 = b$, we have $\min(b_1, b_2) \geq b_1 b_2 / b$. The value $b_1 b_2$ is the number of ordered pairs of sets (S_1, S_2) with S_1, S_2 being two sets from the class $\mathcal{P}_i$ which differ in a (one of them contains a and the other one does not). Let us look at this differently: if $S_1, S_2 \in \mathcal{P}_i$ are two distinct sets, we know that they differ in at least δ elements (by the assumption on $\mathcal{P}$), and therefore the probability that S_1 and S_2 differ in a random element $a \in X \setminus A'$ is at least $\frac{\delta}{n-s+1} \geq \frac{\delta}{n}$. Hence the expected contribution of each pair (S_1, S_2) of distinct sets of $\mathcal{P}_i$ to the quantity b_1b_2 is at least $\frac{\delta}{n}$, and so $\mathbf{E}[b_1b_2] \geq b(b-1)\frac{\delta}{n}$. This further means that the expected contribution of the equivalence class $\mathcal{P}_i$ to the sum of edge weights W_1 is at least $(b-1)\frac{\delta}{n}$. Summing up over all equivalence classes, we find that

$$\mathbf{E}[W_1] \geq \frac{\delta}{n}\sum_{i=1}^{t}(|\mathcal{P}_i|-1) = \frac{\delta}{n}(|\mathcal{P}|-t) \geq \frac{\delta}{n}\left(|\mathcal{P}| - C_2\left(\frac{n}{\delta}\right)^d\right).$$

Combining this with the estimate (5.1), i.e. $s\cdot\mathbf{E}[W_1] = \mathbf{E}[W] \leq 2d_0|\mathcal{P}|$, leads to the inequality

$$2d_0|\mathcal{P}| \geq \frac{s\delta}{n}\left(|\mathcal{P}| - C_2\left(\frac{n}{\delta}\right)^d\right) \geq 4d_0|\mathcal{P}| - 4d_0C_2\left(\frac{n}{\delta}\right)^d.$$

The rabbit is out of the hat: we have $|\mathcal{P}| = O((n/\delta)^d)$ as claimed. □

Bibliography and Remarks. Haussler [Hau95] attributes the idea of the proof of the quantitatively weaker version of the Packing lemma shown above to Dudley. The result was re-discovered by Welzl [Wel88]. The tight bound was first proved in certain geometric cases (e.g., as in Exercise 1) by Chazelle and Welzl [CW89]. The general case is due to Haussler [Hau95]; his proof was simplified by Chazelle in an unpublished manuscript [Cha92] whose presentation we have essentially followed.

Exercises

1. In this exercise, we indicate a simpler proof of the Packing lemma 5.14 for the case when $\mathcal{S}$ is the set system defined by halfplanes on a finite point set in the plane.
   (a)* Let L be a set of n lines in the plane in general position, let x be a point, and let $r < \frac{n}{2}$ be a number. Show that the number of intersections of the lines of L lying at distance at most r from x (that is, intersections v such that the open segment vx is intersected by at most r lines of L) is at least cr^2, with an absolute constant $c > 0$.
   (b) Prove the Packing lemma 5.14 for a set system of the form $\mathcal{H}_2|_P$ for an n-point set P in the plane. If convenient, assume that P is in general position.
   (c)* Extend (a) for hyperplanes in $\mathbf{R}^d$ (there are at least c_dr^d vertices at distance $\leq r$ from any point), and (b) for halfspaces in $\mathbf{R}^d$.

5.4 Matchings with Low Crossing Number

In this section, we establish Theorem 5.4, the upper bound for discrepancy in terms of the dual shatter function. The main tool for the proof are the so-called matchings with low crossing number. First, we consider a particular geometric setting.

5.16 Problem. *Let X be an n-point set in the plane, where the number n of points is even. Moreover, assume for simplicity that X is in general position, with no 3 points collinear. Partition the points of X into pairs (in other words, fix a perfect matching M on the vertex set X), and connect the points in each pair by a straight segment. Let $\kappa(M)$ denote the maximum possible number of these segments that can be intersected by a common line. For a given set X, we take a matching M minimizing $\kappa(M)$. What is the best upper bound for $\kappa(M)$, in terms of n, that can be guaranteed for all n-point sets X?*

For example, in the following picture, the line intersects the 4 thicker segments, and no 5 segments can be intersected by a single line (I hope), and hence $\kappa(M) = 4$ for this particular matching M.

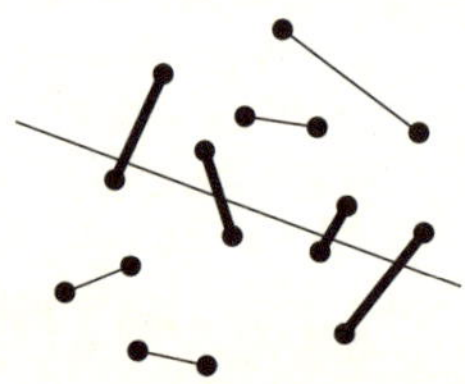

A particular case of a theorem below asserts that we can always get a matching M with $\kappa(M) = O(\sqrt{n})$.

The problem of finding such a good matching can be considered for an arbitrary set system. To this end, we define the notion of a *crossing number.*[1] We formulate the definition for a general graph although we will only need to consider perfect matchings. Let $(X, \mathcal{S})$ be a set system and let $G = (X, E)$ be a graph with vertex set X. We say that a set $S \in \mathcal{S}$ *crosses* an edge $\{u, v\}$ of G if $|S \cap \{u, v\}| = 1$. This is illustrated in the following drawing:

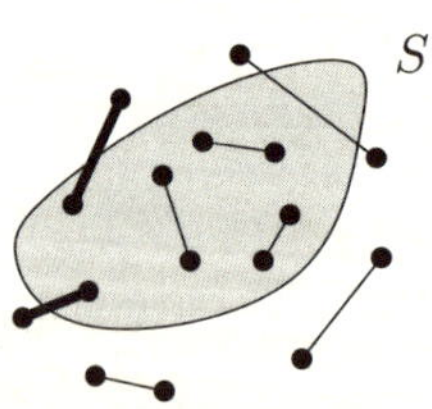

The connecting segments have no geometric meaning here and they just mark the point pairs. The segments for the crossed pairs are again drawn thicker. The *crossing number of G with respect to the set S* is the number of edges of

[1] Not to be confused with the crossing number of a graph G considered in the theory of graph drawing (the minimum number of edge crossings present in a drawing of G in the plane). For the crossing number in the sense of this section, the term *stabbing number* is also used in the literature.

G crossed by S, and the *crossing number of* G is the maximum of crossing numbers of G with respect to all sets of $\mathcal{S}$.

Problem 5.16 above corresponds to the general definition for the case where $\mathcal{S}$ is the set system induced on X by halfplanes. In degenerate cases, such as when a segment is completely contained in the boundary of the considered halfplane, the geometric notion of "crossing" may become unclear but the general set-theoretic definition of crossing can always be applied unambiguously.

Theorem 5.4, the bound for discrepancy in terms of the dual shatter function, can be easily derived from the following result:

5.17 Theorem. *Let $\mathcal{S}$ be a set system on an n-point set X, n even, with $\pi^*_{\mathcal{S}}(m) \leq Cm^d$ for all m, where C and $d > 1$ are constants. Then there exists a perfect matching M on X (i.e. a set of $\frac{n}{2}$ vertex-disjoint edges) whose crossing number is at most $C_1 n^{1-1/d}$, where $C_1 = C_1(C, d)$ is another constant.*

This is a nice and nontrivial result even in the simple geometric case in Problem 5.16 above. Even for this particular case, I know of no proof substantially different from the general one shown below.

Proof of the Dual Shatter Function Bound (Theorem 5.4). Let $(X, \mathcal{S})$ be a set system satisfying the assumptions of Theorem 5.4. Suppose that n is even (if not add one point lying in no set of $\mathcal{S}$). Fix a perfect matching M with edges $\{u_1, v_1\}, \{u_2, v_2\}, \ldots, \{u_k, v_k\}$, $k = \frac{n}{2}$, on X with crossing number $O(n^{1-1/d})$. Define a random coloring $\chi: X \to \{+1, -1\}$ by coloring the points $u_1, \ldots, u_k$ randomly and independently and by setting $\chi(v_i) = -\chi(u_i)$ for all i, as in the picture:

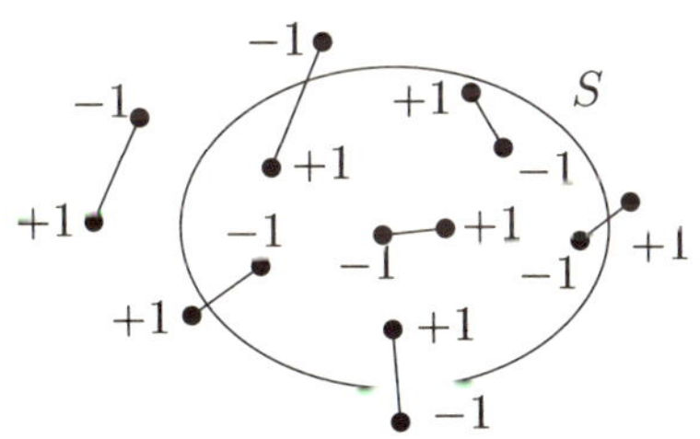

Look at a fixed set $S \in \mathcal{S}$, and classify the edges of M to two types: those with both points inside S or both points outside S, and those crossed by S. The edges of the former type contribute 0 to $\chi(S)$. The contributions of the edges of the latter type to $\chi(S)$ are, by the definition of χ, independent random variables attaining values $+1$ and -1 with equal probability. The number of these variables is $O(n^{1-1/d})$. Thus, the situation is as if we had a random coloring of $|\mathcal{S}|$ sets of size $O(n^{1-1/d})$ each, and the Random coloring lemma 4.1 tells us that $\operatorname{disc}(\chi, \mathcal{S}) = O(n^{1/2-1/2d}\sqrt{\log|\mathcal{S}|})$ with a positive probability. Finally, we have $\log|\mathcal{S}| = O(\log n)$ by the Dual set system lemma 5.10. Theorem 5.4 is proved. □

To prove the Theorem 5.17, we need the following lemma:

5.18 Lemma (Short edge lemma). *Let $\mathcal{S}$ be a set system as in Theorem 5.17. Then for any set (or multiset[2]) $Q \subseteq \mathcal{S}$, there exist points $x, y \in X$ such that the edge $\{x, y\}$ is crossed by at most*

$$C_2 \frac{|Q|}{n^{1/d}}$$

sets of Q, C_2 being a suitable constant.

The proof will be discussed later. In the setting of Problem 5.16 for lines, the lemma implies the following: if X is an n-point set in the plane and we select arbitrary m lines (not passing through any of the points of X, say), then there exist two distinct points of X separated by $O(m/\sqrt{n})$ lines only.

Proof of Theorem 5.17. The basic strategy is to select the edges of the matching M one by one, always taking the "shortest" available edge. The first edge, $\{u_1, v_1\}$, is selected as one crossed by the smallest possible number of sets in $\mathcal{S}$. We could now select $\{u_2, v_2\}$ as a pair of the remaining points such that $\{u_2, v_2\}$ is crossed by the smallest possible number of sets of $\mathcal{S}$, etc. This guarantees a good behavior of the resulting matching on the average, i.e. an average set of $\mathcal{S}$ would cross the right number of edges. However, it might happen that a few exceptional sets of $\mathcal{S}$ would cross many more edges.

The selection strategy is thus improved to penalize the sets of $\mathcal{S}$ that already cross many of the edges selected so far (this "re-weighting strategy" is useful in several proofs in combinatorics). Specifically, suppose that edges $\{u_1, v_1\}, \ldots, \{u_i, v_i\}$ have already been selected. Define the *weight* $w_i(S)$ of a set $S \in \mathcal{S}$ as $2^{\kappa_i(S)}$, where $\kappa_i(S)$ is the number of edges among $\{u_1, v_1\}, \ldots, \{u_i, v_i\}$ crossed by S. In particular, $w_0(S) = 1$ for all $S \in \mathcal{S}$. We select the next edge $\{u_{i+1}, v_{i+1}\}$ as a pair of points among the points of $X_i = X \setminus \{u_1, v_1, \ldots, u_i, v_i\}$ with the total weight of sets crossing $\{u_{i+1}, v_{i+1}\}$ being the smallest possible. We continue in this manner until $\frac{n}{2}$ edges have been selected.

We need to bound the crossing number κ of the resulting matching M. To this end, we estimate the final total weight of all sets of $\mathcal{S}$, i.e. $w_{n/2}(\mathcal{S}) = \sum_{S \in \mathcal{S}} w_{n/2}(S)$. By the definition of $w_{n/2}$, we have $\kappa \leq \log_2 w_{n/2}(\mathcal{S})$.

Let us investigate how $w_{i+1}(\mathcal{S})$ increases compared to $w_i(\mathcal{S})$. Let $\mathcal{S}_{i+1}$ denote the collection of the sets of $\mathcal{S}$ crossing $\{u_{i+1}, v_{i+1}\}$. For the sets of $\mathcal{S}_{i+1}$, the weight increases twice, and for the others it remains unchanged. From this we get

$$w_{i+1}(\mathcal{S}) \leq w_i(\mathcal{S}) - w_i(\mathcal{S}_{i+1}) + 2w_i(\mathcal{S}_{i+1}) = w_i(\mathcal{S}) \left(1 + \frac{w_i(\mathcal{S}_{i+1})}{w_i(\mathcal{S})}\right).$$

[2] Multiset means that Q may contain several copies of the same set $S \in \mathcal{S}$. The cardinality of Q is counted with these multiplicities.

Next, we want to estimate the ratio $w_i(\mathcal{S}_{i+1})/w_i(\mathcal{S})$ using the Short edge lemma. To this end, we define a multiset Q_i of sets. We restrict each set $S \in \mathcal{S}$ to the set X_i, and we add $S \cap X_i$ to Q_i with multiplicity $w_i(S)$. We apply the Short edge lemma 5.18 to this Q_i. This shows that $w_i(\mathcal{S}_{i+1}) \le C_2 w_i(\mathcal{S})/n_i^{1/d}$, where $n_i = |X_i| = n - 2i$. Hence

$$w_{i+1}(\mathcal{S}) \le w_i(\mathcal{S}) \left(1 + \frac{C_2}{(n-2i)^{1/d}}\right),$$

and so

$$w_{n/2}(\mathcal{S}) \le w_0(\mathcal{S}) \prod_{i=0}^{n/2-1} \left(1 + \frac{C_2}{(n-2i)^{1/d}}\right) = |\mathcal{S}| \cdot \prod_{i=0}^{n/2-1} \left(1 + \frac{C_2}{(n-2i)^{1/d}}\right).$$

Taking logarithms and using the inequality $\ln(1+x) \le x$ we obtain

$$\kappa \le \log_2 w_{n/2}(\mathcal{S}) \le \log|\mathcal{S}| + C_2 \sum_{i=0}^{n/2-1} \frac{1}{(n-2i)^{1/d}} \le \log|\mathcal{S}| + C_2 \sum_{j=1}^{n/2} \frac{1}{j^{1/d}}.$$

Bounding the last sum by an integral, we finally obtain $\kappa = O(\log|\mathcal{S}| + n^{1-1/d})$. It remains to apply the Dual set system lemma 5.10 to conclude that the number of sets in $\mathcal{S}$ is polynomial in n. We get $\log|\mathcal{S}| = O(\log n)$ and Theorem 5.17 follows. □

Proof the Short Edge Lemma 5.18. This lemma is a straightforward consequence of the Packing lemma 5.14. First, we form a set system $\mathcal{D}$ dual to Q. We consider the multiset Q as the ground set (if some set appears in Q several times, it is considered with the appropriate multiplicity). For each $x \in X$, we let D_x be the set of all sets of Q containing x, and we put $\mathcal{D} = \{D_x : x \in X\}$.

The symmetric difference $D_x \triangle D_y$ of two sets from $\mathcal{D}$ consists of the sets in Q crossing the pair $\{x, y\}$. We thus want to show that $D_x \triangle D_y$ is small for some $x \ne y$. We may assume $D_x \ne D_y$ for $x \ne y$, for otherwise we are done, and hence $|\mathcal{D}| = |X| = n$.

The primal shatter function $\pi_{\mathcal{D}}$ is certainly no larger than the dual shatter function $\pi_{\mathcal{S}}^*$, and hence $\pi_{\mathcal{D}}(m) \le Cm^d$ by the assumption on $\pi_{\mathcal{S}}^*$. Supposing that any two sets $D_x, D_y \in \mathcal{D}$ have symmetric difference at least δ, the Packing lemma 5.14 implies $n = |\mathcal{D}| = O((|Q|/\delta)^d)$, and therefore $\delta = O(|Q|/n^{1/d})$. □

Remark on Algorithms. The proof given in this section easily leads to a randomized polynomial-time algorithm for finding a coloring as in the dual shatter function bound (Theorem 5.4). A deterministic algorithm can be obtained by the method of conditional probabilities.

Bibliography and Remarks. Matchings with low crossing number were invented by Welzl [Wel88] for the purpose of the *range searching* *problem* in computational geometry. He actually worked with *spanning trees with low crossing number* instead of matchings. Welzl's existence proof was similar to the one presented above, but with a weaker version of the Short edge lemma. The Short edge lemma was subsequently improved by a logarithmic factor, first by Chazelle and Welzl [CW89] in some geometric cases, and then in general as an immediate consequence of the Packing lemma 5.14.

Exercises

1.* Prove that there exist $2n$ points in the plane such that for any perfect matching on them there is a line crossing at least $c\sqrt{n}$ edges. This means that Theorem 5.17 is asymptotically optimal for $d = 2$.
2. Prove that any set system $(X, \mathcal{S})$ as in Theorem 5.17 admits a spanning path with crossing number $O(n^{1-1/d})$. A spanning path is a path connecting all the points of X in some order.
3. (a) Let $S = S_1 + \cdots + S_n$ be the sum of n uniformly distributed independent random 0/1 variables. Calculate the expected value of S^2.
   (b) Let $(X, \mathcal{S})$ be a set system with $|X| = n$ and with $\pi^*_{\mathcal{S}}(m) \leq Cm^d$ ($d > 1$) for all m. Prove that the L_2-discrepancy of $\mathcal{S}$ is $O(n^{1/2-1/2d})$. Show that this result remains valid with an arbitrary measure (weights) on $\mathcal{S}$.

   Remark. This shows that one cannot hope to prove a better lower bound than $n^{1/4}$ for the discrepancy for discs in the plane by methods based on L_2-discrepancy.

   (c)* Generalize (b) for L_p-discrepancy with an arbitrary fixed $p \in [1, \infty)$, again showing $O(n^{1/4})$ upper bound, with the constant of proportionality depending on p.

5.5 Primal Shatter Function and Partial Colorings

Theorem 5.3 asserts the $O(n^{1/2-1/2d})$ bound for the discrepancy of a set system on n points with the primal shatter function bounded by $O(m^d)$. First we prove a weaker bound of $O(n^{1/2-1/2d}(\log n)^{1/2+1/2d})$ under the same conditions, using the Partial coloring lemma 4.13. Then we establish the tight bound by the entropy method. Finally we show one more application of the entropy method.

Proof of the Weaker Bound. Let $(X, \mathcal{S})$ be the considered set system with $\pi_{\mathcal{S}}(m) = O(m^d)$. The idea is to fix a suitable not too large family of "basic sets," and to express each set of $\mathcal{S}$ as an appropriate basic set modified by adding and subtracting suitable small "correction sets." The basic sets will be used as $\mathcal{F}$ in the Partial coloring lemma, and we enforce zero discrepancy for them. The correction sets will play the role of $\mathcal{M}$, and their discrepancy will be as if the coloring were random. Such a partial coloring step is iterated in a standard manner.

Let δ be some yet unspecified parameter bounding the size of the "correction sets." We choose $\mathcal{F} \subseteq \mathcal{S}$ as an inclusion-maximal collection of sets in $\mathcal{S}$ such that any two sets of $\mathcal{F}$ have distance (symmetric difference size) greater than δ. By inclusion-maximality, we infer that for each set $S \in \mathcal{S}$, there exists a set $F = F(S)$ with $|F \triangle S| \leq \delta$, for otherwise we could add S to $\mathcal{F}$. We fix one such $F(S) \in \mathcal{F}$ for each $S \in \mathcal{S}$.

Next, we define the correction sets. Put $\mathcal{A} = \{S \setminus F(S) \colon S \in \mathcal{S}\}$ and $\mathcal{B} = \{F(S) \setminus S \colon S \in \mathcal{S}\}$, and finally $\mathcal{M} = \mathcal{A} \cup \mathcal{B}$. Note that any set $S \in \mathcal{S}$ can be written as

$$S = (F \dot{\cup} A) \setminus B \tag{5.2}$$

for some $F \in \mathcal{F}$ and $A, B \in \mathcal{M}$, where B is completely contained in $F \dot{\cup} A$, as in the following picture:

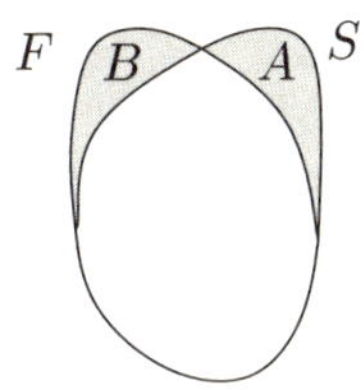

In order to apply the Partial coloring lemma on $\mathcal{F}$ and $\mathcal{M}$, $\mathcal{F}$ must be sufficiently small. The size of $\mathcal{F}$ is bounded according to the Packing lemma 5.14: we have $|\mathcal{F}| = O((n/\delta)^d)$. A simple calculation shows that in order to satisfy the condition $\prod_{F \in \mathcal{F}}(|F| + 1) \leq 2^{(n-1)/5}$ in the Partial coloring lemma, we should set $\delta = C_2 n^{1-1/d} \log^{1/d} n$ for a sufficiently large constant C_2.

We can now apply the Partial coloring lemma, obtaining a no-nonsense partial coloring χ_1 such that the sets of $\mathcal{F}$ have zero discrepancy, while the discrepancy of the sets of $\mathcal{M}$ is bounded by $O(\sqrt{\delta \log |\mathcal{M}|}\,) = O(\sqrt{\delta \log n}\,)$. In view of (5.2), we also have

$$\begin{aligned} |\chi_1(S)| &\leq |\chi_1(F)| + |\chi_1(A)| + |\chi_1(B)| \\ &= O(\sqrt{\delta \log n}) = O(n^{1/2-1/2d} (\log n)^{1/2+1/2d}) \end{aligned}$$

for each $S \in \mathcal{S}$.

To get a (full) coloring with the asserted $O(n^{1/2-1/2d}(\log n)^{1/2+1/2d})$ discrepancy, we iterate the construction described above as usual. This presents no problem here since the shatter function condition is hereditary. We let Y_1 be the set of points colored by χ_1, and let X_2 be the set of the uncolored points. We repeat the argument for the set system $\mathcal{S}$ restricted to X_2, obtaining a partial coloring χ_2, and so on. Note that the auxiliary set systems $\mathcal{F}$ and $\mathcal{M}$ are constructed anew in each partial coloring step, and the maximum size δ of the "correction sets" is getting smaller and smaller as the iterations progress. Having reached a set X_ℓ of size below a suitable constant, we combine all the partial colorings constructed so far into a single full coloring χ; the points of X_ℓ are colored arbitrarily.

We have, for any $S \in \mathcal{S}$,

$$|\chi(S)| \leq \sum_{i=1}^{\ell} |\chi_i(S \cap Y_i)| \leq \sum_{i=1}^{\ell} O\left(n_i^{1/2-1/2d} (\log n_i)^{1/2+1/2d}\right),$$

where $n_i = |X_i| \leq (\frac{9}{10})^{i-1} n$. Since $d > 1$, the summands on the right-hand side decrease geometrically, and we obtain the claimed bound for $\operatorname{disc}(\mathcal{S})$. □

The Tight Bound via Entropy. In the previous proof, we have expressed each $S \in \mathcal{S}$ in the form $S = (F \dot{\cup} A) \setminus B$, a "basic set" with a "correction." For the improved bound, we will use about $\log n$ successive corrections of a smaller and smaller size:

$$S = (\ldots(((A_1 \setminus B_1) \dot{\cup} A_2) \setminus B_2) \dot{\cup} \cdots \dot{\cup} A_k) \setminus B_k, \tag{5.3}$$

where each B_i is subtracted from a set containing it.

Here is the construction providing such decompositions. Let us set $k = \lfloor \log_2 n + 1 \rfloor$, and for each $i = 0, 1, \ldots, k$, let $\mathcal{F}_i$ be an inclusion-maximal subfamily of $\mathcal{S}$ in which every two sets have distance (symmetric difference) greater than $n/2^i$. In particular, we may assume that $\mathcal{F}_0$ consists of the empty set, and we have $\mathcal{F}_k = \mathcal{S}$.

For every set $F \in \mathcal{F}_i$, there exists a set $F' \in \mathcal{F}_{i-1}$ with $|F \triangle F'| \leq n/2^{i-1}$. This follows by inclusion-maximality of $\mathcal{F}_{i-1}$ as in the previous proof. We fix one such F' for each $F \in \mathcal{F}_i$ and we set $A(F) = F \setminus F'$ and $B(F) = F' \setminus F$. We form auxiliary set systems $\mathcal{A}_i = \{A(F)\colon F \in \mathcal{F}_i\}$ and $\mathcal{B}_i = \{B(F)\colon F \in \mathcal{F}_i\}$, for $i = 1, 2, \ldots, k$. Note that each set $F \in \mathcal{F}_i$ can be turned into a set $F' \in \mathcal{F}_{i-1}$ by adding a $B_i \in \mathcal{B}_i$ and subtracting an $A_i \in \mathcal{A}_i$. Since $\mathcal{F}_0 = \{\emptyset\}$ and $\mathcal{F}_k = \mathcal{S}$, we get the decomposition (5.3) for each $S \in \mathcal{S}$ with $A_i \in \mathcal{A}_i$ and $B_i \in \mathcal{B}_i$.

Let us set $\mathcal{M}_i = \mathcal{A}_i \cup \mathcal{B}_i$ and $\mathcal{M} = \bigcup_{i=1}^k \mathcal{M}_i$. For each i, we are going to fix a suitable bound Δ_i for the discrepancy of the sets of $\mathcal{M}_i$, and then we will apply the entropy method (Proposition 4.16) to the auxiliary set system $\mathcal{M}$. This gives us a no-nonsense partial coloring χ_1 on X such that each $\mathcal{M}_i$ has discrepancy at most Δ_i. In view of the decomposition (5.3), the discrepancy of $\mathcal{S}$ under χ_1 is at most $\Delta = 2(\Delta_1 + \Delta_2 + \cdots + \Delta_k)$. If we can manage to get $\Delta = O(n^{1/2-1/2d})$, we can iterate the partial coloring step as in the preceding proof and we get a full coloring with discrepancy $O(n^{1/2-1/2d})$.

What are the restriction on the Δ_i? By the construction, the sizes of the sets in $\mathcal{M}_i$ are at most $s_i = n/2^{i-1}$. By the Packing lemma 5.14, we see that $|\mathcal{F}_i| = O(2^{di})$, and consequently also $|\mathcal{M}_i| = O(2^{di})$. Let us write the constant explicitly as C, i.e. $|\mathcal{M}_i| \leq C \cdot 2^{di}$. The quantitative formulation of the entropy method, Proposition 4.16, tells us we should have

$$\sum_{i=1}^{k} C \cdot 2^{id} h\left(s_i, \Delta_i\right) \le \frac{n}{5}. \tag{5.4}$$

Let us look at a particular value i_0 of the index i, namely the value where the number of sets in $\mathcal{M}_{i_0}$, $C \cdot 2^{i_0 d}$, first exceeds n. The size s_{i_0} of the sets for such i_0 is about $n^{1-1/d}$. The entropy contribution of each of these $\ge n$ sets must be considerably smaller than 1, for otherwise the total entropy would be larger than $\frac{n}{5}$. By inspection of the estimates for $h(s, \Delta)$ in Proposition 4.16, we find that Δ_{i_0} must exceed the square root of the set size, i.e. $n^{1/2-1/2d}$. This is the total discrepancy we would like to get, so we have very little maneuvering room for choosing Δ_{i_0}—only the constant factor can be adjusted.

Fortunately, as i gets larger or smaller than i_0, the room for the choice of Δ_i gets bigger. If $i > i_0$, we have more sets in $\mathcal{M}_i$ but their size decreases, and the entropy contribution of each individual set tends to 0 much faster than the number of sets grows. On the other hand, if i gets smaller than i_0, we have fewer sets; these get bigger but the entropy contribution grows very slowly with the size.

One suitable specific choice for Δ_i is

$$\Delta_i = C_1 n^{1/2-1/2d} \varphi(i)$$

where $\varphi(i) = (1 + |i - i_0|)^{-2}$ and C_1 is a sufficiently large constant. As a function of i, $\varphi(i)$ is a "hill" with peak at i_0. Since the sum $\sum_{i=-\infty}^{\infty} \varphi(i) \le 2\sum_{j=1}^{\infty} j^{-2}$ converges, we have $\sum_{i=1}^{k} \Delta_i = O(\Delta_{i_0}) = O(n^{/12-1/2d})$ and the overall discrepancy bound is as desired.

Let $H_i = |\mathcal{M}_i| \cdot h(s_i, \Delta_i)$ denote the total entropy contribution for the sets of $\mathcal{M}_i$. Invoking Proposition 4.16, we obtain

$$\begin{aligned} H_i &\le K \cdot C \cdot 2^{id} e^{-\Delta_i^2/4s_i} \log_2\left(2 + \frac{\sqrt{s_i}}{\Delta_i}\right) \\ &= K \cdot C \cdot 2^{id} e^{-\lambda_i^2/4} \log_2\left(2 + \frac{1}{\lambda_i}\right), \end{aligned}$$

where $\lambda_i = \Delta_i/\sqrt{s_i} = C_1 \varphi(i) 2^{(i-1)/2}/n^{1/2d}$. For $i = i_0$, we calculate that λ_{i_0} is a constant which can be made as large as we wish by setting C_1 sufficiently large. For other values of i, λ_i behaves roughly like $2^{(i-i_0)/2}\lambda_{i_0}$ (the influence of the $\varphi(i)$ factor is negligible). Therefore for i growing above i_0, the value of H_i decreases superexponentially fast. On the other hand, for $i < i_0$, $h(s_i, \Delta_i)$ behaves roughly like $\log(1/\lambda_i) \approx (i_0 - i)/2$, and we have $H_i \approx 2^{-d(i_0-i)}(i_0 - i)\lambda_{i_0}$, which decreases exponentially with i. Altogether we derive that the total entropy bound (5.4) can be made an arbitrarily small fraction of n by choosing C_1 sufficiently large. I believe it is better for the reader to complete this rough argument by a detailed proof by himself/herself, in case of interest, rather than reading a tedious precise calculation. This concludes the proof of Theorem 5.3. □

A Discrepancy Bound in Terms of Degree: Entropy Again. We demonstrate another application of the entropy method. Let $(X, \mathcal{S})$ be a set system and and let t be its maximum degree (i.e. the maximum, over all $x \in X$, of the number of sets of $\mathcal{S}$ containing x). The Beck–Fiala theorem 4.3 claims $\operatorname{disc}(\mathcal{S}) \leq 2t - 1$, and it is not known how to improve this bound substantially in terms of t alone. However, if we also allow a moderate dependence on $|X|$, we can do better for a wide range of values of n and t:

5.19 Theorem. *Let $\mathcal{S}$ be a set system on an n-point set X, and let t be the maximum degree of $\mathcal{S}$. Then* $\operatorname{disc}(\mathcal{S}) = O(\sqrt{t} \cdot \log n)$.

A still better bound, of $O(\sqrt{t \log n})$, was recently obtained by geometric methods; see the remarks to Section 4.3.

Sketch of Proof. We prove that under the conditions of the theorem, there exists a no-nonsense partial coloring with discrepancy $O(\sqrt{t})$. The final bound then follows by a standard iteration of the partial coloring step.

We note that if the maximum degree is t, then the sum of the sizes of the sets in $\mathcal{S}$ is at most nt. Let $\mathcal{S}_i \subseteq \mathcal{S}$ consist of the sets of $\mathcal{S}$ of size between 2^i and 2^{i+1}; the degree condition thus gives $|\mathcal{S}_i| \leq nt/2^i$. Set $\Delta = C\sqrt{t}$ for a sufficiently large absolute constant C. A calculation similar to the one in the previous proof but simpler (the $\varphi(i)$ factor is not present, for instance) shows that

$$\sum_{i=0}^{\log_2 n} |\mathcal{S}_i| \cdot h(2^{i+1}, \Delta) \leq \frac{n}{5}\,.$$

and the existence of the no-nonsense partial coloring follows from Proposition 4.16. □

Bibliography and Remarks. The proof of the weaker bound for discrepancy from the primal shatter function follows [MWW93]. The tight bound is from [Mat95], but here the proof looks simpler because a big part of the work has already been done in previous sections.

Theorem 5.19 was proved by Srinivasan [Sri97], who improved a slightly weaker bound of $O(\sqrt{t \log t} \log n)$ due to Beck and Spencer (see [Spe87]). Srinivasan's proof is different from the one presented above; it uses a relation of the problem to the k-permutation problem (see Exercise 4.5.5) noted by Bohus [Boh90]. The best known $O(\sqrt{t \log n})$ bound is a consequence of a recent result of Banaszczyk[Ban98]; see the remarks to Section 4.3.

Exercises

1. Complete the calculation in the proof of Theorem 5.19 and/or in the proof of Theorem 5.3.

2. (Tusnády's problem revisited) Use the entropy method to prove an $O(\log^{5/2} n)$ bound for Tusnády's problem (combinatorial discrepancy for axis-parallel rectangles), improving Theorem 4.14 slightly. Consider the set system $\mathcal{F}$ as in the proof of that theorem, but with $t = 1$, and partition the sets by sizes, prescribing a suitable discrepancy bound Δ_i for each size 2^i.
3.* (The k-permutation problem revisited) Let $\mathcal{P}_k$ be a set system on $\{1, 2, \ldots, n\}$ defined by k permutations as in Exercise 4.5.5. By the entropy method, prove $\mathrm{disc}(\mathcal{P}_k) = O(\sqrt{k} \cdot \log n)$, with the the constant of proportionality independent of k. Use canonical intervals along each permutation.
4. (Arithmetic progressions revisited) Let $\mathcal{A}_n$ be the set system of all arithmetic progressions on the set $\{1, 2, \ldots, n\}$ as in Exercise 4.5.7.
(a)* Using the entropy method and a suitable decomposition of arithmetic progressions into canonical intervals, show that $\mathcal{A}_n$ has a no-nonsense partial coloring with discrepancy $O(n^{1/4})$.
(b)* Show that $\mathcal{A}_n$ restricted to an arbitrary m-point subset X of $\{1, 2, \ldots, n\}$ has a no-nonsense partial coloring with discrepancy $O(n^{1/4})$ as well, and hence that $\mathrm{disc}(\mathcal{A}_n) = O(n^{1/4} \log n)$.
(c)** Can you improve the $O(n^{1/4} \log n)$ upper bound from (b)? A bound of $O(n^{1/4} \log\log n)$ is not too difficult, while the tight $O(n^{1/4})$ bound requires additional consideration of the structure of arithmetic progressions restricted to a subset of $\{1, 2, \ldots, n\}$; see [MS96].

6. Lower Bounds

In this chapter, we finally begin with the mathematically most fascinating results in geometric discrepancy theory: the lower bounds (we have already seen some lower bounds in Chapter 4 but not in a geometric setting). So far we have not answered the basic question, Problem 1.1, namely whether the discrepancy for axis-parallel rectangles must grow to infinity as $n \to \infty$. An answer is given in Section 6.1, where we prove that $D(n, \mathcal{R}_2)$ is at least of the order $\sqrt{\log n}$. Note that, in order to establish a such a result, we have to show that for any n-point set P in the unit square, some axis-parallel rectangle exists with a suitably high discrepancy. So we have to take into account all possible sets P simultaneously, although we have no idea what they can look like. The proof is a two-page gem due to Roth, based on a cleverly constructed system of orthogonal functions on the unit square. In dimension d, the same method gives $D(n, \mathcal{R}_d) = \Omega((\log n)^{(d-1)/2})$.

In Section 6.2, we present a proof of a great result of Schmidt: an asymptotically tight lower bound for the discrepancy for axis-parallel rectangles in the plane, $D(n, \mathcal{R}_2) = \Omega(\log n)$. Orthogonal functions from Roth's proof are employed again, but they are combined in a more sophisticated manner. Unlike to the previous section, the proof only works in the plane. Obtaining an improvement over Roth's lower bound in dimension d seems considerably more difficult, and so far the success has been moderate.

After these two intellectually challenging proofs, the reader can take a breath while studying Section 6.3. There, we derive a lower bound by a simple (but clever) reduction from the results already proved. Namely, we obtain an $\Omega(\log n)$ estimate for the Lebesgue-measure discrepancy for axis-parallel squares within the unit square.

Next, we tackle lower bounds for discrepancy for objects with rotation allowed. With the current knowledge, the simplest case is the combinatorial discrepancy for halfplanes in the plane. In Section 6.4, we show a proof formally similar to Roth's method from Section 6.1, and Section 6.5 explains a conceptually somewhat different approach to the same result (although leading to almost identical calculations). This method involves no auxiliary functions but it replaces the given point set by the union of several of its slightly shifted copies.

In Section 6.6, we give a tight lower bound for the Lebesgue-measure discrepancy for halfplanes, based on the point-replication method from Section 6.5. As we know, combinatorial lower bounds can be derived from Lebesgue-measure ones (Proposition 1.8), and so the separate treatment of the combinatorial case in the previous two sections is, strictly speaking, unnecessary. But it is a good introduction to the Lebesgue-measure discrepancy setting which, perhaps surprisingly, turns out to be more challenging than the combinatorial case. In the proof, we have to deal with a substantial new phenomenon in the estimates, involving the nonnegativity of a rather complicated function. This nonnegativity can be proved via elementary calculus, but it also follows from an interesting theory of so-called positive definite functions. We present a very small fragment of this theory in Section 6.7. First we derive the nonnegativity result we need for the discrepancy lower bound, and then, for reader's interest, we describe a nice application in a geometric problem (unrelated to discrepancy) concerning isometric embeddings of metric spaces.

6.1 Axis-Parallel Rectangles: L_2-Discrepancy

In this section, we explain the arguably simplest known lower-bound proof in geometric discrepancy, showing that the discrepancy $D(n, \mathcal{R}_2)$ for axis-parallel rectangles is at least of the order $\sqrt{\log n}$ and so, in particular, it grows to infinity as $n \to \infty$.

First, we need to recall that the L_2-discrepancy of a point set $P \subset [0,1]^d$ for corners is

$$D_2(P, \mathcal{C}_d) = \sqrt{\int_{[0,1]^d} D(P, C_x)^2 \,\mathrm{d}x},$$

where $C_x = \prod_{i=1}^d [0, x_i)$ is the corner defined by x and $D(P, C_x) = n \cdot \mathrm{vol}(C_x) - |P \cap C_x|$.

The L_2-discrepancy for corners (and, consequently, the worst-case discrepancy for axis-parallel boxes) can be estimated as follows:

6.1 Theorem (Roth's lower bound for corners). *For any fixed d, the L_2-discrepancy for corners in dimension d satisfies*

$$D_2(n, \mathcal{C}_d) = \Omega(\log^{(d-1)/2} n).$$

Therefore, we also have $D(n, \mathcal{R}_d) = \Omega(\log^{(d-1)/2} n)$ for the worst-case discrepancy for axis-parallel boxes.

As we know from Section 2.2, the lower bound in this theorem for the L_2-discrepancy is asymptotically the best possible. On the other hand, the consequent lower bound for the worst-case discrepancy is not tight and better lower bounds are known (although for dimension $d \geq 3$, the improvement established so far is very slight).

We demonstrate the proof of the theorem only in the plane. For higher dimensions, the notation and calculations become more complicated, but no essential new idea is required, and the proof for an arbitrary dimension is left as Exercise 1.

Proof of Theorem 6.1 for d = 2. Let $P \subset [0,1]^2$ be an arbitrary n-point set in the unit square, fixed throughout the proof. We use the shorthand $D(x)$ for the (signed) discrepancy of the corner C_x, i.e. $D(x) = D(P, C_x) = n \cdot \mathrm{vol}(C_x) - |P \cap C_x|$. Our goal is to lower-bound the integral

$$\int_{[0,1]^2} D(x)^2 \, \mathrm{d}x.$$

We proceed as follows. We choose a suitable auxiliary function $F\colon [0,1]^2 \to \mathbf{R}$, and we use the Cauchy–Schwarz inequality in integral form, which in our case says that

$$\int FD \leq \sqrt{\int F^2} \sqrt{\int D^2}$$

for any square-integrable functions F and D. For simpler notation, we leave out the domain of integration, which is $[0,1]^2$, and the integration variable. We thus have

$$D_2(P, \mathcal{C}_2) = \sqrt{\int D^2} \geq \frac{\int FD}{\sqrt{\int F^2}}.$$

The heart of the proof is in the choice of the auxiliary function F. This function will be "small on the average," meaning that $\int F^2 = O(\log n)$, but the integral $\int FD$ will be "large," of the order $\log n$. The function F depends on the set P, and it "collects" its discrepancy in a suitable sense.

Let us choose an integer m such that $2n \leq 2^m < 4n$, and let us put $N = 2^m$. For $j = 0, 1, \ldots, m$, we define functions $f_j\colon [0,1]^2 \to \{-1, 0, 1\}$. In order to do this, we subdivide the unit square into a grid of 2^m small rectangles. We use 2^j rectangles vertically and 2^{m-j} rectangles horizontally; the following drawing illustrates the definition for $n = 4$, $m = 3$, and $j = 1$:

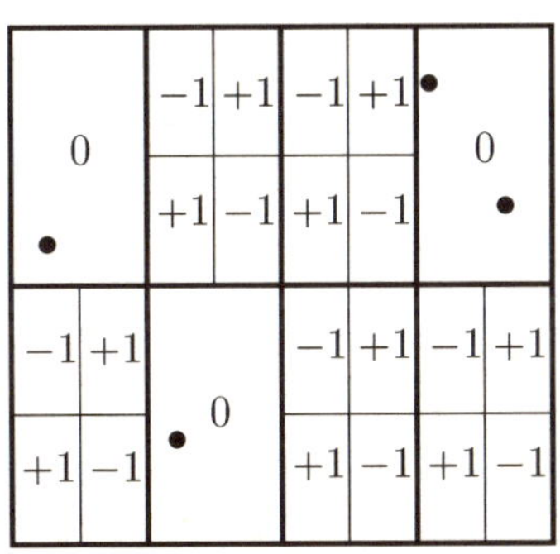

If a small rectangle R in this grid contains at least one point of P, we define the value of f_j as 0 everywhere on R. On the other hand, each empty R is further subdivided into four congruent quadrants. The function f_j is equal to 1 in the upper right and lower left quadrants, and it equals -1 in the upper left and lower right quadrants.

An important property of the functions f_j is their *orthogonality*: for any $i < j$ we have $\int f_i f_j = 0$. Indeed, by overlaying the grid of $2^i \times 2^{m-i}$ rectangles used in the definition of f_i and the $2^j \times 2^{m-j}$ grid appearing in the definition of f_j, we obtain a $2^j \times 2^{m-i}$ grid of smaller rectangles. It is easily checked that on each of these rectangles, either the product $f_i f_j$ is identically 0, or it looks as is indicated in the following schematic drawings:

$$\begin{array}{|c|c|}\hline -1 & +1 \\ \hline +1 & -1 \\ \hline \end{array} \quad \text{or} \quad \begin{array}{|c|c|}\hline +1 & -1 \\ \hline -1 & +1 \\ \hline \end{array}.$$

In any case, the integral of $f_i f_j$ over each rectangle of the combined grid is 0.

We now define the function F by setting $F = f_0 + f_1 + \cdots + f_m$. First, using the orthogonality of the functions f_j, we estimate

$$\int F^2 = \sum_{i,j=0}^{m} \int f_i f_j = \sum_{i=0}^{m} \int f_i^2 \leq \sum_{i=0}^{m} 1 \leq \log_2 n + 3.$$

It remains to prove $\int FD = \Omega(\log n)$. To this end, it is enough to bound $\int f_j D$ from below by a positive constant $c > 0$ for each $j = 0, 1, \ldots, m$. Let us consider the function $f_j D$ on one of the small rectangles R appearing in the definition of f_j. We are only interested in empty rectangles R (containing no points of P), because we have $f_j = 0$ on the other rectangles. Since there are $N \geq 2n$ rectangles R in total, and at most n of them may contain points of P, there exist at least n empty rectangles R, and thus it suffices to prove that $\int_R f_j D = \Omega(\frac{1}{n})$ for each empty rectangle R. Let R_{LL} denote the *l*ower *l*eft quadrant of the rectangle R, and similarly for R_{LR}, R_{UL}, R_{UR}, as in this picture:

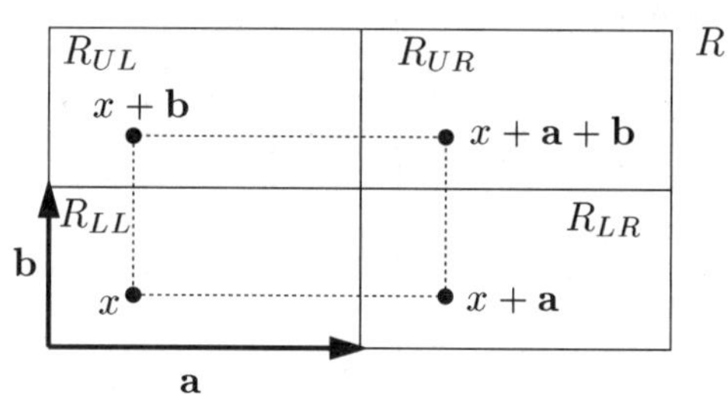

Let $\mathbf{a}$ and $\mathbf{b}$ be the vectors defined by the horizontal and vertical sides of R_{LL}, as in the drawing, and let $a = 2^{-j-1}$ and $b = 2^{j-m-1}$ be their lengths.

We rewrite

$$\int_R f_j D = \int_{R_{LL}} D - \int_{R_{LR}} D - \int_{R_{UL}} D + \int_{R_{UR}} D$$
$$= \int_{R_{LL}} \Big[D(x) - D(x+\mathbf{a}) - D(x+\mathbf{b}) + D(x+\mathbf{a}+\mathbf{b}) \Big] \, dx.$$

We expand the expression under the integration sign using the definition of the function $D(x)$; we obtain

$$n\Big[\mathrm{vol}(C_x) - \mathrm{vol}(C_{x+\mathbf{a}}) - \mathrm{vol}(C_{x+\mathbf{b}}) + \mathrm{vol}(C_{x+\mathbf{a}+\mathbf{b}})\Big]$$
$$-\Big[|P\cap C_x| - |P\cap C_{x+\mathbf{a}}| - |P\cap C_{x+\mathbf{b}}| + |P\cap C_{x+\mathbf{a}+\mathbf{b}}|\Big].$$

A short consideration reveals that the combination of the areas in the first square brackets equals the area of the rectangle $[x_1, x_1+a) \times [x_2, x_2+b)$, which is the same as the area of R_{LL}, i.e. $ab = 2^{-m-2}$. Similarly one finds that the contribution of any point $p \in P$ to the expression in the second pair of square brackets is nonzero only if the point p lies in the rectangle $[x_1, x_1+a) \times [x_2, x_2+b)$, but there are no points in this rectangle (since it is contained in the empty rectangle R). Hence

$$\int_R f_j D = \int_{R_{LL}} n \cdot \mathrm{vol}(R_{LL}) = n \cdot \mathrm{vol}(R_{LL})^2 = \frac{n}{2^{2m+4}} = \Omega\left(\tfrac{1}{n}\right).$$

This finishes the proof of Theorem 6.1 in dimension 2. □

Bibliography and Remarks. Theorem 6.1 and the presented proof are due to Roth [Rot54] (the paper only treats the planar case but the d-dimensional generalization is straightforward). Other proofs of Theorem 6.1 are possible using Fourier transform methods, which give the result even for the family of all axis-parallel cubes in $[0,1]^d$ instead of all axis-parallel boxes (see Section 7.2), but Roth's approach is much simpler.

Exercises

1.* Prove Theorem 6.1 for an arbitrary dimension d (you may want to try the case $d = 3$ first). Estimate the constant of proportionality obtained at the leading term of the resulting bound for $D_2(n, \mathcal{C}_d)$.

6.2 Axis-Parallel Rectangles: the Tight Bound

Here we prove the asymptotically optimal lower bound for the discrepancy for axis-parallel rectangles (or for corners):

6.2 Theorem (Schmidt's lower bound for corners). *For any n-point set $P \subset [0,1]^2$, there is an axis-parallel rectangle R with discrepancy $|D(P,R)| \geq c \log n$, for a suitable constant $c > 0$. That is, $D(n, \mathcal{R}_2) = \Omega(\log n)$ (and also $D(n, \mathcal{C}_2) = \Omega(\log n)$).*

Recall that there are sets whose L_2-discrepancy for corners is $O(\sqrt{\log n})$ (Section 2.2). For such sets, the worst-case discrepancy $\Omega(\log n)$ must be caused by a very small fraction of "very bad" corners. This, of course, makes Theorem 6.2 the more difficult to prove.

Schmidt's result improves Roth's lower bound in the plane by a factor of $\sqrt{\log n}$. A similar improvement in higher dimensions turns out to be much more challenging (although it is widely believed that it should be possible). The current best lower bound for any fixed dimension $d \geq 3$ is $\Omega((\log n)^{(d-1)/2}(\log\log n/\log\log\log n)^{1/(2d-2)})$, which is just a little bit bigger than Roth's bound.

Proof of Theorem 6.2. We use the ideas and notation from the above proof of Theorem 6.1. In particular, we take over the definition of the function $D(x)$, of the numbers m, N, and of the functions f_j without any modification. However, we employ another auxiliary function G. This time we base the lower bound on the obvious inequality

$$\int DG \leq (\sup|D|) \int |G|;$$

thus, we will estimate

$$D(P, \mathcal{C}_2) = \sup_{x \in [0,1]^2} |D(x)| \geq \frac{\int DG}{\int |G|}.$$

The function G is

$$G = (1 + cf_0)(1 + cf_1) \cdots (1 + cf_m) - 1,$$

where $c \in (0,1)$ is a certain small constant (later on we will see how small c must be chosen).[1] After multiplying out the parentheses, the function G has the form $G_1 + G_2 + \cdots + G_m$, where

$$G_k = c^k \sum_{0 \leq j_1 < j_2 < \cdots < j_k \leq m} f_{j_1} f_{j_2} \cdots f_{j_k}.$$

[1] For the peace of mind of a reader wondering how anyone could ever come up with such a proof, we remark that functions defined by products of this type were used in analysis earlier, under the name *Riesz product*.

First we consider a product $f_{j_1} f_{j_2} \cdots f_{j_k}$, as we did in the proof of Theorem 6.1 in the special case $k = 2$. By overlaying the rectangular grids from the definition of the functions f_{j_i}, we obtain a $2^{j_k} \times 2^{m-j_1}$ grid of rectangles. By induction on k, one can check that on each of these rectangles, the product $f_{j_1} f_{j_2} \cdots f_{j_k}$ either is identically 0 (this happens whenever the rectangle contains a point of P, but it may also happen for some empty rectangles of the combined grid), or it has the form

$$\begin{array}{|c|c|}\hline -1 & +1 \\ \hline +1 & -1 \\ \hline \end{array} \quad \text{or} \quad \begin{array}{|c|c|}\hline +1 & -1 \\ \hline -1 & +1 \\ \hline \end{array} \tag{6.1}$$

there. In particular, we have $\int f_{j_1} f_{j_2} \cdots f_{j_k} = 0$ (a "generalized orthogonality"), and consequently $\int G_k = 0$ for all $k > 0$.

We estimate

$$\int |G| \le \int 1 + \int (1 + cf_0) \cdots (1 + cf_m) = 1 + 1 + \sum_{k=1}^{m} \int G_k = 2.$$

It remains to bound the integral of the function $DG = D \cdot (G_1 + G_2 + \cdots + G_m)$ from below. The function G_1 is just the F from the proof of Theorem 6.1 multiplied by c, and so we already know that $\int DG_1 \ge cc_0 \log n$, where $c_0 > 0$ is an absolute constant (independent of c). It remains to show that the integrals of the remaining terms, $G_2 D, \ldots, G_m D$, are all considerably smaller in absolute value.

We know that on each rectangle R of the $2^{j_k} \times 2^{m-j_1}$ grid, the product $f_{j_1} f_{j_2} \cdots f_{j_k}$ either is 0 or looks like (6.1); in the latter case, R contains no points of P. The integral over the rectangle R of the product of the function D with a function of the form (6.1) was calculated above in the proof of Theorem 6.1. Its absolute value equals $\frac{1}{16} n \cdot \mathrm{vol}(R)^2$, and so the integral over all rectangles R together is at most $O(n \cdot \mathrm{vol}(R))$ in absolute value (note that all the rectangles R have the same area). For convenience, let us put $q = j_k - j_1$. We have $n \cdot \mathrm{vol}(R) = n 2^{-j_k} 2^{j_1 - m} = O(2^{-q})$. The rest of the proof is a summation of all terms in a suitable order:

$$\begin{aligned} \sum_{k=2}^{m} \left| \int G_k D \right| &\le \sum_{k=2}^{m} c^k \sum_{0 \le j_1 < \cdots < j_k \le m} \left| \int f_{j_1} \cdots f_{j_k} D \right| \\ &\le \sum_{k=2}^{m} c^k \sum_{0 \le j_1 < \cdots < j_k \le m} O(2^{j_k - j_1}) \\ &\le \sum_{k=2}^{m} c^k \sum_{j_1=0}^{m-k+1} \sum_{q=k-1}^{m-j_1} \sum_{j_1 < j_2 < \cdots < j_{k-1} < j_1 + q} O(2^{-q}) . \end{aligned}$$

In the innermost sum, the indices $j_2, \ldots, j_{k-1}$ can be chosen, among the $q-1$ numbers lying between j_1 and $j_1 + q$, in $\binom{q-1}{k-2}$ ways. To simplify the formulas, we can let the indices j_1 and q in the sums run up to m. Then we change the order of summation. We thus upper-bound the previous sum by

$$O(1) \cdot \sum_{k=2}^{m} c^k \sum_{j_1=0}^{m} \sum_{q=k-1}^{m} 2^{-q} \binom{q-1}{k-2} = O(1) \cdot \sum_{j_1=0}^{m} \sum_{q=1}^{m} 2^{-q} \sum_{k=2}^{q+1} \binom{q-1}{k-2} c^k.$$

The innermost sum equals

$$c^2 \sum_{k=2}^{q+1} \binom{q-1}{k-2} c^{k-2} = c^2 \sum_{j=0}^{q-1} \binom{q-1}{j} c^j = c^2 (1+c)^{q-1}.$$

From the second innermost sum, we obtain

$$\sum_{q=1}^{m} 2^{-q} \sum_{k=2}^{q+1} \binom{q-1}{k-2} c^k = c^2 \sum_{q=1}^{m} (1+c)^{q-1} 2^{-q} = \frac{c^2}{2} \sum_{\ell=0}^{m-1} \left(\frac{1+c}{2}\right)^{\ell} = O(c^2)$$

(assuming $c < \frac{1}{2}$, say). Finally the whole sum over j_1 is $O(mc^2) = O(c^2 \log n)$, where the constant of proportionality hidden in the $O(.)$ notation does not depend on c. Thus, for a sufficiently small but fixed c we have $\int GD \geq \int G_1 D - \sum_{k=2}^{m} \left| \int G_k D \right| \geq cc_0 \log n - O(c^2 \log n) = \Omega(\log n)$. Schmidt's lower bound is proved. □

Bibliography and Remarks. Theorem 6.2 was first proved by Schmidt [Sch72], and the (different) proof presented above was found by Halász [Hal81]. In the same paper, Halász also proved the lower bound of $\Omega(\sqrt{\log n})$ for the L_1-discrepancy (Exercise 1). Another version of the proof of Schmidt's lower bound, due to Liardet, is reproduced in [DT97].

Previously, it was known that for any fixed $p > 1$ and for any fixed $d \geq 2$, the L_p-discrepancy for corners in $\mathbf{R}^d$ is at least $\Omega(\log^{(d-1)/2} n)$, with the constant of proportionality depending on p and d. For $p \geq 2$, this follows from Roth's theorem discussed in the preceding section, and for $p \in (1, 2)$ it was proved by Schmidt [Sch77a] (see Exercise 3). As we know from Section 2.2, this bound is tight.

The (worst-case) discrepancy for axis-parallel boxes in $\mathbf{R}^d$ for $d > 2$ remains an intriguing open problem (called "the great open problem" by Beck and Chen [BC87]). The first progress since Roth's $\Omega(\log^{(d-1)/2})$ lower bound and the Halton–Hammersley $O(\log^{d-1} n)$ upper bound in the 1950s was Beck's mathematically very impressive but quantitatively small improvement of the lower bound in $\mathbf{R}^3$, from $\Omega(\log n)$ to $\Omega(\log n (\log \log n)^{1/8-\varepsilon})$ for an arbitrarily small $\varepsilon > 0$ [Bec89c]. Recently Baker [Bak] simplified Beck's method and generalized it to an arbitrary dimension d. He proved the lower bound of

$\Omega((\log n)^{(d-1)/2}(\log\log n/\log\log\log n)^{1/(2d-2)})$ mentioned in the text. Also the problem of determining the L_1-discrepancy in higher dimensions remains, to my knowledge, wide open.

Let us mention one interesting question of Erdős [Erd64] related to the result of this section (strengthening it in a certain sense). If $x_1, x_2, \ldots$ is any infinite sequence in $[0,1)$, does there exist an $a \in [0,1)$ such that $\limsup_{n\to\infty} |D(\{x_1, x_2, \ldots, x_n\}, [0,a))| = \infty$?

The answer is yes, and the correct order of magnitude of the discrepancy is $\log n$ [Sch77b]. In fact, the discrepancy is at least $\Omega(\log n)$ for almost all a. This was proved by Sós [Sós76] for the $(\{n\alpha\})$ sequences and by Tijdeman and Wagner [TW80] for arbitrary sequences (also see [BC87] or [DT97] for more information).

Exercises

1. Modify the method presented in the text to prove a lower bound for the L_1-discrepancy of corners: $D_1(n, \mathcal{C}_2) = \Omega(\sqrt{\log n})$. Use the auxiliary function $H = \left(\prod_{j=0}^{m}(1+\gamma f_j)\right) - 1$, where $\gamma = \mathrm{i}c/\sqrt{\log n}$, i standing for the imaginary unit. Apply the inequality $\left|\int HD\right| \leq (\sup|H|)\int|D|$, valid for any (complex and measurable) functions H, D.
2. (Orthogonality of $\mathbf{r}$-functions) For an integer $r \geq 0$, the one-dimensional *Rademacher function* R_r is defined by $R_r(x) = (-1)^{\lfloor 2^{r+1}x\rfloor}$, and for a nonnegative integer vector $\mathbf{r} = (r_1, r_2, \ldots, r_d)$ and for $x \in \mathbf{R}^d$, we put $R_{\mathbf{r}}(x) = \prod_{k=1}^{d} R_{r_k}(x_k)$. An $\mathbf{r}$-function is any function $f\colon [0,1)^d \to \{-1,0,1\}$ such that for any binary canonical box[2] B of size $2^{-r_1}\times 2^{-r_2}\times \cdots \times 2^{-r_d}$, f restricted to B equals either 0, or $R_{\mathbf{r}}$, or $-R_{\mathbf{r}}$.
(a) Let $\mathbf{r}^{(1)}, \mathbf{r}^{(2)}, \ldots, \mathbf{r}^{(\ell)}$ be nonnegative integer vectors, and let f_j be an $\mathbf{r}^{(j)}$-function, $j = 1, 2, \ldots, \ell$. Prove that if the following condition does *not* hold then $\int_{[0,1]^d} f_1(x)f_2(x)\cdots f_\ell(x)\,\mathrm{d}x = 0$: for each k, each number occurs an even number of times in the the sequence $(r_k^{(1)}, r_k^{(2)}, \ldots, r_k^{(\ell)})$.
(b)* Give an example of a collection $\mathbf{r}^{(1)}, \ldots \mathbf{r}^{(\ell)}$ of distinct nonzero 3-dimensional vectors, all of them with the same sum of coordinates, such that the product of the corresponding Rademacher functions $R_{\mathbf{r}^{(j)}}$ has nonzero integral over $[0,1)^3$. (This indicates one source of problems with generalizing the lower bound to higher dimensions.)
3. (L_p-discrepancy lower bound)
(a)* Let Y be a set of d-component nonnegative integer vectors, each of them having the sum of components equal to m. For each $\mathbf{r} \in Y$, let $f_{\mathbf{r}}$ be an $\mathbf{r}$-function (as in Exercise 2). Prove the following inequality for an arbitrary integer $t \geq 1$ (Schmidt's lemma):

---
[2] We recall that a binary canonical box is a Cartesian product of binary canonical intervals, and a binary canonical interval has the form $[k/2^q, (k+1)/2^q)$.

$$\int_{[0,1]^d} \left(\sum_{\mathbf{r} \in Y} f_{\mathbf{r}}(x) \right)^{2t} \mathrm{d}x \le (2t)^{t(d-1)} (m+1)^{t(d-1)}.$$

(b) Let $d = 2$, let $D(x)$ be the discrepancy function for an n-point set $P \subset [0,1]^2$ as in Section 6.1, and let F be the auxiliary function as in that section. Use (a) and Hölder's inequality to conclude that $D_p(P, \mathcal{C}_2) = \Omega(\sqrt{\log n})$ for any fixed $p > 1$.
(c) Generalize (b) to an arbitrary fixed dimension $d \ge 2$, heading for the bound $\Omega(\log^{(d-1)/2} n)$ (do Exercise 6.1.1 first).

6.3 A Reduction: Squares from Rectangles

What is the discrepancy for axis-parallel squares? First of all, if we allow for arbitrary squares in the plane, we can obtain any corner as the intersection of a square with the unit square, so the discrepancy for all axis-parallel squares is, according to our definition, clearly of the order $\log n$ by Theorem 6.2. A much more interesting question is obtained if we restrict ourselves to the axis-parallel squares completely contained in the unit square. There is a surprising elementary argument proving that the discrepancy is of the same order of magnitude as that for rectangles. We haste to remark that no such direct reduction is known in higher dimensions relating axis-parallel cubes to axis-parallel boxes.

6.3 Theorem. *The Lebesgue-measure discrepancy for axis-parallel squares contained in* $[0,1]^2$ *is at least* $\Omega(\log n)$.

Proof. Let P be a fixed n-point set in $[0,1]^2$, set $M = D(P, \mathcal{R}_2)$, and let $R_0 \in \mathcal{R}_2$ be a rectangle with $|D(P, R_0)| = M$. By Theorem 6.2, we know $M = \Omega(\log n)$. Moreover, by decreasing M by an arbitrarily small amount, we may assume that R_0 is chosen in such a way that no point of P lies on its boundary or on the boundaries of the finitely many auxiliary rectangles constructed in the sequel.

Let Δ be the largest absolute value of the discrepancy of an axis-parallel square contained in $[0,1]^2$. For contradiction, let us suppose that Δ is much smaller than M.

First, assume that the rectangle R_0 can be completed to a square $Q_0 \subseteq [0,1]^2$ as in Fig. 6.1(a), and that the longer side of R_0 is at least twice as long as the shorter one (later we show how to eliminate these assumptions). Moreover, suppose that $D(P, R_0) = +M$ (the case $D(P, R_0) = -M$ is handled symmetrically). In the rest of this proof, let us write $D(R)$ for $D(P, R)$, for any rectangle R.

Divide the square Q_0 into two squares, Q_1 and Q_2, and two rectangles, using the upper horizontal side of R_0 as one of the dividing segments; see

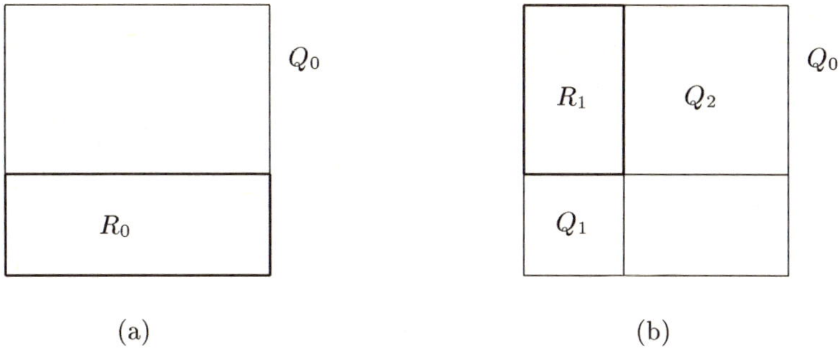

Fig. 6.1. Deriving the discrepancy of squares I.

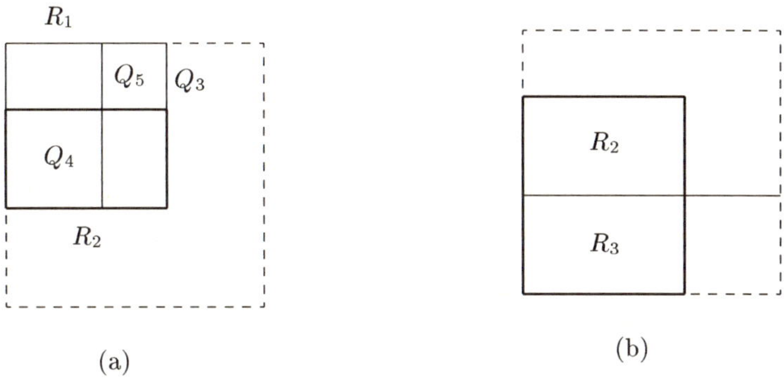

Fig. 6.2. Deriving the discrepancy of squares II.

Fig. 6.1(b). The rectangle R_1, marked by thick lines in the picture, can be expressed as $R_1 = (Q_0 \setminus R_0) \setminus Q_2$, and hence $D(R_1) = D(Q_0) - D(R_0) - D(Q_2) \leq -M + 2\Delta$. So the (signed) discrepancy of R_1 is about the same as that of R_0 but with opposite sign. Complete R_1 to another square Q_3 (Fig. 6.2(a)), and subdivide Q_3 in the way we subdivided Q_0, obtaining two smaller squares Q_4 and Q_5. Let R_2 be the rectangle (marked thick) $R_2 = ((Q_3 \setminus R_1) \setminus Q_5) \cup Q_4$. We then derive $D(R_2) \geq M - 5\Delta$. Consider the rectangle R_3 arising from R_0 by cutting off a square on the right (Fig. 6.2(b)); we have $D(R_3) \geq M - \Delta$. Finally, for the rectangle $R_4 = R_3 \dot{\cup} R_2$ we have $D(R_4) \geq M - \Delta + M - 5\Delta = 2M - 6\Delta$. Since M was assumed to be the maximum possible discrepancy (in absolute value), we must have $|D(R_4)| \leq M$, and so $2M - 6\Delta \leq M$. Therefore $\Delta \geq \frac{M}{6}$.

It remains to show how to get rid of our restrictive assumptions on the initial rectangle R_0. First of all, if its side ratio is smaller than 2, we can start with its complement in the square Q_0 (with discrepancy at least $M - \Delta$). It remains to handle the case when both the possible squares extending R_0 (the one going up and the one going down—that is, assuming that R_0 is longer in the horizontal direction) reach out of the unit square. This is left as an exercise. □

Bibliography and Remarks. The lower bound $\Omega(\log n)$ for axis-parallel squares was first established by Halász (unpublished; the proof is reproduced in [BC87]) by modifying the method used for rectangles. A surprising elementary argument similar to the one given above was found by Ruzsa [Ruz93]. This was motivated by a question of Laczkovich: can the discrepancy of any set for axis-parallel cubes be much smaller than that for axis-parallel boxes? Ruzsa's result shows that the answer is no in the plane, but the problem remains open in higher dimensions. For an arbitrary dimension d, there is a direct lower bound of $\Omega(\log^{(d-1)/2} n)$ for axis-parallel cubes due to Beck (see Section 7.2).

Exercises

1. Complete the proof of Theorem 6.3, i.e. given a rectangle R_0 with discrepancy M, find a rectangle R_0' with discrepancy at least $M - O(\Delta)$ whose completion to a square (as in Fig. 6.1(a)) is contained in $[0,1]^2$.
2. Not everything has discrepancy going to infinity!
(a) Let $R = [0,a) \times [0,b)$ be a fixed axis-parallel rectangle, and let $\mathcal{T}$ denote the family of all parallel translates of R that are completely contained in $[0,1]^2$. Show that $D(n,\mathcal{T}) = O(1)$ (for any fixed R).
(b)** Let T be an arbitrary fixed triangle and let $\mathcal{T}$ be the family of all parallel translates of T that are completely contained in $[0,1]^2$. Show $D(n,\mathcal{T}) = O(1)$ (it may be a good idea to start with some particular triangles).
The result in (b) is due to G. Wagner (unpublished) and it was communicated to me by W. L. Chen.

6.4 Halfplanes: Combinatorial Discrepancy

The discrepancy lower bounds derived so far, concerning axis-parallel objects, have polylogarithmic orders of magnitude. Now we start dealing with halfplanes, one of the simplest classes of geometric objects invariant under rotation and translation, and the lower bound we are aiming at is of a much larger order: $n^{1/4}$. We prove

6.4 Theorem (Alexander's lower bound for halfplanes). *For each n, there exists an n-point set P in the plane whose combinatorial discrepancy for halfplanes satisfies* $\operatorname{disc}(P,\mathcal{H}_2) = \Omega(n^{1/4})$. *That is, for any red-blue coloring*

of P, there exists a halfplane in which one color outnumbers the other by at least $\Omega(n^{1/4})$.

The proof, similar to many other lower-bound proofs in discrepancy theory, actually bounds the L_2-discrepancy. We thus have to define an appropriate measure on the set of halfplanes. Although various more or less reasonable choices may come to mind, it turns out that there is an essentially unique "nicest" measure. This is a classical tool and object of study in geometry, with many applications.

Invariant Measure on Lines. We introduce a measure ν on the set of all lines in the plane. We ignore the set of lines passing through the origin (it has measure 0). Any other line ℓ is uniquely determined by its point x closest to the origin. We express x in polar coordinates. This defines a mapping Ψ, assigning to each line ℓ a pair $(r, \varphi) \in \mathbf{R}^2$, where r and φ are the polar coordinates of x (with $0 \le \varphi < 2\pi$):

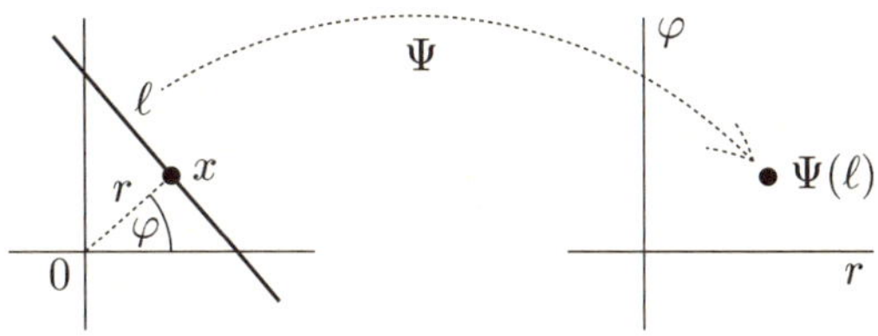

Now if L is a set of lines, we define $\nu(L)$ as the Lebesgue measure of the set $\Psi(L)$ in the (r, φ) plane, where r and φ are interpreted as *Cartesian* coordinates. For instance, the set $\{(r, \varphi)\colon 0 \le r \le 1, 0 \le \varphi < 2\pi\}$ is a rectangle, not a disc!

The first pleasant property of the measure ν is *motion invariance*: if L' arises from L by a rigid motion (i.e., without changing the relative position of the lines) then $\nu(L') = \nu(L)$. This is not difficult to see. Clearly, the measure does not change by rotating L around the origin, since this corresponds to a parallel translation of $\Psi(L)$ in the φ-direction in the (r, φ)-plane. Any rigid motion can be obtained by composing such rotations with translations parallel to the x-axis, so let us check how a translation of L by a vector $(a, 0)$ works in the (r, φ)-plane. In the following drawing, we see that φ remains fixed and r changes to $r + a\cos\varphi$:

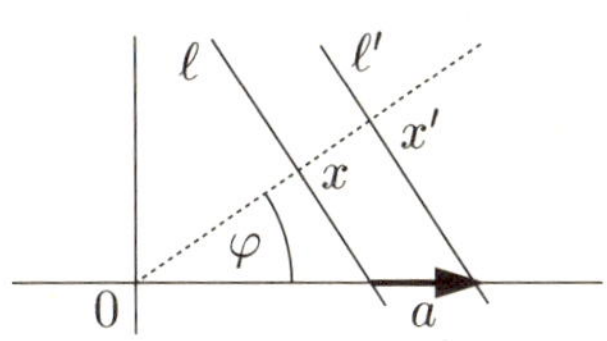

Therefore, each slice of $\Psi(L')$ by a line parallel to the r-axis has the same one-dimensional measure as the corresponding slice of $\Psi(L)$, and so the two-dimensional measures are the same as well.

Let us remark that up to scaling, our measure is the only "reasonable" motion-invariant measure on the lines in the plane (similar to the Lebesgue measure being the only motion-invariant measure on the points of the plane). There are elegant, although less elementary, ways of deriving the concrete formula for ν from the motion-invariance requirement (the Haar measure on the group of rigid motions is used as an intermediate step).

Here is another remarkable property of the measure ν:

6.5 Theorem (Perimeter formula). *For any convex set $K \subseteq \mathbf{R}^2$, the ν-measure of the set of all lines intersecting K equals the perimeter of K.*

We actually need a very special case of Theorem 6.5 only, namely for K being a segment (then the measure should be twice the length of the segment). This is quite easy (Exercise 1).

A Side Remark. Let us note Theorem 6.5 immediately implies that if $K \subseteq K'$ are both convex, then the perimeter of K is no larger than the perimeter of K'. This result is not as easy to prove without such an aid as it might seem.

Finite Differencing. We need one more detour, this time leading us into calculus. The *forward differencing operator* Δ_h acts on a real function f as follows:

$$\Delta_h f(x) = f(x+h) - f(x).$$

Its t-fold iteration can be expressed as

$$\Delta_h^t f(x) = \sum_{i=0}^{t} (-1)^{t-i} \binom{t}{i} f(x+ih). \tag{6.2}$$

We will need to estimate the tth difference using derivatives. For h small, $\frac{\Delta_h f(x)}{h}$ approximates the derivative $f'(x)$; indeed, the Mean value theorem gives $\Delta_h f(x_0) = hf'(\xi)$ for some $\xi \in (x_0, x_0+h)$ if f is differentiable in (x_0, x_0+h). For a larger order t, intuitively, $\frac{\Delta_h^t(x)}{h^t}$ should approximate the tth derivative $f^{(t)}(x)$. Precisely, we have the following generalization of the Mean value theorem:

$$\Delta_h^t f(x_0) = h^t f^{(t)}(\xi) \qquad \text{for some } \xi \in (x_0, x_0+th) \tag{6.3}$$

provided that f is t-times continuously differentiable on the interval (x_0, x_0+th). The proof is left as Exercise 3. For the purposes of this section, we will only need that

$$\Delta_h^t f(x_0) = O\Big(h^t \sup_{\xi \in (x_0, x_0+th)} \big|f^{(t)}(\xi)\big|\Big),$$

with the constant of proportionality depending on t only, which is somewhat more straightforward to prove.

Proof of Theorem 6.4. The set P with large combinatorial discrepancy for halfplanes can be taken as the $\sqrt{n} \times \sqrt{n}$ regular grid placed within the square $S = [0, \frac{1}{4}]^2$ (the side $\frac{1}{4}$ is chosen so that the perimeter is 1 and, consequently, ν is a probability measure on the set of lines intersecting S). We will assume that $\sqrt{n}$ is integral and even, since the modifications for the general case are easy and would just complicate the notation. In such case, we put $P = \left(\left(\frac{1}{8\sqrt{n}}, \frac{1}{8\sqrt{n}}\right) + \frac{1}{4\sqrt{n}}\mathbf{Z}^2\right) \cap S$, as in the following picture[3] for $n = 16$:

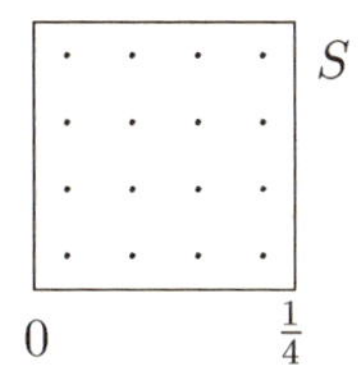

Let $\mathcal{U}$ denote the set of all upper halfplanes (ones lying above their boundary line) whose boundary lines intersect the square S. Ignoring vertical lines (which have measure 0), the lines are in a bijective correspondence with their upper halfplanes, and so ν can be regarded as a probability measure on the set $\mathcal{U}$ of upper halfplanes.

Let $\chi: P \to \{-1, +1\}$ be any given coloring. To establish Theorem 6.4, we are going to prove $\mathrm{disc}_{2,\nu}(\chi, P, \mathcal{U}) = \Omega(n^{1/4})$; recall that $\mathrm{disc}_{2,\nu}(\chi, P, \mathcal{U})$ is defined as

$$\left(\int_{\mathcal{U}} \chi(P \cap \gamma)^2 \, \mathrm{d}\nu(\gamma)\right)^{1/2}.$$

In this formula, we integrate the discrepancy over all halfplanes $\gamma \in \mathcal{U}$. Thus, we are going to deal with various auxiliary functions defined on $\mathcal{U}$. (Note that in Roth's proof in Section 6.1, we have been working with functions defined on the set of all corners, although this was somewhat obscured by the one-to-one correspondence of corners with the points of the unit square.) First, we formally introduce a function $D: \mathcal{U} \to \mathbf{R}$ expressing the discrepancy of a given halfplane, by setting $D(\gamma) = \chi(P \cap \gamma)$. This function is the sum of the contributions of individual points, namely $D(\gamma) = \sum_{p \in P} \chi(p) I_p(\gamma)$, where $I_p: \mathcal{U} \to \{0, 1\}$ is the indicator function defined by

$$I_p(\gamma) = \begin{cases} 1 & \text{for } p \in \gamma \\ 0 & \text{for } p \notin \gamma. \end{cases}$$

[3] Let us remark that the same proof goes through for more general sets P. Let δ and δ' be, respectively, the largest and the smallest distance between any pair of distinct points in a set P. We say that P is *dense* if the ratio δ/δ' is less than $C\sqrt{n}$, for some constant $C > 0$. (Clearly, this is asymptotically the smallest ratio one can have.) The lower bound in Theorem 6.4 holds for any dense set P in the plane. On the other hand, in order to get a lower bound for the combinatorial discrepancy for halfplanes, *some* requirement on P is necessary, since if the points of P are in convex position, then there is a coloring with discrepancy at most 2.

Next, we introduce functions $f_p: \mathcal{U} \to \mathbf{R}$, $p \in P$. These play the role of the orthogonal functions f_i from Roth's proof in Section 6.1, but they are not really orthogonal, at least not exactly so. Instead, we will show that they are near-orthogonal, meaning that the cross-terms $\int f_p f_q$ for $p \neq q$ are sufficiently small compared to the quadratic terms $\int f_p^2$.

Set $w = cn^{-1/2}$, where $c > 0$ is a sufficiently small positive constant, and let $\mathbf{w}$ be the vertical vector $(0, w)$. For each point p, we consider the auxiliary points $p + i\mathbf{w}$, for $i = -2, -1, \ldots, 2$, as in the following picture:

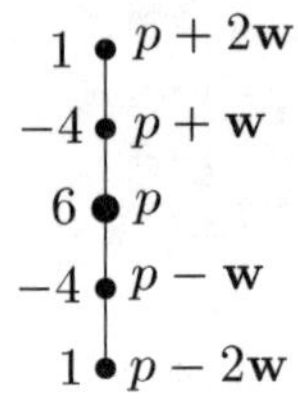

The point $p+i\mathbf{w}$ is assigned the weight $(-1)^i\binom{4}{i+2}$; these weights are written near the points. The value of the function f_p for an upper halfplane γ is defined as the sum of weights of the points among $p + i\mathbf{w}$, $i = -2, -1, \ldots, 2$, that are contained in γ. Written by a formula,

$$f_p = I_{p-2\mathbf{w}} - 4I_{p-\mathbf{w}} + 6I_p - 4I_{p+\mathbf{w}} + I_{p+2\mathbf{w}}. \tag{6.4}$$

The weights of the points come from the fourth-order differencing formula, and the role they play in the proof will become apparent later. Finally, we set $F = \sum_{p \in P} \chi(p) f_p$.

As in Section 6.1, we will use the inequality $\sqrt{\int D^2} \geq \int FD / \sqrt{\int F^2}$ (again omitting the integration domain, which is $\mathcal{U}$, and the integration variable γ). This time we estimate the numerator first.

We have

$$\begin{aligned} \int FD &= \int \Bigl(\sum_{p \in P} \chi(p) f_p\Bigr)\Bigl(\sum_{q \in P} \chi(q) I_q\Bigr) \\ &= \sum_{p \in P} \chi(p)^2 \int f_p I_p + \sum_{p,q \in P,\, p \neq q} \int \chi(p)\chi(q) f_p I_q \\ &\geq \sum_p \int f_p I_p - \sum_{p \neq q} \Bigl| \int f_p I_q \Bigr|. \end{aligned} \tag{6.5}$$

Before immersing into the calculations, let us present an informal view of the proof. The functions I_p and f_p are defined on the domain $\mathcal{U}$, which is the set of all the upper halfplanes intersecting the square S, but for simplicity, we will consider them defined on the set of all upper halfplanes in this informal outline (it turns out that the restriction to the domain $\mathcal{U}$ does not really matter for the calculation). These functions can be illustrated graphically in the (r, φ)-plane used in the definition of the invariant measure ν on lines (recall that each line not passing through the origin corresponds to a point in the (r, φ)-plane). The set of all lines lying below a point p is delimited by parts of a sine curve in the (r, φ)-plane, as in the following picture:

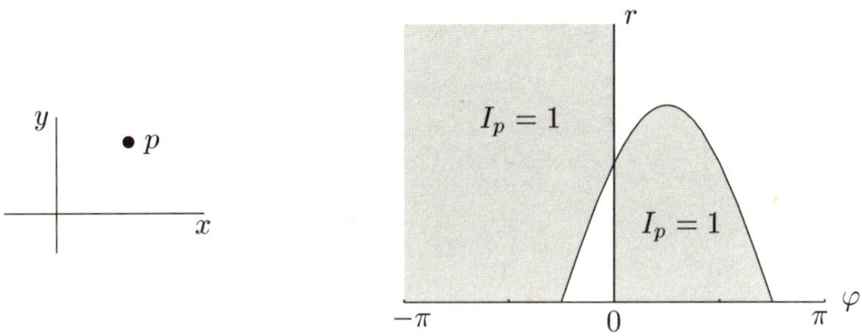

The function I_p is 1 in the gray areas and 0 in the white areas. The function f_p is a linear combination of several functions $I_{p+i\mathbf{w}}$. In the picture, f_p is only nonzero in a narrow band along the curve delimiting the region $I_p = 1$. The following picture shows a magnified small region near this boundary, with the values of f_p in the four bands:

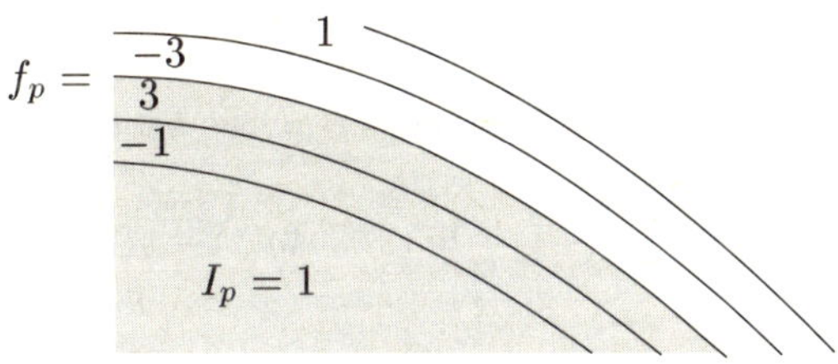

The reason for $\int f_p I_p$ being "large" is that within the region $I_p = 1$, f_p is $+3$ in one of the bands and only -1 in the other band, so there is a significant positive excess. On the other hand, if we plot f_p and I_q for two points p and q lying sufficiently far apart, we get a picture like this:

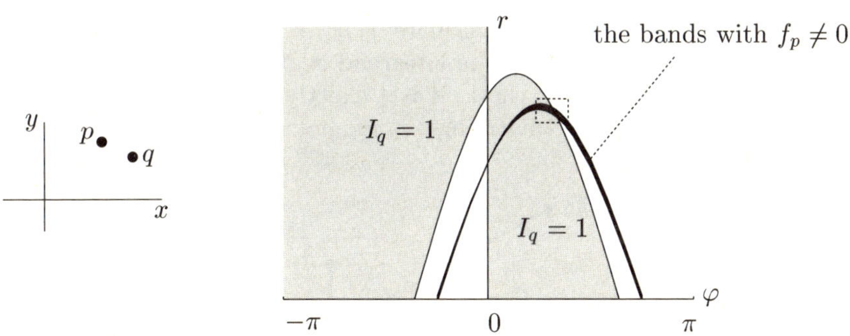

Here is a detail of the marked area where the boundary of the region $I_q = 1$ crosses the bands with $f_p \neq 0$:

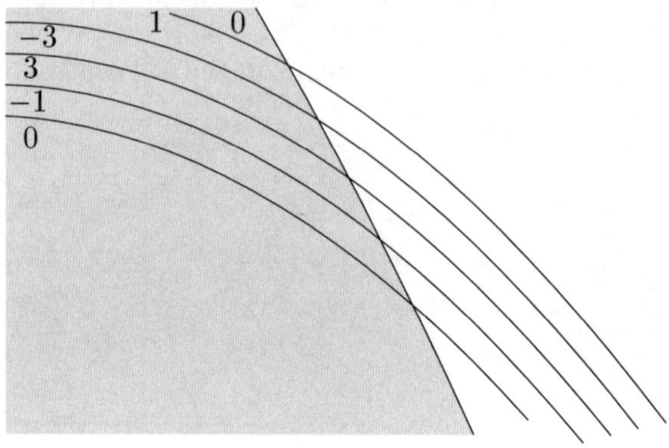

In such a situation, the contribution of the four bands to $\int f_p I_q$ very nearly cancels out, and so $\int f_p I_q$ is very small. One can say that f_p is "focused on" I_p, following the boundary and making a significant contribution, but any other I_q gets "blurred" with respect to f_p. Of course, this fairy-tale has to be substantiated by an actual calculation, which we do next.

A formula and an estimate for $\int f_p I_q$ are provided by the following lemma, in which the finite differencing enters the stage:

6.6 Lemma. (i) *For any two points $p, q \in S$, we have*

$$\int f_p I_q = -g(-2) + 4g(-1) - 6g(0) + 4g(1) - g(2) = -\Delta_1^4 g(-2),$$

where $g(x) = g_{p,q,\mathbf{w}}(x) = \|p - q - x\mathbf{w}\|$.

(ii) *For p, q with $\|p - q\| \geq 4w$, we have*

$$\left| \int f_p I_q \right| = O\left(\frac{w^4}{\|p - q\|^3} \right).$$

Proof of Part (i). First we calculate $\int (I_{p_1} - I_{p_2}) I_q$, where $p_1, p_2 \in S$ and p_1 lies vertically above p_2. The integrand is always either 0 or 1 (since $I_{p_1} \geq I_{p_2}$), and as is easy to check, it is 1 exactly for the upper halfplanes whose boundary intersects the triangle $p_1 q p_2$ but doesn't intersect the side $p_1 q$:

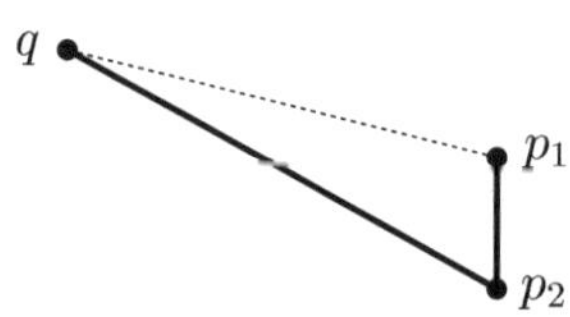

Using the perimeter formula (Theorem 6.5), we derive

$$\int (I_{p_1} - I_{p_2}) I_q = \|p_1 - p_2\| + \|p_2 - q\| - \|p_1 - q\|. \qquad (6.6)$$

The definition (6.4) of f_p can we rewritten to

$$f_p = -(I_{p-\mathbf{w}} - I_{p-2\mathbf{w}}) + 3(I_p - I_{p-\mathbf{w}}) - 3(I_{p+\mathbf{w}} - I_p) + (I_{p+2\mathbf{w}} - I_{p+\mathbf{w}}).$$

Substituting the right-hand side for f_p into $\int f_p I_q$ and using (6.6), we obtain part (i) of the lemma.

Proof of Part (ii). If α denotes the angle of the vectors $\mathbf{w}$ and $q-p$ and if $\delta = \|p-q\|$ as in the picture,

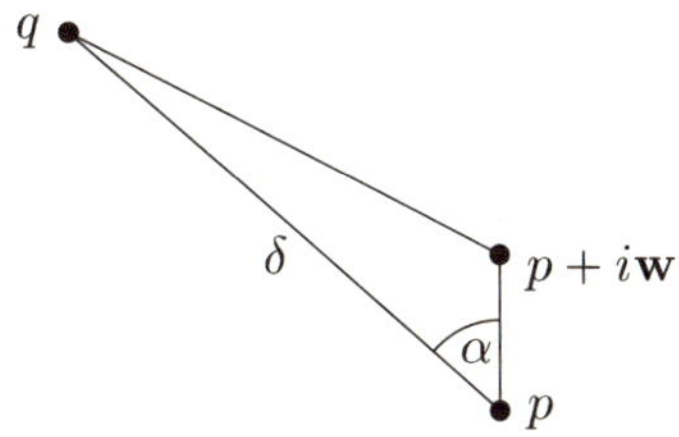

then the cosine theorem yields $\|p - q + i\mathbf{w}\| = \sqrt{\delta^2 + i^2 w^2 - 2\delta i w \cos\alpha}$. With this notation, the quantity $\left|\int f_p I_q\right|$ depends on the three parameters w, δ, and α, and we would like to bound it by $O(w^4/\delta^3)$ uniformly for all α. First, we reduce this to an essentially univariate problem (with α as an extra, and harmless, parameter) by a suitable change of variable. Namely, we put $h = \frac{w}{\delta}$. Note that the assumption $\|p-q\| \geq 4w$ implies $h \leq \frac{1}{4}$. We obtain $\|p + i\mathbf{w} - q\| = \delta \cdot g_1(ih)$, where $g_1(x) = \sqrt{1 + x^2 - 2x\cos\alpha}$, and $\int f_p I_q = -\delta \cdot \Delta_h^4 g_1(-2h)$.

Now it suffices to prove $\left|\Delta_h^4 g_1(-2h)\right| \leq Ch^4$ for all $h \in [0, \frac{1}{4}]$, with a constant C independent of α. A brute-force approach is to calculate the Taylor formula at $h = 0$ with terms up to h^3 and a remainder term of the order $O(h^4)$. It turns out that all the terms up to h^3 have zero coefficients and that the $O(h^4)$ term is bounded uniformly for all α. The Taylor formula calculation is a bit laborious but possible, and it even becomes easy using a computer algebra system.

A proof without any calculation uses the knowledge about finite differencing presented at the beginning of the section. It is easy to see that the function $g_1(x)$ has a continuous fourth derivative on $[-\frac{1}{2}, \frac{1}{2}]$ for all α, and that this derivative depends on α continuously. By (6.3), we have

$$\Delta_h^4 g_1(-2h) = h^4 g_1^{(4)}(\xi)$$

for some $\xi \in (-2h, 2h)$. Since $|g_1^{(4)}(x)|$ is a continuous function of x and α on the compact set $[-\frac{1}{2}, \frac{1}{2}] \times [0, 2\pi]$, it is bounded by some constant C there, and $\left|\Delta_h^4 g_1(-2h)\right| \leq Ch^4$ follows. Part (ii) of Lemma 6.6 is proved. □

End of the Proof of Theorem 6.4. From the explicit formula in part (i) of Lemma 6.6, we calculate $\int f_p I_p = 4w$, and hence $\sum_{p\in P} \int f_p I_p = 4nw$.

Next, we want to show that $\sum_{p\neq q} \left|\int f_p I_q\right|$ is considerably smaller than $4nw$. To compute this sum, we fix p, and we divide the points q into groups. The kth group is

$$G_k = \left\{ q \in P\colon \|p-q\| \in \left[\frac{k}{4\sqrt{n}}, \frac{k+1}{4\sqrt{n}}\right) \right\}$$

$k = 1, 2, \ldots$, as in the following picture:

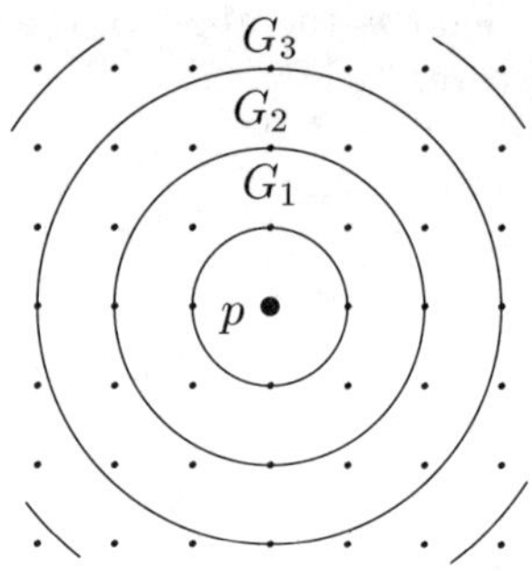

We observe that $|G_k| = O(k)$, as can be seen by the following standard volume argument. For each point $q \in G_k$, consider the disc of radius $\frac{1}{8\sqrt{n}}$, say, centered at q. These discs do not overlap and they are all contained in the annulus with center p, inner radius $\frac{k-1}{4\sqrt{n}}$, and outer radius $\frac{k+2}{4\sqrt{n}}$. The area of such annulus is $O(k/n)$ and hence it can only contain $O(k)$ disjoint discs of area $\Omega(\frac{1}{n})$ each.

Since $\|p-q\| \geq 4w$ for any two distinct points $p, q \in P$, the estimate $\left|\int f_p I_q\right| = O(w^4/\|p-q\|^3)$ in Lemma 6.6(ii) applies, and so

$$\sum_{q\in P\setminus\{p\}} \left|\int I_p f_q\right| \leq \sum_{k=1}^{\infty} |G_k| \frac{O(w^4)}{(k/4\sqrt{n})^3} = O(n^{3/2}w^4) \sum_{k=1}^{\infty} \frac{1}{k^2} = O(n^{3/2}w^4).$$

By the formula (6.5) for $\int FD$, we obtain

$$\int FD \geq 4wn - C_1 n^{5/2} w^4 = \sqrt{n} \cdot (4c - C_1 c^4)$$

with some absolute constant C_1. By choosing c sufficiently small in terms of C_1, we get $\int FD = \Omega(\sqrt{n})$.

It remains to bound $\int F^2 = \sum_{p,q\in P} \int f_p f_q$. By substituting for f_q from (6.4), we have

$$\int F^2 \leq \sum_{i=-2}^{2} O(1) \cdot \sum_{p,q\in P} \left|\int f_p I_{q+i\mathbf{w}}\right|.$$

For $i = 0$, we have already calculated that the inner sum over $p, q \in P$ is $O(\sqrt{n})$. The sums for the other i can be estimated in almost exactly the same way; we leave the details to the reader.

Altogether we have derived $\int F^2 = O(\sqrt{n})$, and so

$$\sqrt{\int D^2} \geq \frac{\int FD}{\sqrt{\int F^2}} = \Omega(n^{1/4}).$$

Theorem 6.4 is proved. □

Higher Dimensions. It is not difficult to extend the above lower-bound proof to halfspaces in an arbitrary fixed dimension d. First, we need a motion-invariant measure ν on the set of all hyperplanes in $\mathbf{R}^d$. The measure introduced for lines in this section can be generalized as follows. A hyperplane h not containing the origin is again characterized by the points x closest to the origin. The point x can be determined by its distance to the origin, $r = \|x\|$, and by the point $\xi = \frac{x}{r}$ of the unit sphere S^{d-1}. For such a hyperplane h, we put $\Psi(h) = (r, \xi)$. The range of the mapping Ψ is thus $[0, \infty) \times S^{d-1}$. The ν-measure of a set of hyperplanes is given by the measure of its Ψ-image, where on $[0, \infty)$ we take the usual (Lebesgue) measure, and on the sphere S^{d-1} we also take the "usual" measure (surface area for $d = 3$), but we scale it by a suitable constant factor. Such a measure on hyperplanes is invariant under rigid motions; this can be verified in the same way as we did for lines. Therefore, the measure of the set of all hyperplanes intersecting any given segment is proportional to its length. Here we fix the scaling so that this measure equals twice the length as was the case in the plane (this is different from the scaling usually used in the integral-geometry literature).

To prove the d-dimensional analogue of Theorem 6.4, we let P be a d-dimensional $n^{1/d} \times \cdots \times n^{1/d}$ grid placed in a suitable cube. The side of this cube is chosen in such a way that the total measure of the hyperplanes intersecting the cube is 1. The functions f_p are defined analogously to the planar case, but the finite differencing formula of order $d+2$ is taken as a basis. This order is just enough to make the sum of the terms $\int f_p I_q$ sufficiently small. All this leads to the asymptotically tight discrepancy bound $\Omega(n^{1/2-1/2d})$ for halfspaces in $\mathbf{R}^d$.

Bibliography and Remarks. Theorem 6.4 was proved by Alexander [Ale90], using some ideas from earlier work of his and Stolarsky (e.g. Stolarsky [Sto73]; see [Ale90] for more references). Some of the precursors of this result will be mentioned in the remarks to Section 7.1, where we also give a brief overview of lower bounds for various classes of geometric figures. A new presentation of Alexander's result was given by Chazelle et al. [CMS95]; the main new ingredient in this paper is an explicit use of finite differencing. This method will be explained below in Section 6.5. The proof in the present section is another modification, due to the author this book, of the basic ideas of Alexander [Ale90] and of Chazelle et al. [CMS95]. Being formulated in terms of suitable near-orthogonal functions, the proof becomes perhaps more natural for a reader familiar with Roth's proof for axis-parallel rectangles.

More on the motion-invariant measure for lines and related subjects can be found in Santaló's book [San76] or, with a more advanced and more demanding presentation, in Schneider and Wieacker [SW93]. The perimeter formula (Theorem 6.5) is sometimes called *Crofton's formula* in the literature. But there are several other Crofton's formulas, and also the perimeter formula goes back at least to Cauchy [Cau50], so here we use the more neutral name for the formula.

In dimensions $d > 2$, the measure of hyperplanes intersecting a given convex body K is not related to the surface area of K anymore. It is proportional to the expected width of K in a randomly chosen direction (the width in direction x is the distance of the two supporting hyperplanes perpendicular to x), and for bodies with a smooth boundary, it is also proportional to the mean curvature (see [San76]).

Exercises

1. (a) Prove that for any segment s in $\mathbf{R}^2$, the ν-measure of the set of lines intersecting s equals twice the length of s.
(b) Prove the perimeter formula (Theorem 6.5) with K being a convex polygon.
(c)* Prove Theorem 6.5.
2. Consider the definition of the motion-invariant measure ν on the set of all planes in $\mathbf{R}^3$.
(a) Prove in detail that the measure of the set of planes intersecting any given segment is proportional to its length.
(b) Calculate the total measure that should be assigned to the unit sphere S^2 in the definition of ν so that the measure of the planes intersecting a segment equals twice its length.
3. (a) Show that $\Delta_h^t p(x)$ is identically 0 for any polynomial $p(x)$ of degree smaller than t.
(b)* Prove that for any fixed t and any function f, t-times continuously differentiable on $(x_0, x_0 + th)$, we have
$$|\Delta_h^t f(x_0)| \leq C_t h^t \sup_{\xi \in (x_0, x_0+th)} |f^{(t)}(\xi)|,$$
for some suitable constant C_t independent of f. (This is somewhat easier than (6.3) and suffices for the proof in this section.)
(c)** Prove the formula (6.3).

4.* Go through the proof of the d-dimensional lower bound sketched at the end of the section carefully, including all the calculations. Note that the elementary calculus approach to (the higher-dimensional analogue of) Lemma 6.6 becomes unmanageable for high dimensions, while the finite-differencing argument works unchanged.

6.5 Combinatorial Discrepancy for Halfplanes Revisited

In this section, we present another proof of Theorem 6.4, closer in spirit to the original proof of Alexander. Although the approach is conceptually somewhat different, it leads to nearly identical calculations and estimates. The finite differencing effect is achieved not by employing auxiliary functions as in the preceding proof, but by replacing the considered point set by several slightly shifted copies of it, each copy being assigned an appropriate weight.

From the previous section, we will need the material concerning the motion-invariant measure ν on lines (and on upper halfplanes) and some basically self-contained parts of the proof given there. A significant part of the notation and definitions are also taken over. In particular, we let $S = [0, \frac{1}{4}]^2$ and we write $\mathcal{U}$ for the set of upper halfplanes with boundary intersecting S. For a point $p \in S$ and a halfplane $\gamma \in \mathcal{U}$, $I_p(\gamma) = 1$ if $p \in \gamma$ and $I_p(\gamma) = 0$ otherwise.

The Alexander–Stolarsky Formula. Our first additional tool is an elegant formula expressing the L_2-discrepancy of a point set in S using certain signed distance sums. In the sequel, we will also need it for *generalized colorings*. By a generalized coloring of a point set P, we mean an arbitrary mapping $\chi: P \to \mathbf{R}$. We also recall that $\mathrm{disc}_{2,\nu}(P,\mathcal{U},\chi) = \left(\int_{\mathcal{U}} \chi(P\cap\gamma)^2 \,\mathrm{d}\nu(\gamma)\right)^{1/2}$.

6.7 Lemma (Alexander–Stolarsky formula). *Let P be a finite point set in the square S and let $\chi: P \to \mathbf{R}$ be a mapping (generalized coloring) such that (important assumption!) $\chi(P) = \sum_{p\in P} \chi(p) = 0$. Then we have*

$$\mathrm{disc}_{2,\nu}(P,\mathcal{U},\chi)^2 = -\sum_{p,q\in P} \chi(p)\chi(q)\|p-q\|.$$

Proof. Since $\chi(P) = 0$, we have $\chi(P\cap\gamma) = -\chi(P\cap(\mathbf{R}^2\setminus\gamma))$ for any halfplane γ. We rewrite

$$\begin{aligned}\chi(P\cap\gamma)^2 &= -\chi(P\cap\gamma)\chi(P\cap(\mathbf{R}^2\setminus\gamma))\\ &= -\sum_{p,q\in P} \chi(p)\chi(q)I_p(\gamma)[1-I_q(\gamma)].\end{aligned}$$

For $p = q$, the term in the above sum is 0. If $\{p, q\}$ is an unordered pair of distinct points of P, the sum contains two terms involving p and q: $\chi(p)\chi(q)I_p(\gamma)[1 - I_q(\gamma)]$ and $\chi(p)\chi(q)I_q(\gamma)[1 - I_p(\gamma)]$. The first term contributes 1 for the halfplanes γ containing p but not q, and the second term contributes 1 for the γ containing q but not p. Hence the sum of the two terms is 1 exactly if the boundary line of γ intersects the segment pq, and the ν-measure of the set of these γ is $2\|p - q\|$. Therefore,

$$\begin{aligned}\operatorname{disc}_{2,\nu}(P,\mathcal{U},\chi)^2 &= \int_{\mathcal{U}} \chi(P \cap \gamma)^2 \, d\nu(\gamma) \\ &= - \sum_{\{p,q\} \subseteq P,\, p \neq q} \chi(p)\chi(q) \cdot 2\|p - q\| \\ &= - \sum_{p,q \in P} \chi(p)\chi(q)\|p - q\|.\end{aligned}$$

□

As was advertised above, the proof of Theorem 6.4 will be based on finite differencing, and the finite differencing effect will be achieved by considering several slightly shifted copies of the given set P. First we check that such a replication of the point set cannot decrease the L_2-discrepancy by much.

6.8 Lemma (Replication lemma). *Let P be a finite point set in the square S, and let $\chi: P \to \mathbf{R}$ be a (generalized) coloring with $\chi(P) = 0$. Let $\mathbf{w}_1, \mathbf{w}_2, \ldots, \mathbf{w}_k$ be translation vectors and let $c_1, c_2, \ldots, c_k$ be real weights, where both k and $c_1, \ldots, c_k$ are considered as constants. Put $P_i = P + \mathbf{w}_i$ and define a (multi)set $\tilde{P}$ as the disjoint union $P_1 \dot\cup P_2 \dot\cup \cdots \dot\cup P_k$. Suppose that no points have been shifted outside S, that is, $\tilde{P} \subseteq S$. Finally, define a generalized coloring $\tilde{\chi}: P \to \mathbf{R}$ by setting $\tilde{\chi}(p + \mathbf{w}_i) = c_i\chi(p)$ for each $p \in P$ and each $i = 1, 2, \ldots, k$; see a picture:*

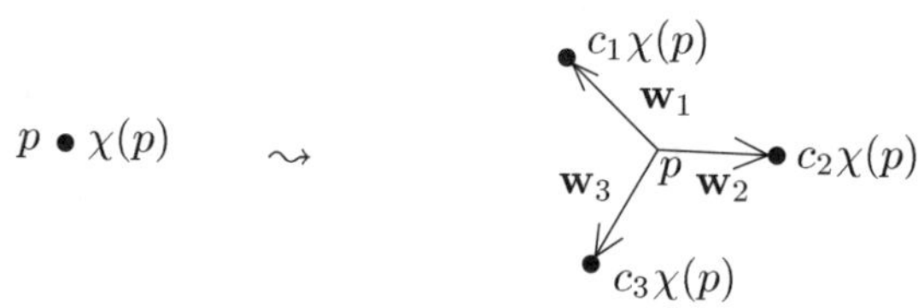

Then the L_2-discrepancy of P under χ has at least the same order of magnitude as that of $\tilde{P}$ under $\tilde{\chi}$:

$$\operatorname{disc}_{2,\nu}(P,\mathcal{U},\chi) = \Omega(\operatorname{disc}_{2,\nu}(\tilde{P},\mathcal{U},\tilde{\chi})).$$

Proof. By the Alexander–Stolarsky formula, if $\chi(P) = 0$, then the L_2-discrepancy of P under χ does not change by translating P provided that the translated set remains within the square S. That is, if χ_i is the coloring of P_i defined by $\chi_i(p + \mathbf{w}_i) = \chi(p)$ then

$$\begin{aligned}\operatorname{disc}_{2,\nu}(P,\mathcal{U},\chi)^2 &= -\sum_{p,q\in P}\chi(p)\chi(q)\|p-q\|\\ &= -\sum_{p,q\in P_i}\chi_i(p)\chi_i(q)\|p-q\| = \operatorname{disc}_{2,\nu}(P_i,\mathcal{U},\chi_i)^2.\end{aligned}$$

By the Cauchy–Schwarz inequality, we obtain

$$\tilde{\chi}(\tilde{P}\cap\gamma)^2 = \left(\sum_{i=1}^{k} c_k\chi_i(P_i\cap\gamma)\right)^2 \le \left(\sum_{i=1}^{k} c_k^2\right)\left(\sum_{i=1}^{k}\chi_i(P_i\cap\gamma)^2\right).$$

Consequently,

$$\begin{aligned}\operatorname{disc}_{2,\nu}(\tilde{P},\mathcal{U},\tilde{\chi})^2 &\le \left(\sum_{i=1}^{k} c_k^2\right)\left(\sum_{i=1}^{k}\operatorname{disc}_{2,\nu}(P_i,\mathcal{U},\chi_i)^2\right)\\ &= k\left(\sum_{i=1}^{k} c_k^2\right)\operatorname{disc}_{2,\nu}(P,\mathcal{U},\chi)^2.\end{aligned}$$

□

Second Proof of Theorem 6.4. Many parts of the proof are almost identical to the proof in the previous section. We again assume that n is a perfect square and we take the same $\sqrt{n}\times\sqrt{n}$ grid in the square S for P. We also put $w = cn^{-1/2}$ and $\mathbf{w} = (0, w)$.

Let $\chi\colon P \to \{-1,+1\}$ be a fixed coloring. In order to apply the fine tools just developed, we would need that $\chi(P) = 0$, which in general need not be the case. This is easy to rectify, however. Suppose, for instance, that the points of P colored by 1 outnumber the points colored by -1 by some $\Delta > 0$. If $\Delta > cn^{1/4}$ for a suitable positive constant c then any halfplane containing the whole P has large enough discrepancy, so we may disregard such a coloring χ. Otherwise, recolor some $\frac{\Delta}{2}$ points of P colored by 1. If we now show that the discrepancy of this modified χ is at least 2Δ, then the original discrepancy was at least Δ.

Thus, assume $\chi(P) = 0$. From the set P, we pass to a set $\tilde{P}$ and a generalized coloring $\tilde{\chi}$ by replicating each point $p \in P$ three times, as in the drawing:

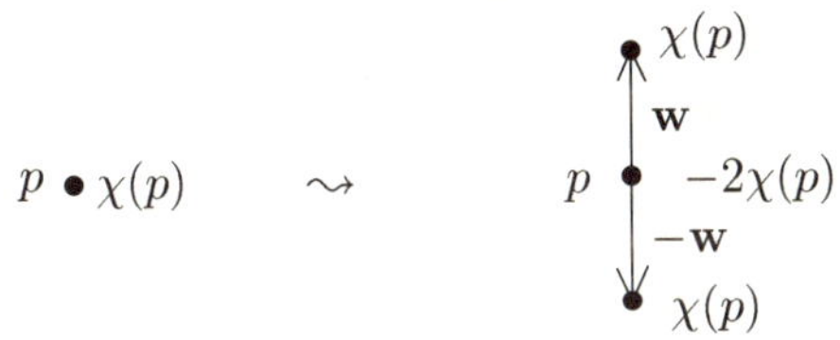

This is as in the Replication lemma 6.8 with $k = 3$, $\mathbf{w}_1 = -\mathbf{w}$, $\mathbf{w}_2 = 0$, $\mathbf{w}_3 = \mathbf{w}$, $c_1 = c_3 = 1$, and $c_2 = -2$. This time, the coefficients c_1, c_2 and

c_3 are taken from the finite-differencing formula of order 2. (In this proof method, the differencing order will effectively be doubled, and so we end up with the fourth-order differencing as in the former proof.)

By the Replication lemma 6.8, it suffices to prove $\mathrm{disc}_{2,\nu}(\tilde{P},\mathcal{U},\tilde{\chi}) = \Omega(n^{1/4})$. We use the Alexander–Stolarsky formula for expressing the L_2-discrepancy of $\tilde{P}$ under $\tilde{\chi}$, and we group together the terms coming from copies of the same original point $p \in P$:

$$\begin{aligned}\mathrm{disc}_{2,\nu}(\tilde{P},\mathcal{U},\tilde{\chi})^2 &= -\sum_{\tilde{p},\tilde{q}\in\tilde{P}} \tilde{\chi}(\tilde{p})\tilde{\chi}(\tilde{q})\,\|\tilde{p}-\tilde{q}\| \\ &= \sum_{p,q\in P} \chi(p)\chi(q)J(p,q),\end{aligned}$$

where

$$J(p,q) = -\|p-q-2\mathbf{w}\|+4\|p-q-\mathbf{w}\|-6\|p-q\|+4\|p-q+\mathbf{w}\|-\|p-q+2\mathbf{w}\|.$$

This formula for $J(p,q)$ is perhaps best explained by the following picture:

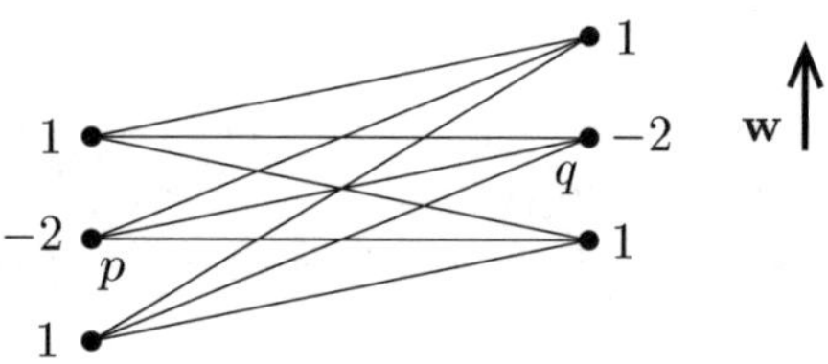

For a pair of original points $p, q \in P$, we are summing the distances drawn there, and the weight of a distance is the product of the weights of its endpoints. (The calculation of $J(p,q)$ hides the passage from the second-order differencing implicit in the coefficients used in the definition of $\tilde{\chi}$ to the fourth-order differencing. The reader may want to contemplate how $J(p,q)$ corresponds to two successive applications of the second-order differencing operator Δ_1^2 to the function $g(x) = \|p-q+x\mathbf{w}\|$.)

As it happens, the quantity $J(p,q)$ is identical to $\int f_p I_q$ considered in the previous proof, as can be seen by comparing the above formula for $J(p,q)$ with the formula for $\int I_p f_q$ in Lemma 6.6. Hence, exactly the same calculations yield

$$\mathrm{disc}_{2,\nu}(\tilde{P},\mathcal{U},\tilde{\chi})^2 \ge \sum_{p\in P} J(p,p) - \sum_{p,q\in P,\, p\neq q} |J(p,q)| = \Omega(\sqrt{n}).$$

Theorem 6.4 is proved once again. □

Remarks. Here is an intuitive view of the presented proof. The set $\tilde{P}$ is composed of triples of close points, and the weights of the points in each triple compensate. If a halfplane crosses some triple (i.e. contains some of its points but not all of them) then this triple contributes 1 to the "imbalance" of this halfplane. A random halfplane is likely to cross about $wn \approx \sqrt{n}$ triples (by the Perimeter formula 6.5). The proof shows that the contributions of these triples behave somewhat like independent random variables (the randomness comes from the halfplane!); we can interpret $J(p,q)$ as the covariance of the contribution of the triple coming from the point p and the triple coming from the point q. The proof shows that the covariances are sufficiently small compared to the expected contributions of the individual triples.

In the proof in higher dimension, the set $\tilde{P}$ is formed according to the finite differencing formula of order $t = \lceil (d+2)/2 \rceil$, and this leads to $J(p,q) = -\Delta_1^{2t} g(x)$ with $g(x) = \|p - q + xw\|$.

Bibliography and Remarks. This section follows the presentation of Chazelle et al. [CMS95] quite closely. That paper contains still another version of the proof, due to Chazelle, with an interesting relation to the famous "Buffon needle" experiment.

6.6 Halfplanes: the Lebesgue-Measure Discrepancy

In this section we prove

6.9 Theorem. *For all $n \geq 1$, the discrepancy of any n-point set P in the unit square for halfplanes satisfies $D(P, \mathcal{H}_2) = \Omega(n^{1/4})$.*

The basic approach to the proof is the same as in the previous section (point replication), but if we did the replication in exactly the same way as before, an essential part of the forthcoming proof would fail! The key new idea is to step out of the plane and to shift the replicated points in the 3-dimensional space in the direction perpendicular to the original plane.

Proof. Let $P \subseteq [0,1]^2$ be an arbitrary n-point set, fixed throughout the proof. By the conventions introduced in Section 1.2, $D(P, A)$ stands for the "signed discrepancy" of a set A. Since P is fixed, we omit it from the notation; that is, we put

$$D(A) = n \cdot \mathrm{vol}_{\square}(A) - |P \cap A|$$

for a (Lebesgue-measurable) set $A \subseteq \mathbf{R}^2$. We thus want to prove that there exists a halfplane γ with $|D(\gamma)| = \Omega(n^{1/4})$.

In this proof (as well as in some subsequent ones), it is useful to view the function $A \mapsto D(A)$ as a signed measure (a signed measure is a countably additive real function defined on a σ-algebra of sets; it is like a measure but without the requirement of nonnegativity). Part of this signed measure is concentrated on the point set P, and another part is a continuous measure on the unit square. In particular, we can integrate various functions according to D; we have

$$\int_X f(x)\,\mathrm{d}D(x) = n \cdot \int_X f(x)\,\mathrm{d}x - \sum_{p \in P} f(p)$$

(a reader not feeling at ease with signed measures can just regard the left-hand side as a textual abbreviation of the right-hand side).

Similar to the previous section, we are going to replicate the set P and the corresponding signed measure D, but here we also increase the dimension of the ambient space. First, we think of the square $[0,1]^2$ as lying in the $z=0$ plane in $\mathbf{R}^3$, and we regard D as a signed measure in the 3-dimensional space (concentrated in the $z=0$ plane).

Instead of the discrepancy for 2-dimensional halfplanes, we will consider the discrepancy for 3-dimensional halfspaces. At the end of Section 6.4, we introduced the motion-invariant measure ν on the set of all planes in $\mathbf{R}^3$, and we remarked that the ν-measure of all planes intersecting a segment s is twice the length of s. Let $\mathcal{U}_3$ be the set of all upper halfspaces in $\mathbf{R}^3$ whose bounding planes intersect the cube $[0,1]^2 \times [-\frac{1}{2}, \frac{1}{2}]$. Obviously, since the cube has 12 edges, we have $\nu(\mathcal{U}_3) \le 24$ (we do not need the exact ν-measure of $\mathcal{U}_3$ although it is possible to calculate it). We will work with L_2-discrepancy averaged over the halfspaces of $\mathcal{U}_3$. The measure ν restricted to $\mathcal{U}_3$ is not a probability measure, but it could be rescaled by a constant factor to become one. Note that if we find a halfspace $\gamma \in \mathcal{U}_3$ with the discrepancy $|D(\gamma)|$ large, then intersecting this γ with the $z=0$ plane yields a halfplane with the same large discrepancy.

We now describe the replication of D. We fix $w = cn^{-1/2}$ as in the previous proofs, and this time we let $\mathbf{w} = (0,0,w)$. We replicate the 2-dimensional signed measure D, by shifting one copy from the $z=0$ plane by $\mathbf{w}$ upwards and one copy by $-\mathbf{w}$ downwards, and we assign the middle copy in the $z=0$ plane the weight -2, as the following picture indicates:

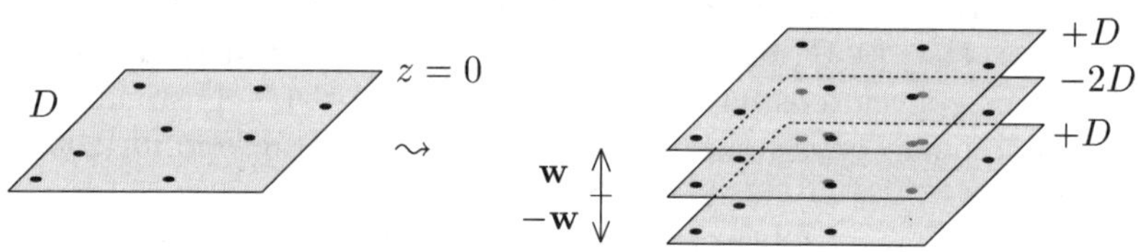

Thus, we formally define

$$\tilde{D}(A) = D(A + \mathbf{w}) - 2D(A) + D(A - \mathbf{w})$$

for a set $A \subseteq \mathbf{R}^3$ (we will use this definition for halfspaces only).

In the proof, we will show that the quadratic average over $\gamma \in \mathcal{U}_3$ of $\tilde{D}(\gamma)$ is $\Omega(n^{1/4})$, and analogous to the preceding section, this will allow us to conclude that there exists a halfspace γ with $|D(\gamma)| = \Omega(n^{1/4})$. To this end, we need analogies of the Alexander–Stolarsky formula 6.7 and of the Replication lemma 6.8.

For the Alexander–Stolarsky formula, the counterpart of the generalized coloring χ in Lemma 6.7 is the signed measure $\tilde{D}$ here. Namely, we have

$$\int_{\mathcal{U}_3} \tilde{D}(\gamma)^2 \, \mathrm{d}\nu(\gamma) = -\int_{\mathbf{R}^3}\int_{\mathbf{R}^3} \|\tilde{p}-\tilde{q}\| \, \mathrm{d}\tilde{D}(\tilde{p}) \, \mathrm{d}\tilde{D}(\tilde{q}). \tag{6.7}$$

The proof is exactly the same as before, with summation replaced by integration (note that we have $\tilde{D}(\mathbf{R}^3) = 0$). Using the definition of $\tilde{D}$, (6.7) can also be rewritten, as in the previous section, to

$$\int_{\mathcal{U}_3} \tilde{D}(\gamma)^2 \, \mathrm{d}\nu(\gamma) = \int_{\mathbf{R}^2}\int_{\mathbf{R}^2} J(p,q) \, \mathrm{d}D(p) \, \mathrm{d}D(q),$$

where $J(p,q)$ is formally the same as before:

$$\begin{aligned} J(p,q) = {} & -\|p-q-2\mathbf{w}\| + 4\|p-q-\mathbf{w}\| - 6\|p-q\| \\ & +4\|p-q+\mathbf{w}\| - \|p-q+2\mathbf{w}\|. \end{aligned} \tag{6.8}$$

The crucial advantage over the previous section is that the vectors $q-p$ and $\mathbf{w}$ are now always orthogonal.

The analogy of the Replication lemma 6.8, again with the same proof, reads

$$\int_{\mathcal{U}_3} D(\gamma)^2 \, \mathrm{d}\nu(\gamma) = \Omega\left(\int_{\mathcal{U}_3} \tilde{D}(\gamma)^2 \, \mathrm{d}\nu(\gamma)\right) \tag{6.9}$$

and so for proving Theorem 6.9, it suffices to show that

$$\int_{\mathbf{R}^2}\int_{\mathbf{R}^2} J(p,q) \, \mathrm{d}D(p) \, \mathrm{d}D(q) = \Omega(n^{1/2}).$$

By the definition of the signed measure D, we can rewrite

$$\int_{\mathbf{R}^2}\int_{\mathbf{R}^2} J(p,q) \, \mathrm{d}D(p) \, \mathrm{d}D(q) = E_{\lambda\lambda} + E_{\lambda P} + E_{P2} + E_{PP}, \tag{6.10}$$

where

$$\begin{aligned} E_{\lambda\lambda} &= n^2 \cdot \int_{[0,1]^2}\int_{[0,1]^2} J(x,y) \, \mathrm{d}x \, \mathrm{d}y, \\ E_{\lambda P} &= -n \cdot \sum_{p\in P} \int_{[0,1]^2} J(p,x) \, \mathrm{d}x, \\ E_{P2} &= \sum_{p\in P} J(p,p), \text{ and} \\ E_{PP} &= \sum_{\substack{p,q\in P \\ p\neq q}} J(p,q) \end{aligned}$$

(here the λ in the subscripts refers to the Lebesgue measure).

We estimate these terms one by one. As for E_{P2}, $J(p,p)$ always equals $4w$, so $E_{P2} = 4nw = 4cn^{1/2}$. This is the substantial positive contribution we need for lower-bounding the L_2-discrepancy, and it remains to show that the other terms cannot cancel it out.

The magnitude of term $E_{\lambda P}$ can be bounded using the estimates for $J(p,q)$ derived earlier. We have

$$|E_{\lambda P}| \le n \sum_{p \in P} \int_{[0,1]^2} |J(p,x)| \, \mathrm{d}x \le n \sum_{p \in P} \int_{\mathbf{R}^2} |J(p,x)| \, \mathrm{d}x.$$

Obviously, the integral in the last expression does not depend on p. For the x with $\|p-x\| \le 4w$, we use the estimate $|J(p,x)| = O(w)$, which immediately follows from the definition (6.8) of $J(p,q)$. For $\|p-x\| > 4w$, we apply the bound $|J(p,q)| = O(w^4/\|p-q\|^3)$ from Lemma 6.6(ii). Therefore,

$$\begin{aligned} \int_{\mathbf{R}^2} |J(p,x)| \, \mathrm{d}x &\le \int_{\|p-x\|\le 4w} O(w) \, \mathrm{d}x + \int_{\|p-x\|>4w} O\left(\frac{w^4}{\|p-x\|^3}\right) \mathrm{d}x \\ &= O(w^3) + O(w^4) \cdot \int_{4w}^{\infty} \frac{2\pi r}{r^3} \, \mathrm{d}r \\ &= O(w^3). \end{aligned}$$

Consequently, $|E_{\lambda P}| = O(w^3 n^2) = O(c^3 n^{1/2})$, and so for a small enough c, $|E_{\lambda P}|$ is much smaller than E_{P2}.

Now we get to the term E_{PP}, and this is where the situation becomes more interesting. Formally, the sum $\sum_{p,q \in P,\, p \ne q} J(p,q)$ looks the same as the one we handled successfully in the previous section. But there the set P was chosen at our will, but now we have no control over its distribution! This is really a problem; for example, the reader may want to check that if all the points of P coincide, or lie in a tiny cluster, then E_{PP} is as big as $\Omega(n^{3/2})$. Of course, one can immediately object that if all points are tightly clustered then P cannot have low discrepancy for halfplanes. Indeed, low discrepancy for halfplanes means some kind of uniform distribution. But it is not at all obvious how one should derive a good enough estimate for the order of magnitude of E_{PP} even under the low-discrepancy assumption, since a low-discrepancy set might still possibly contain small local clusters of points that kill the summation.

The following lemma saves the situation, showing that the term E_{PP}, even though possibly large, is always nonnegative, and consequently harmless in (6.10). (The same result also gives us $E_{\lambda\lambda} \ge 0$ for free.)

6.10 Lemma (Nonnegativity lemma). *For any two points p, q in the $z = 0$ plane, we have $J(p,q) \ge 0$.*

Let us remark that the nonnegativity fails in the setting of the previous section, when the vector $\mathbf{w}$ lies in the plane and need not be orthogonal to $q - p$ (Exercise 5).

For the planar case we are dealing with, this lemma can be proved by an ad hoc approach using elementary calculus. The function $J(p,q)$ depends on w and on $\delta = \|p-q\|$. Using the substitution $h = \frac{w}{\delta}$ as in Section 6.4, we can calculate from (6.8) that

$$J(p,q) = 2\delta \cdot \left(3 - 4\sqrt{1+h^2} + \sqrt{1+4h^2}\right).$$

Having plotted a graph of the function in parentheses, one may start believing that it is indeed nonnegative for all $h > 0$. This can be verified by elementary calculus, and this concludes the proof of Theorem 6.9. □

The conclusion of the previous proof cannot really satisfy a curious reader. First of all, if we try to generalize the result for halfspaces in an arbitrary dimension d, everything else goes through easily, but it is not obvious how to prove the higher-dimensional analogue of the nonnegativity lemma—at least the approach via elementary calculus seems inadequate. And second, it is not clear why such a property should hold: is it purely by chance, or is it a manifestation of some interesting general phenomenon?

The first problem, the higher-dimensional generalization, can be dealt with by changing the definition of the replicated signed measure $\tilde{D}$, which leads to an expression whose non-negativity for an arbitrary dimension follows by an easy calculation. This route is indicated in Exercise 3 below. But in this method the nonnegativity phenomenon also looks like a miraculous coincidence. In the next section, we thus show another proof of the Nonnegativity lemma 6.10, which provides some insight into this miracle and shows connections of the Nonnegativity lemma to geometric properties of the Euclidean space and to other interesting results.

Bibliography and Remarks. Here we again follow Alexander's ideas [Ale90], with some additional twists from [CMS95]. In Alexander [Ale90], the Nonnegativity lemma 6.10 was proved for dimension 2 only (using elementary calculus). In [Ale91] it was established in general, by methods presented in the next section. An alternative, more elementary way of obtaining the bound for higher dimensions, outlined in Exercise 3 below, is due to Chazelle et al. [CMS95].

In [Ale91], Alexander states the results of his method in a considerable geometric generality (instead of the unit cube, he considers domains on convex surfaces). Actually, the method depends very little on the specific properties of the Lebesgue measure. It turns out that the d-dimensional Lebesgue measure can be replaced by an arbitrary probability measure μ in $\mathbf{R}^d$ with bounded support, and the lower bound depends on the μ-volume of balls—see Exercise 2 for a precise formulation. (We mostly discuss the combinatorial and Lebesgue-measure discrepancy in this book, but here the method works in such a simple way for arbitrary measures that this generalization is perhaps worth mentioning.)

Another interesting generalization was investigated by Rogers [Rog94]. He considers an $m \times m$ chessboard, consisting of $n = m^2$ unit squares, where each of these squares is colored red or blue. By extending Alexander's method, he proves that there is a halfplane for which the difference of red and blue areas contained in it is at least $\Omega(n^{1/4})$.

Exercises

1. Go through the proof of the Alexander–Stolarsky formula and of the Replication lemma, and modify them to obtain (6.7) and (6.9).
2. (A lower bound for approximating arbitrary measures by finite sets)
(a)* Let μ be a probability measure with support contained in $[0,1]^2$, such that all halfplanes are μ-measurable, and suppose that $\alpha \in (1,2]$ and C are constants such that $\mu(B(x,r)) \leq Cr^\alpha$ holds for all points $x \in \mathbf{R}^2$ and all radii $r > 0$ (recall that $B(x,r)$ denotes the ball of radius r centered at x). For an n-point set $P \subset [0,1]^2$, define the signed measure D by $D(A) = n \cdot \mu(A) - |P \cap A|$. By generalizing the proof given in this section, show that there exists a halfplane γ with $|D(\gamma)| = \Omega(n^{1/2-1/2\alpha})$, where the constant of proportionality depends on α. (A d-dimensional analogue can be proved in a very similar way.)
(b)* Show that it is sufficient to assume the condition $\mu(B(x,r)) \leq Cr^\alpha$ only for $r = c_1 n^{-1/\alpha}$ with a suitable sufficiently small constant $c_1 > 0$.
Remark. Nontrivial examples of measures μ with $0 < \alpha < d$ are provided by various fractal sets. One of them is the *Sierpiński Carpet*, which arises by subdividing the unit square into a 3×3 square grid, deleting the middle square, and repeating this construction recursively in each of the 8 remaining squares. Here $\alpha = \log 8/\log 3 = 1.89\ldots$ coincides with the Hausdorff dimension. Also for many other fractals, the assumption in this exercise is satisfied for α being the Hausdorff dimension. Unfortunately, a matching upper bound for the discrepancy is not known. The question of determining the discrepancy for halfplanes with respect to fractal measures was communicated to me by Robert Tichy.
3. Complete the following outline of another version of the proof of Theorem 6.9. (This is perhaps the shortest and most elementary way known.) Embed $\mathbf{R}^2$ into $\mathbf{R}^4$ as the plane $x_3 = x_4 = 0$. With $\mathbf{w}_1 = (0,0,w,0)$ and $\mathbf{w}_2 = (0,0,0,w)$, set $\tilde{D}(A) = \sum_{i_1,i_2=0}^{1} (-1)^{i_1+i_2} D(A - i_1\mathbf{w}_1 - i_2\mathbf{w}_2)$. Give a formula for $\int_{\mathcal{U}_4} \tilde{D}(\gamma)^2 \,d\nu(\gamma)$ analogous to (6.10), with an appropriate version of $J(p,q)$, where $\mathcal{U}_4$ is the set of the upper halfspaces intersecting the cube $[0,1]^2 \times [-\frac{1}{2},\frac{1}{2}]^2$ and ν is the measure on $\mathcal{U}_4$ induced by the translation-invariant measure on the hyperplanes in $\mathbf{R}^4$. Check that the estimates for E_{P2} and $E_{\lambda P}$ go through, and establish the nonnegativity of $J(p,q)$ using the formula (6.3) on page 184. *Generalize the proof to an arbitrary dimension d.

4. Calculate the measure of the set of all planes intersecting the unit cube $[0,1]^3$ (how many times bigger it is than the measure of the planes intersecting a unit-length segment?).
5. Show that the function $J(p,q)$ defined in Section 6.4 need not be nonnegative if we do not assume that $\mathbf{w}$ and $q-p$ are orthogonal vectors. Thus, adding an extra dimension is crucial for the proof method in the current section.

6.7 A Glimpse of Positive Definite Functions

In this section, we derive the Nonnegativity lemma 6.10 as a simple consequence of a theory concerning the so-called positive definite functions. We present only a small part of this theory, in a very special setting just sufficient for our application. In the second part of the section, we leave our main theme of discrepancy and look at some problems on isometric embeddings that motivated the development of the techniques explained in the first part.

At the very beginning, we recall the definition of the one-dimensional Fourier transform. Here we will need to know very little about it besides the definition, but in the subsequent chapter we will meet the Fourier transform again (in a higher-dimensional setting) and we will use it for proving discrepancy lower bounds and other interesting results.

The Fourier Transform. Let f be a real or complex function of one real variable with $f \in L_1(\mathbf{R})$, which means that the Lebesgue integral $\int_{-\infty}^{\infty} |f(x)|\,\mathrm{d}x$ exists and is finite. The *Fourier transform* of f, denoted by $\hat{f}$, is a function defined by

$$\hat{f}(\xi) = \frac{1}{\sqrt{2\pi}} \int_{-\infty}^{\infty} f(x)\, e^{-\mathrm{i}x\xi}\,\mathrm{d}x,$$

where i stands for the imaginary unit. The Fourier transform of a real function is, in general, a complex-valued function of the real variable ξ. This transform enjoys a number of interesting properties, and we will encounter some of them later. Here we only mention the fact (not needed in this section) that if also $\hat{f} \in L_1(\mathbf{R})$, then f can be recovered from $\hat{f}$ by means of the following inversion formula:

$$f(x) = \frac{1}{\sqrt{2\pi}} \int_{-\infty}^{\infty} \hat{f}(\xi)\, e^{\mathrm{i}x\xi}\,\mathrm{d}\xi.$$

The proof is not as simple as one might suspect at first sight, since if one tries to verify the formula by a direct substitution of the definition of $\hat{f}$, one gets the divergent integral $\int_{-\infty}^{\infty} e^{\mathrm{i}(\xi_1-\xi_2)x}\mathrm{d}x$. The proof is usually done by a limit argument; see Rudin [Rud74] for one.

Positive Definite Functions. A function $f\colon \mathbf{R} \to \mathbf{R}$ is called *positive definite* if we have, for any numbers $x_1,\ldots,x_n \in \mathbf{R}$ and $\tau_1,\ldots,\tau_n \in \mathbf{R}$,

$$\sum_{j,k=1}^{n} \tau_j \tau_k f(x_j - x_k) \geq 0.$$ [4]

Let us note that this concept can be related to the perhaps better known notion of positive definite matrices. Namely, for $x_1, x_2, \ldots, x_n \in \mathbf{R}$, define the $n \times n$ matrix $A = A(f; x_1, x_2, \ldots, x_n)$ by setting $a_{jk} = f(x_j - x_k)$. Then f being positive definite means that A is a positive semidefinite matrix for any choice of $x_1, x_2, \ldots, x_n$; that is, $\tau^T A \tau \geq 0$ for any real column vector $\tau \in \mathbf{R}^n$. (In analogy with the terminology for matrices, it would perhaps be more appropriate to say "a positive semidefinite function" instead of "a positive definite function," but we stick to the traditional terminology.)

Nontrivial examples of positive definite functions are not obvious. The following lemma gives a method for producing some using the Fourier transform.

6.11 Lemma. *Suppose that a real function $\varphi = \hat{\psi}$ is the Fourier transform of a nonnegative real function $\psi \in L_1(\mathbf{R})$. Then φ is positive definite.*

Proof. Let $x_1, \ldots, x_n \in \mathbf{R}$ and $\tau_1, \ldots, \tau_n$ be given. We calculate

$$\begin{aligned} \sum_{j,k=1}^{n} \tau_j \tau_k \varphi(x_j - x_k) &= \frac{1}{\sqrt{2\pi}} \int_{-\infty}^{\infty} \psi(\xi) \sum_{j,k} \tau_j \tau_k e^{-\mathrm{i}(x_j - x_k)\xi} \, \mathrm{d}\xi \\ &= \frac{1}{\sqrt{2\pi}} \int_{-\infty}^{\infty} \psi(\xi) \Big(\sum_j \tau_j e^{-\mathrm{i}x_j\xi} \Big) \Big(\sum_j \tau_j e^{\mathrm{i}x_j\xi} \Big) \, \mathrm{d}\xi. \end{aligned}$$

Since all the x_j and τ_j are real, the sum in the second pair of parentheses is the complex conjugate of the one in the first pair of parentheses, and we obtain

$$\sum_{j,k=1}^{n} \tau_j \tau_k \varphi(x_j - x_k) = \frac{1}{\sqrt{2\pi}} \int_{-\infty}^{\infty} \psi(\xi) \Big| \sum_j \tau_j e^{\mathrm{i}x_j\xi} \Big|^2 \, \mathrm{d}\xi \geq 0.$$

□

The following lemma provides one of the most important examples of a positive definite function.

6.12 Lemma. *The function $h(x) = e^{-x^2}$ is positive definite.*

[4] This definition is appropriate if we wish to stay in the domain of real numbers. Sometimes it is useful to consider complex-valued functions as well; then positive definiteness means $\sum_{j,k=1}^{n} \tau_j \overline{\tau_k} f(x_j - x_k) \geq 0$ for all complex $\tau_1, \ldots, \tau_n$, where the bar over τ_k denotes complex conjugate. Also, the definition can be formulated for a function $f\colon G \to \mathbf{R}$, where G is an (additively written) commutative group, in which case the x_j are chosen in G while the τ_j remain real or complex.

Sketch of Proof. We claim that h is the Fourier transform of the function $e^{-x^2/4}$ (and thus the previous lemma applies). We have

$$\begin{aligned}\int_{-\infty}^{\infty} e^{-x^2/4} \cdot e^{-\mathrm{i}x\xi}\, \mathrm{d}x &= \int_{-\infty}^{\infty} e^{-x^2/4-\mathrm{i}x\xi}\, \mathrm{d}x \\ &= e^{-\xi^2} \cdot \int_{-\infty}^{\infty} e^{-(x/2+\mathrm{i}\xi)^2}\, \mathrm{d}x.\end{aligned}$$

Since we are integrating in the complex plane, it is not really honest to say that the last integral is transformed by the substitution $y = (x/2 + \mathrm{i}\xi)$ to the classical integral $\int_{-\infty}^{\infty} e^{-y^2}\, \mathrm{d}y = \sqrt{\pi}$. Nevertheless, such a transformation is possible using some simple considerations with Cauchy's theorem about holomorphic functions. In any case, for showing the positive definiteness of h, we do not really need the value of the integral $\int_{-\infty}^{\infty} e^{-(x/2+\mathrm{i}\xi)^2}\, \mathrm{d}x$, we only need that it is a positive constant independent of ξ, and for those familiar with the basics of complex analysis, this is quite easy to check. □

Next, we introduce a concept similar to positive definite functions, which is already intimately related to the Nonnegativity lemma from the previous section. Namely, we say that a function $g\colon \mathbf{R} \to \mathbf{R}$ is *of negative type* if we have, for any n points $x_1, \ldots, x_n \in \mathbf{R}$ and any real numbers $\tau_1, \ldots, \tau_n$ with $\tau_1 + \cdots + \tau_n = 0$ (this last condition should not be overlooked!),

$$\sum_{j,k=1}^{n} \tau_j \tau_k g(x_j - x_k) \leq 0.$$

6.13 Observation. *If $f(x)$ is a positive definite function then the function $g(x) = 1 - f(x)$ is of negative type.*

Proof. With $\tau_1 + \cdots + \tau_n = 0$, we have

$$\sum_{j,k=1}^{n} \tau_j \tau_k g(x_j - x_k) = \left(\sum_{j=1}^{n} \tau_j\right)^2 - \sum_{j,k=1}^{n} \tau_j \tau_k f(x_j - x_k) \leq 0.$$

□

In order to establish the Nonnegativity lemma 6.10, we prove the following statement:

6.14 Proposition. *The function $x \mapsto \sqrt{1 + x^2}$ is of negative type.*

To see the relevance of this result for the Nonnegativity lemma, we recall that the function $J(p, q)$ in that lemma arose as a signed sum of the distances drawn in the following picture (also see Section 6.4):

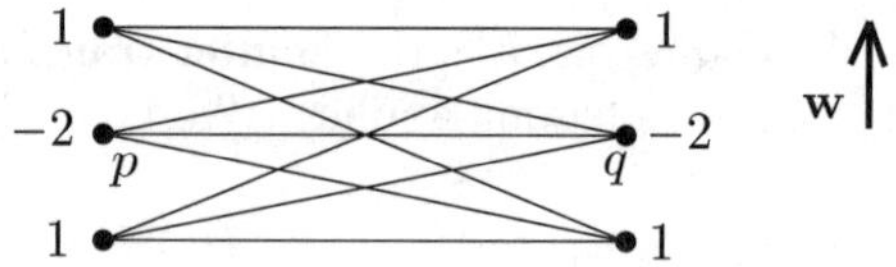

Namely, we have

$$J(p,q) = -\sum_{j,k=1}^{3} \tau_j\tau_k \cdot \|p - q - (j-k)\mathbf{w}\|$$
$$= -\|p-q\| \cdot \sum_{j,k=1}^{3} \tau_j\tau_k\sqrt{1+h^2(j-k)^2}$$

with $h = w/\|p-q\|$, $\tau_1 = \tau_3 = 1$ and $\tau_2 = -2$. The way Proposition 6.14 implies $J(p,q) \geq 0$ for all p, q should now be clear. The nonnegativity of the higher-dimensional version of $J(p,q)$ follows from the proposition in the same way.

For the proof of Proposition 6.14, we need

6.15 Lemma. *If $f\colon \mathbf{R} \to [0,\infty)$ is a nonnegative function such that the function $g_\lambda(x) = e^{-\lambda^2 f^2(x)}$ is positive definite for each value of the parameter $\lambda \in \mathbf{R}$ then f is of negative type.*

Since e^{-x^2} is positive definite by Lemma 6.12, also $e^{-\lambda^2(1+x^2)}$ is obviously positive definite, and Proposition 6.14 follows from Lemma 6.15 with $f(x) = \sqrt{1+x^2}$.

Proof of Lemma 6.15. Let us enjoy some magic of real analysis. We first re-express the considered function f as a suitable integral. Set $\varphi(x) = (1-e^{-x^2})/x^2$. As is easy to check, the integral $\int_0^\infty \varphi(x)\,\mathrm{d}x$ converges to some positive constant C (whose value is not important for us). Now let t be any fixed positive real number and let λ be a new variable. By the substitution $x = t\lambda$ we obtain

$$C = \int_0^\infty \varphi(x)\,\mathrm{d}x = \int_0^\infty \varphi(t\lambda)t\,\mathrm{d}\lambda = \frac{1}{t}\int_0^\infty \frac{1-e^{-t^2\lambda^2}}{\lambda^2}\,\mathrm{d}\lambda,$$

and therefore

$$t = \frac{1}{C}\int_0^\infty \frac{1-e^{-t^2\lambda^2}}{\lambda^2}\,\mathrm{d}\lambda.$$

Using this for $t = f(x)$, we find the integral representation

$$f(x) = \frac{1}{C}\int_0^\infty \frac{1-e^{-\lambda^2 f(x)^2}}{\lambda^2}\,\mathrm{d}\lambda.$$

If we prove that the integrand is, for each value of λ, a function of x of negative type, we can infer that f is of negative type too (this can easily be seen from the definition of a function of negative type). But the integrand is of negative type because of the assumption that $e^{-\lambda^2 f^2}$ is positive definite and by Observation 6.13. This concludes the alternative proof of the Nonnegativity lemma 6.10. □

An Application Concerning Isometric Embeddings. Although it may be unnecessary for most readers, we first recall that a *metric space* is a pair (X, ρ), where X is a set and $\rho: X \times X \to \mathbf{R}$ is a nonnegative real function, the *metric*, satisfying $\rho(x, y) = \rho(y, x)$, $\rho(x, y) = 0$ if and only if $x = y$, and $\rho(x, y) + \rho(y, z) \geq \rho(x, z)$. A mapping $f: X \to Y$ is an *isometric embedding* of a metric space (X, ρ) into a metric space (Y, σ) if $\sigma(f(x), f(y)) = \rho(x, y)$ for all $x, y \in X$. Such an isometric embedding can be thought of as finding an exact copy of (X, ρ) in (Y, σ).

Questions about isometric embeddability of various metric spaces are important in several branches of analysis and also in applications. As a sample, let us look at the following question. Consider the real line $\mathbf{R}$ with the metric $\rho(x, y) = \sqrt{|x - y|}$ (one can check that it is really a metric). Can it be embedded into a Euclidean space equipped with the usual Euclidean metric? Intuitively, the embedding must locally "stretch" the line (since x, y lying very close together have $\rho(x, y)$ much bigger than $|x - y|$) but, on larger scales, the image of the line must be somehow "wound together." It turns out that an isometric embedding into $\mathbf{R}^d$ is not possible for any d (Exercise 1), but there exists an isometric embedding into the separable Hilbert space ℓ_2. We recall that ℓ_2 is defined as the space of all infinite sequences $x = (x_1, x_2, \ldots)$ with $x_1, x_2, \ldots \in \mathbf{R}$ and $\sum_{i=1}^{\infty} x_i^2 < \infty$, and with the distance of two such sequences $x, y \in \ell_2$ given by $\|x - y\| = \left(\sum_{i=1}^{\infty} (x_i - y_i)^2\right)^{1/2}$. For our purposes, ℓ_2 can be thought of as a kind of limit of the Euclidean spaces $\mathbf{R}^d$ for $d \to \infty$. The existence of an isometric embedding of the line with the metric ρ defined above into ℓ_2 can be proved very elegantly via positive definite functions. The following more general result holds:

6.16 Theorem (Schoenberg). *Let $0 < \gamma < 1$ and let ρ be the distance function on $\mathbf{R}^d$ given by $\rho(x, y) = \|x - y\|^\gamma$. Then the metric space $(\mathbf{R}^d, \rho)$ can be isometrically embedded into ℓ_2.*

We will only prove the particular case with $d = 1$ and $\gamma = \frac{1}{2}$ mentioned above the theorem, but it is not difficult to extend the proof method shown below to arbitrary d and $\gamma \in (0, 1)$.

Before we begin with the proof, we generalize the definition of a positive definite function. A bivariate function $K: X \times X \to \mathbf{R}$, where X is some set, is called a *(real) positive definite kernel* if

$$\sum_{j,k=1}^{n} \tau_j \tau_k K(x_j, x_k) \geq 0$$

holds for any choice of elements $x_1, x_2, \ldots, x_n \in X$ and all $\tau_1, \ldots, \tau_n \in \mathbf{R}$.[5]

If (X, ρ) is a metric space and $f: \mathbf{R} \to \mathbf{R}$ is a real function, then f is called *positive definite on* (X, ρ) if $(x, y) \mapsto f(\rho(x, y))$ is a positive definite kernel on X. A kernel of negative type and a function of negative type on a metric space are defined analogously.

For proving the promised special case of Theorem 6.16, we first derive the following criterion for isometric embeddability into ℓ_2:

6.17 Proposition (Schoenberg). *A separable*[6] *metric space* (X, ρ) *admits an isometric embedding into the separable Hilbert space* ℓ_2 *if and only if the function* $x \mapsto x^2$ *is of negative type on* (X, ρ).

As a first step towards proving this proposition, we need a "compactness" lemma, saying that the embeddability is determined by embeddability of finite subsets.

6.18 Lemma (Menger). *Let* X *be a separable metric space such that any* n*-point subspace of* X *can be isometrically embedded into* $\mathbf{R}^n$, $n = 1, 2, \ldots$. *Then* X *can be isometrically embedded into* ℓ_2.

We leave the proof of this lemma for Exercise 2.

For the proof of Proposition 6.17, it will be notationally convenient to consider $n+1$ points $x_0, x_1, \ldots, x_n \in X$. Also, for brevity, we write ρ_{jk} for $\rho(x_j, x_k)$. The condition in Proposition 6.17 that the function $x \mapsto x^2$ be of negative type on (X, ρ) can be written as follows:

$$\sum_{j,k=0}^{n} \tau_j \tau_k \rho_{jk}^2 \leq 0 \tag{6.11}$$

holds for all real $\tau_0, \tau_1, \ldots, \tau_n$ with $\sum_{j=0}^{n} \tau_j = 0$.

This condition can be neatly reformulated using the so-called *Gram matrix*, which is a generally useful tool in distance-related problems in Euclidean spaces. We thus make a small detour to introduce this concept. For given n vectors $v_1, v_2, \ldots, v_n$ in $\mathbf{R}^d$, the Gram matrix is the $n \times n$ matrix G with elements $g_{jk} = \langle v_j, v_k \rangle$, with $\langle .,. \rangle$ denoting the usual scalar product in $\mathbf{R}^d$. Since we have $\langle x, y \rangle = \frac{1}{2}(\|x\|^2 + \|y\|^2 - \|x - y\|^2)$ by the cosine theorem, the Gram matrix can be expressed solely in terms of distances, namely we have $g_{jk} = \frac{1}{2}(\|v_j\|^2 + \|v_k\|^2 - \|v_j - v_k\|^2)$.

Returning to our embedding condition, we define, for the given $n+1$ points $x_0, x_1, \ldots, x_n \in X$, the $n \times n$ matrix $G = (g_{jk})_{j,k=1}^{n}$ by setting

---

[5] Reproducing kernels on Hilbert spaces mentioned in Section 1.4 are examples of positive definite kernels, as is not difficult to check. Indeed, positive definite kernels in the just defined sense are sometimes called "reproducing kernels" in the literature.

[6] A metric space (X, ρ) is *separable* if it has a countable dense subset, where a subset $A \subseteq X$ is called *dense* if for each point $x \in X$, there are points of A arbitrary close to x.

$$g_{jk} = \frac{1}{2}\left(\rho_{j0}^2 + \rho_{0k}^2 - \rho_{jk}^2\right). \tag{6.12}$$

Whenever $f\colon \{x_0, x_1, \ldots, x_n\} \to \mathbf{R}^n$ is an isometric embedding, this G is the Gram matrix of the vectors $f(x_1) - f(x_0)$, $f(x_2) - f(x_0), \ldots, f(x_n) - f(x_0)$.

Next, we note that the condition (6.11) holds for all $\tau_0, \ldots, \tau_n$ if and only if the matrix G is positive semidefinite, i.e. $\sum_{j,k=1}^n \tau_j \tau_k g_{jk} \geq 0$ for all $\tau_1, \tau_2, \ldots, \tau_n \in \mathbf{R}$. This follows by substituting $-(\tau_1 + \tau_2 + \cdots + \tau_n)$ for τ_0 into (6.11) and rearranging.

Hence, proving Proposition 6.17 amounts to showing that the points $x_0, x_1, \ldots, x_n \in X$ can be isometrically embedded into $\mathbf{R}^n$ if and only if the matrix G given by (6.12) is positive semidefinite.[7] We prove only the "if" direction, which gives us the "if" direction in Proposition 6.17, the one we need for establishing Theorem 6.16. The "only if" part is left as Exercise 3.

Now we need the fact that any positive semidefinite and symmetric matrix G can be written in the form $G = M^T DM$, where M is a nonsingular $n \times n$ matrix and D is a diagonal matrix whose diagonal consists of r ones followed by $n - r$ zeros, with r standing for the rank of G. This fact is usually mentioned in a more general setting, in connection with the Sylvester law of inertia for quadratic forms. What we need here can be proved by a simple diagonalization algorithm resembling the Gauss elimination but done symmetrically on both rows and columns so that the matrix stays symmetric. We should perhaps stress that M is not required to be orthogonal (and, in general, it cannot be), so the matters are simpler than when dealing with diagonalizations of the form $G = M^{-1}DM$.

Having a matrix $M = (m_{jk})_{j,k=1}^n$ with $G = M^T DM$ at our disposal for the matrix G considered above, we define points $y_0, y_1, \ldots, y_n \in \mathbf{R}^r$ by setting $y_0 = 0$ and $y_j = (m_{1j}, m_{2j}, \ldots, m_{rj})$, $j = 1, 2, \ldots, n$. We claim that this configuration of $n + 1$ points in $\mathbf{R}^r$ is isometric to $x_0, x_1, \ldots, x_n \in X$. From the identity $G = M^T DM$, we see that the Gram matrix of the vectors $y_1, \ldots, y_n$ is just G, and this implies $\|y_j - y_k\| = \rho_{jk}$ for all $j, k = 0, 1, \ldots, n$. This proves the "if" part of Proposition 6.17. □

The proof of Theorem 6.16 for $d = 1$ and $\gamma = \frac{1}{2}$ is now immediate. To prove that $\mathbf{R}$ with the metric $(x, y) \mapsto \sqrt{|x - y|}$ embeds isometrically into ℓ_2, it is enough to show, by Proposition 6.17, that the function $x \mapsto |x|$ is of negative type. But this is a direct consequence of Lemma 6.12 and Lemma 6.15. □

---

[7] Thus, the scalar product on the Hilbert space is an example of a positive definite kernel. In some sense, it is "the" example, since a theorem of Moore says that any positive definite kernel $K(x, y)$ on a set X can be represented as $K(x, y) = \langle Tx, Ty \rangle$, where T is a mapping of X into a Hilbert space (of a sufficiently large cardinality, so not necessarily the separable one). See, e.g., Aharoni et al. [AMM85] for a proof (in the complex case) and references.

Bibliography and Remarks. Alexander [Ale91] showed that results and methods of Schoenberg [Sch37], [Sch38] concerning positive definite functions provide the nonnegativity results needed for the discrepancy lower-bound proof. Here we have only discussed particular cases of Schoenberg's and Alexander's results.

The material concerning isometric embeddings into the separable Hilbert space is mostly taken from Schoenberg's papers [Sch37], [Sch38]; the reader can also find references to previous work on the subject (by Wilson, von Neumann, Menger, and others) there. Schoenberg's results are more general than Theorem 6.16, also involving metric spaces arising by transformations of the Euclidean metric by functions other than $x \mapsto x^\gamma$. In [Sch38], Schoenberg proved another version of Proposition 6.17, namely that a separable metric space (X, ρ) is embeddable into the separable Hilbert space if and only if the functions $x \mapsto e^{-\lambda x^2}$ are positive definite on (X, ρ) for all $\lambda > 0$. This criterion has been generalized by Bretagnolle et al. [BDCK66] to embedding into the function spaces $L_p(0,1)$, $1 \leq p \leq 2$. For instance, they show that for $p \in [1, 2]$, a separable Banach space X can be isometrically embedded into $L_p(0,1)$ if and only if $x \mapsto e^{-|x|^p}$ is a positive definite function on X. A more recent application of methods involving positive definite functions to embeddability questions was presented by Aharoni et al. [AMM85], who consider questions about uniformly continuous embeddability in Banach spaces. This paper also lists various interesting properties of positive definite functions. A recent book concerning isometric embeddings, and embeddings into ℓ_1 in particular, is Deza and Laurent [DL97].

Exercises

1.* For a natural number d, consider the metric space $(\{0, 1, 2, \ldots, d+1\}, \rho)$ with $\rho(i, j) = \sqrt{|i - j|}$. Prove that this metric space cannot be isometrically embedded into $\mathbf{R}^d$ with the usual Euclidean metric. Use the Gram matrix.
2. Prove Lemma 6.18. You may use the fact that if A, B are two finite sets in $\mathbf{R}^n$ and $f \colon A \to B$ is a bijective isometric mapping then f can be extended to an isometry $\bar{f} \colon \mathbf{R}^n \to \mathbf{R}^n$.
3.* Show that the Gram matrix of arbitrary n vectors $v_1, v_2, \ldots, v_n \in \mathbf{R}^d$ is positive semidefinite. (Together with the considerations made in the text above, this establishes the "only if" part of Proposition 6.17.)
4. (a)* Prove that the function $x \mapsto e^{-x^2}$ is positive definite on the metric space $\mathbf{R}^d$ with the usual Euclidean metric, for any $d \geq 1$. Generalize Lemma 6.11 suitably to functions on $\mathbf{R}^d$.
   (b) Use (a) to prove Theorem 6.16 with $d \geq 1$ arbitrary and $\gamma = \frac{1}{2}$.

5. (a) Prove the following analogue of Lemma 6.15: Let $\alpha \in (0,2)$. If $f: \mathbf{R} \to [0,\infty)$ is a nonnegative function such that the function $g_\lambda(x) = e^{-\lambda^2 f^2(x)}$ is positive definite for each value of the parameter $\lambda \in \mathbf{R}$ then f^α is of negative type. Where does the proof fail for $\alpha \geq 2$?
(b) Use (a) to prove Theorem 6.16 for $d = 1$ and $\gamma \in (0,1)$ arbitrary.

7. More Lower Bounds and the Fourier Transform

In the previous chapters, we have seen several approaches to lower bounds in combinatorial and geometric discrepancy. Here we are going to discuss another, very powerful method developed by Beck, based on the Fourier transform. Although one can argue that, deep down, this method is actually related to eigenvalues and proofs using orthogonal or near-orthogonal functions, proofs via the Fourier transform certainly look different, being less geometric and more akin to classical harmonic analysis. For many results obtained by this method, such as the tight lower bound for the discrepancy for discs of a single fixed radius, no other proofs are known.

We are going to demonstrate the method on two examples (belonging to the technically simplest ones). In Section 7.1, we estimate the discrepancy for arbitrarily rotated squares, and in Section 7.2, we show an $\Omega((\log n)^{(d-1)/2})$ lower bound for axis-parallel cubes in $\mathbf{R}^d$, an analogue of Roth's bound for axis-parallel boxes. These two examples by far do not exhaust all the significant ideas in this area. There are numerous other and more general results, and they are by no means routine generalizations; in many of them, nice new tricks appear. But since these proofs are usually technically somewhat demanding and appear difficult to present convincingly in one or two lectures, we prefer to refer to the existing literature for more proofs via the Fourier transform method (Beck [Bec88b] can be particularly recommended as further reading).

Instead, we present another application of harmonic analysis, in the Euclidean Ramsey theory (which is a field not so remote from discrepancy theory). In Section 7.3, we reproduce Bourgain's proof of a theorem of Katznelson and Weiss, saying that any set of positive upper density in the plane contains a pair of points of any prescribed sufficiently large distance.

7.1 Arbitrarily Rotated Squares

Let $\mathcal{Q}$ denote the family of all squares in the plane (with arbitrary orientations). Since we study the discrepancy of point sets in the unit square, and since the intersection of any halfplane with the unit square can be simulated by the intersection with a large enough square, the discrepancy for $\mathcal{Q}$ is at least $\Omega(n^{1/4})$ by Theorem 6.9. We re-derive this result here, and we also get extra information, namely about the discrepancy for "small" squares. For a real number R, let $\mathcal{Q}_R$ denote the family of all squares with side in the interval $[R, 2R]$.

7.1 Theorem. *For any $R \in [\frac{1}{\sqrt{n}}, \frac{1}{2}]$ we have $D(n, \mathcal{Q}_R) \geq cn^{1/4}\sqrt{R}$ for a constant $c > 0$.*

For the proof, we need some preparation.

Fourier Transform in the Plane. Let $L_1(\mathbf{R}^2)$ denote the set of all measurable real or complex functions such that the integral of $|f|$ over $\mathbf{R}^2$ is finite. For a function $f \in L_1(\mathbf{R}^2)$, the Fourier transform $\hat{f}$ is a function on the plane, generally complex-valued even if f is real, defined by

$$\hat{f}(\xi) = \frac{1}{2\pi}\int_{\mathbf{R}^2} f(x)\, e^{-\mathrm{i}\langle x,\xi\rangle}\, \mathrm{d}x,$$

where $\langle x, \xi\rangle = x_1\xi_1 + x_2\xi_2$ is the scalar product.[1] For completeness, we should also mention the inversion formula, although we do not need it in this section. In this case, it reads

$$f(x) = \frac{1}{2\pi}\int_{\mathbf{R}^2} \hat{f}(\xi)\, e^{\mathrm{i}\langle x,\xi\rangle}\, \mathrm{d}\xi \tag{7.1}$$

What we do need is the Parseval–Plancherel theorem. For a real or complex function on $\mathbf{R}^2$, let us write $\|f\|_2 = \left(\int_{\mathbf{R}^2} |f(x)|^2\, \mathrm{d}x\right)^{1/2}$, and let $L_2(\mathbf{R}^2)$be the set of all complex measurable functions on $\mathbf{R}^2$ with $\|f\|_2$ finite.

7.2 Theorem (Parseval–Plancherel theorem). *Under suitable assumptions, the Fourier transform preserves the norm $\|.\|_2$. Namely, if $f \in L_1(\mathbf{R}^2) \cap L_2(\mathbf{R}^2)$ then also $\hat{f} \in L_2(\mathbf{R}^2)$ and $\|f\|_2 = \|\hat{f}\|_2$.*

---

[1] To put the Fourier series expansion and the one-dimensional Fourier transform we have encountered earlier and the two-dimensional transform treated here into a wider perspective, it might be useful to mention a general setting. Let G be a locally compact Abelian topological group. Let G^* stand for the group of all *continuous characters* of G, i.e. continuous homomorphisms γ mapping G into the multiplicative group of complex numbers of absolute value 1. For instance, for G being the real numbers with addition, G^* consists of all maps $x \mapsto e^{-iyx}$, $y \in \mathbf{R}$, and hence G^* can be identified with G—this is the case of the one-dimensional Fourier transform. If G is the interval $[0, 1)$ with addition modulo 1, G^* is isomorphic to $(\mathbf{Z}, +)$, and the Fourier transform is the usual Fourier series of a periodic function.

The Fourier transform of a complex function f with domain G is a complex function with domain G^* defined by

$$\hat{f}(\gamma) = \int_G f(g)\gamma(g)\, \mathrm{d}\mu(g),$$

where μ is a translation-invariant (Haar) measure on G, usually normalized suitably. Of course, the defining integral need not exist for all f.

The proof is not quite straightforward. A proof of an analogous statement for the one-dimensional Fourier transform can be found in [Rud74], and a version (very similar) for $\mathbf{R}^d$ is in [Rud91]. Also see Exercise 2.

We will also make use of the *convolution theorem*. For $f, g \in L_1(\mathbf{R}^2)$, the *convolution* $f * g$ is defined by $(f * g)(y) = \frac{1}{2\pi}\int_{\mathbf{R}^2} f(x)g(y-x)\,\mathrm{d}x$, and the theorem claims that $\widehat{f * g} = \hat{f} \cdot \hat{g}$, i.e. the Fourier transform converts the convolution into a pointwise product. We will actually use this in a situation when one of the functions is not an "honest function" but rather a so-called distribution (like the Dirac delta function). In order to avoid introducing the somewhat sophisticated notion of distributions (which are not really necessary here), we will re-derive the particular convolution result we need.

Two Lemmas for Translates of a Set. Let $P \subset [0,1]^2$ be an n-point set which we consider fixed throughout the proof, and let A be a bounded Lebesgue-measurable set. For definiteness, suppose that A is contained in the unit disc $B(0,1)$ centered at the origin. Later in the proof, A will be a square centered at the origin, but here we prove some statements in general in order to emphasize where the "squareness" really enters into play.

Let $A + x$ stand for A translated by the vector x, and let $\Delta_A(x) = D(P, A+x) = n \cdot \mathrm{vol}_\Box(A+x) - |(A+x) \cap P|$ be the discrepancy of P for $A + x$. Note that $\Delta_A(x) = 0$ as soon as $A + x$ does not intersect the unit square, which is certainly the case for $x \notin B(0,3)$. We are going to investigate the L_2-norm $\|\Delta_A\|_2 = \left(\int_{\mathbf{R}^2} \Delta_A^2(x)\,\mathrm{d}x\right)^{1/2}$. This is not quite the kind of L_2-discrepancy we have been working with earlier since the measure used for the integration is not a probability measure. But since we can restrict the integration domain to $B(0,3)$, it follows that if we prove $\|\Delta_A\|_2 \geq M$ for some number M then there is a translated copy $A + x$ of A with discrepancy at least a constant multiple of M.

The following lemma by no means provides a large discrepancy lower bound, but nevertheless, it is an important ingredient in the proof. It elaborates on the observation that any set of area between δ and $1-\delta$ in the unit square has discrepancy at least δ (where $0 < \delta \leq \frac{1}{2}$).

7.3 Lemma (Trivial discrepancy lemma). *Let* $A \subseteq B(0, \frac{1}{4})$ *be a measurable set with* $\frac{1}{5n} \leq \mathrm{vol}(A) \leq \frac{4}{5n}$. *Then we have* $\|\Delta_A\|_2 \geq c_1$ *for a constant* $c_1 > 0$ *(independent of A).*

Proof. For all x in the disc $B((\frac{1}{2},\frac{1}{2}),\frac{1}{4})$, we have $A + x \subset [0,1]^2$. But if $A + x \subseteq [0,1]^2$ then $|\Delta_A(x)| \geq \frac{1}{5}$, since either $P \cap (A+x) = \emptyset$ and then $\Delta_A(x) = n \cdot \mathrm{vol}(A+x) \geq \frac{1}{5}$, or $P \cap (A+x) \neq \emptyset$ and then $\Delta_A(x) \leq n \cdot \mathrm{vol}(A+x) - 1 \leq -\frac{1}{5}$. Hence $\int_{\mathbf{R}^2} \Delta_A^2 \geq \frac{1}{25}\mathrm{vol}(B((\frac{1}{2},\frac{1}{2}),\frac{1}{4})) = c_1^2$. □

Next, we approach one of the key ideas of the proof of Theorem 7.1. To formulate it, it is convenient to think of the discrepancy $D(P,X)$ for a set X as a signed measure of X, as in Section 6.6. That is, for a (Lebesgue-measurable) set $X \subseteq \mathbf{R}^2$, and considering P fixed, we set $D(X) = D(P,X) =$

$n \cdot \mathrm{vol}_\Box(X) - |P \cap X|$. We also recall how a function $f\colon \mathbf{R}^2 \to \mathbf{R}$ is integrated according to D:

$$\int_{\mathbf{R}^2} f(x)\,\mathrm{d}D(x) = n \cdot \int_{[0,1]^2} f(x)\,\mathrm{d}x - \sum_{p \in P} f(p).$$

In particular, we can define the Fourier transform of D by setting

$$\hat{D}(\xi) = \frac{1}{2\pi} \int_{\mathbf{R}^2} e^{-\mathrm{i}\langle y,\xi\rangle}\,\mathrm{d}D(y).$$

Denoting the characteristic function of the set A by I_A, we have

7.4 Lemma (Point component/shape component separation). *Let $A \subseteq \mathbf{R}^2$ be a measurable and bounded set. Then we have*

$$2\pi\|\Delta_A\|_2 = \|\hat{D} \cdot \hat{I}_{-A}\|_2 = \left(\int_{\mathbf{R}^2} |\hat{D}(\xi)|^2 \cdot |\hat{I}_{-A}(\xi)|^2\,\mathrm{d}\xi\right)^{1/2}.$$

Before proving this lemma, which is easy, let us explain its name and meaning a little. The L_2-norm of the function Δ_A is expressed using an integral of a product of two real functions, $|\hat{D}|^2$ and $|\hat{I}_{-A}|^2$. The first of them is fully determined by the set P and has nothing to do with A, so we can call it the *point component* of the discrepancy. The other function, $|\hat{I}_{-A}|^2$, depends on A but not on the point distribution, and we suggest to call it the *shape component*. In a lower bound proof, we usually have little control over the point component, but the shape component can sometimes be expressed explicitly or estimated suitably. In a moment we will learn more about this, but first we prove the lemma.

Proof of Lemma 7.4. We have

$$\Delta_A(x) = D(A+x) = \int_{\mathbf{R}^2} I_{A+x}(y)\,\mathrm{d}D(y) = \int_{\mathbf{R}^2} I_{-A}(x-y)\,\mathrm{d}D(y),$$

and so

$$\begin{aligned}
\hat{\Delta}_A(\xi) &= \frac{1}{2\pi} \int_{\mathbf{R}^2} \Delta_A(x) e^{-\mathrm{i}\langle x,\xi\rangle}\,\mathrm{d}x \\
&= \frac{1}{2\pi} \int_{\mathbf{R}^2} \int_{\mathbf{R}^2} I_{-A}(x-y) e^{-\mathrm{i}\langle x,\xi\rangle}\,\mathrm{d}D(y)\,\mathrm{d}x \\
&= \frac{1}{2\pi} \int_{\mathbf{R}^2} \int_{\mathbf{R}^2} I_{-A}(x-y) e^{-\mathrm{i}\langle x,\xi\rangle}\,\mathrm{d}x\,\mathrm{d}D(y)
\end{aligned}$$

(exchanging the order of integration is allowed by Fubini's theorem). This is further rewritten to

$$\begin{aligned}
&\int_{\mathbf{R}^2} \left(\frac{1}{2\pi} \int_{\mathbf{R}^2} I_{-A}(x-y) e^{-\mathrm{i}\langle x-y,\xi\rangle}\, \mathrm{d}x \right) e^{-\mathrm{i}\langle y,\xi\rangle}\, \mathrm{d}D(y) \\
&= \int_{\mathbf{R}^2} \hat{I}_{-A}(\xi)\, e^{-\mathrm{i}\langle y,\xi\rangle}\, \mathrm{d}D(y) \\
&= 2\pi \cdot \hat{I}_{-A}(\xi) \cdot \hat{D}(\xi).
\end{aligned}$$

The lemma now follows from the Parseval–Plancherel theorem.[2] □

Back to Squares. Let us write $Q(r,\vartheta)$ for the square of side $2r$ centered at the origin such that the angle of its side with the x-axis is ϑ, where $0 \le \vartheta \le \frac{\pi}{4}$:

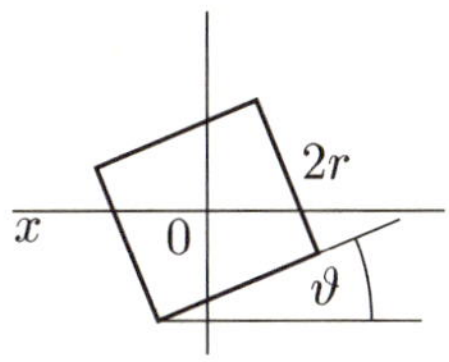

From now on, we are going to use $Q(r,\vartheta)$ in the role of the set A. The area of $Q(r,\vartheta)$ is $4r^2$, and so the Trivial discrepancy lemma 7.3 tells us that for $\frac{1}{2\sqrt{5n}} \le r \le \frac{1}{\sqrt{5n}}$, we have $\|\Delta_{Q(r,\vartheta)}\|_2 \ge c_1$. And from Lemma 7.4, we get

$$4\pi^2 \|\Delta_{Q(r,\vartheta)}\|_2^2 = \int_{\mathbf{R}^2} |\hat{D}(\xi)|^2 \cdot g_{r,\vartheta}(\xi)\, \mathrm{d}\xi \tag{7.2}$$

with $g_{r,\vartheta}(\xi) = \left|\hat{I}_{Q(r,\vartheta)}(\xi)\right|^2$.

Here is the basic strategy of the proof of Theorem 7.1. Suppose for a moment we could show that the function $g_{r,\vartheta}(\xi)$ increases significantly enough by increasing the argument r; say that we had, in some improbable TV world, $g_{ar,\vartheta}(\xi) \ge a \cdot g_{r,\vartheta}(\xi)$ for all factors $a \ge 1$ and all r, ϑ, and ξ. Then, by increasing the side of the considered square a times, the integrand on the r.h.s. of (7.2) would grow at least a times at each point, and so the squared norm $\|\Delta_{Q(r,\vartheta)}\|_2^2$ would increase at least a times as well (note that $\hat{D}$ stays the same). Starting with the side length $r = \frac{1}{\sqrt{5n}}$ and setting $a = R$, we would get that the L_2-discrepancy for squares of side R is $\Omega(n^{1/4}\sqrt{R})$.

Why is this scenario too optimistic? One answer is, because the conclusion is simply not true: so far we were dealing with the family of all translates of a single square (the side was fixed, and the argument did not use ϑ in any way), and the discrepancy for translates of a single square is at most $O(\log n)$.

2 This was the promised hidden application of the convolution theorem. A formal problem here is that there is no function φ, in the usual sense, representing our signed measure D, i.e. such that $D(X) = \int_X \varphi(x)\, \mathrm{d}x$, because a part of D is concentrated in the points of P. To use the convolution theorem explicitly, we would need to introduce distributions.

Another, perhaps more intuitive explanation in terms of the behavior of the function $g_{r,\vartheta}(\xi)$ will be given in remarks below.

To repair the failure of the above argument, we will not try to bound the ratio $g_{ar,\vartheta}(\xi)/g_{r,\vartheta}(\xi)$, but we will first take averages of both the numerator and denominator over ϑ and over a suitable range of r. We put

$$G_R(\xi) = \operatorname*{ave}_{r\in(R/2,R)} \operatorname*{ave}_{\vartheta} g_{r,\vartheta}(\xi) = \frac{2}{R}\int_{R/2}^{R} \frac{4}{\pi}\int_0^{\pi/4} g_{r,\vartheta}(\xi)\,\mathrm{d}\vartheta\,\mathrm{d}r.$$

(Depending on personal taste, one may also think of choosing $r \in (\frac{R}{2}, R)$ and $\vartheta \in (0, \frac{\pi}{4})$ uniformly at random, and write the expectation operator $\mathbf{E}[.]$ instead of the averages.) And this way, everything works!

7.5 Lemma (Amplification lemma). *For any $R > 0$ and $a \geq 1$, we have*

$$G_{aR}(\xi) \geq c_2 a \cdot G_R(\xi),$$

for all $\xi \in \mathbf{R}^2$ and an absolute constant $c_2 > 0$.

Before proving this lemma, let us finish the proof of Theorem 7.1. It suffices to modify the failed scenario above a little, by involving Fubini's theorem at a suitable moment. Let $R_0 = \frac{1}{\sqrt{5n}}$ and $R \in [R_0, \frac{1}{2}]$. We have

$$\begin{aligned}
\operatorname*{ave}_{r\in[\frac{R}{2},R)} \operatorname*{ave}_{\vartheta} \|\Delta_{Q(r,\vartheta)}\|_2^2 &= \operatorname*{ave}_{r\in(\frac{R}{2},R)} \operatorname*{ave}_{\vartheta} \int_{\mathbf{R}^2} |\hat{D}(\xi)|^2 g_{r,\vartheta}(\xi)\,\mathrm{d}\xi \\
&= \int_{\mathbf{R}^2} |\hat{D}(\xi)|^2\, G_R(\xi)\,\mathrm{d}\xi && \text{(Fubini)} \\
&\geq c_2 \frac{R}{R_0} \cdot \textstyle\int_{\mathbf{R}^2} |\hat{D}(\xi)|^2\, G_{R_0}(\xi)\,\mathrm{d}\xi && \text{(Lemma 7.5)} \\
&= c_2 R\sqrt{5n} \operatorname*{ave}_{r\in(\frac{R_0}{2},R_0)} \operatorname*{ave}_{\vartheta} \|\Delta_{Q(r,\vartheta)}\|_2^2 && \text{(Fubini)} \\
&= \Omega\left(R\sqrt{n}\right) && \text{(Lemma 7.3).}
\end{aligned}$$

Therefore, we have $D(P, \mathcal{Q}_R) = \Omega(n^{1/4}\sqrt{R})$. This proves Theorem 7.1. □

Proof of the Amplification Lemma 7.5. This is an exercise in calculus, but not a trivial one. First we determine the Fourier transform of $I_{Q(r,\vartheta)}$. By symmetry, it suffices to calculate it for $\vartheta = 0$ and then rotate. By definition, we obtain

$$\begin{aligned}
\hat{I}_{Q(r,0)}(\xi) &= \frac{1}{2\pi}\int_{-r}^{r}\int_{-r}^{r} e^{-\mathrm{i}(x_1\xi_1+x_2\xi_2)}\,\mathrm{d}x_1\,\mathrm{d}x_2 \\
&= \frac{1}{2\pi}\left(\int_{-r}^{r} e^{-\mathrm{i}x_1\xi_1}\,\mathrm{d}x_1\right)\left(\int_{-r}^{r} e^{-\mathrm{i}x_2\xi_2}\,\mathrm{d}x_2\right),
\end{aligned}$$

where $\xi = (\xi_1, \xi_2)$. With some effort, one can evaluate the integral, obtaining

$$\hat{I}_{Q(r,0)}(\xi) = \frac{1}{2\pi} \cdot \frac{4\sin(\xi_1 r)\sin(\xi_2 r)}{\xi_1 \xi_2}.$$

To express $\hat{I}_{Q(r,\vartheta)}(\xi)$, we rotate the system of coordinates by the angle $-\vartheta$ (note that the Fourier transform commutes with rotations of the plane around the origin). Namely, we have $\hat{I}_{Q(r,\vartheta)}(\xi) = \hat{I}_{Q(r,0)}(\xi')$, where $(\xi'_1, \xi'_2) = (\xi_1 \cos\vartheta + \xi_2 \sin\vartheta, -\xi_1 \sin\vartheta + \xi_2 \cos\vartheta)$. Since we average over all directions ϑ in the definition of $G_R(\xi)$, it suffices to prove the Amplification lemma with $\xi_1 > 0, \xi_2 = 0$. For such a $\xi = (\xi_1, 0)$ we have

$$G_R(\xi) = \frac{4}{\pi^2} \operatorname*{ave}_{r\in(R/2,R)} \operatorname*{ave}_{\vartheta} \frac{\sin^2(r\xi_1 \cos\vartheta)\sin^2(r\xi_1 \sin\vartheta)}{\xi_1^4 \cos^2\vartheta \sin^2\vartheta}.$$

For the considered range $0 \le \vartheta \le \frac{\pi}{4}$, we have $\sin\vartheta \approx \vartheta$ and $\cos\vartheta \approx 1$, where $f \approx g$ means that f is bounded by constant multiples of g from both above an below. So we can simplify a little:

$$G_R(\xi) \approx \operatorname*{ave}_{\vartheta} \operatorname*{ave}_{r\in(R/2,R)} \frac{\sin^2(r\xi_1 \cos\vartheta)\sin^2(r\xi_1 \sin\vartheta)}{\xi_1^4 \vartheta^2}. \tag{7.3}$$

We distinguish two cases. First, let $\xi_1 \le \frac{1}{R}$. Then we have, for any $r \in (\frac{R}{2}, R)$ and any $\vartheta \in [0, \frac{\pi}{4}]$, $\sin(r\xi_1 \cos\vartheta) \approx R\xi_1$ and $\sin(r\xi_1 \sin\vartheta) \approx R\xi_1\vartheta$. From (7.3), we calculate

$$G_R(\xi) \approx \operatorname*{ave}_{\vartheta} \frac{(R\xi_1)^2 (R\xi_1\vartheta)^2}{\xi_1^4 \vartheta^2} = R^4.$$

Next, let $\xi_1 > \frac{1}{R}$. We split the integration interval $(0, \frac{\pi}{4})$ for ϑ in (7.3) into two subintervals, $0 \le \vartheta \le \frac{1}{R\xi_1}$ and $\frac{1}{R\xi_1} \le \vartheta \le \frac{\pi}{4}$. For the second subinterval, we only calculate an upper bound on the integral, and for this we use the simple estimates $|\sin(r\xi_1 \sin\vartheta)| \le 1$, $|\sin(r\xi_1 \cos\vartheta)| \le 1$. For the first subinterval $0 \le \vartheta \le \frac{1}{R\xi_1}$, we have $\sin(r\xi_1 \sin\vartheta) \approx R\xi_1\vartheta$ as before, but moreover, we claim that

$$\operatorname*{ave}_{r\in(R/2,R)} \sin^2(r\xi_1 \cos\vartheta) \approx 1.$$

for any $\vartheta \in [0, \frac{\pi}{4}]$. The upper bound is clear since the sine function is ≤ 1. For a lower bound, we observe that as r runs from $\frac{R}{2}$ to R, the argument of the sine function, $r\xi_1 \cos\vartheta$, varies over an interval of length at least $\frac{1}{4}$ (say), therefore over at least a fixed fraction of the period of the sine function. (This is where the averaging over r becomes important in the proof.) Substituting these estimates into (7.3), we find

$$
\begin{aligned}
G_R(\xi) &\approx \int_0^{\pi/4} \operatorname*{ave}_{r\in(R/2,R)} \frac{\sin^2(r\xi_1\cos\vartheta)\sin^2(r\xi_1\sin\vartheta)}{\xi_1^4\vartheta^2}\,\mathrm{d}\vartheta \\
&\approx \int_0^{1/R\xi_1} \frac{R^2}{\xi_1^2}\cdot \operatorname*{ave}_{r\in(R/2,R)} \sin^2(r\xi_1\cos\vartheta)\,\mathrm{d}\vartheta + O\left(\int_{1/R\xi_1}^{\pi/4} \frac{1}{\xi_1^4\vartheta^2}\,\mathrm{d}\vartheta\right) \\
&\approx \left[\frac{\vartheta R^2}{\xi_1^2}\right]_{\vartheta=0}^{1/R\xi_1} + O\left(\left[-\frac{1}{\vartheta\xi_1^4}\right]_{\vartheta=1/R\xi_1}^{\pi/4}\right) \approx \frac{R}{\xi_1^3}.
\end{aligned}
$$

Recall that this is valid for $\xi = (\xi_1, 0)$ and $\xi_1 > \frac{1}{R}$. Combining this with the estimate $G_R(\xi) \approx R^4$ for $\xi_1 \le \frac{1}{R}$ derived earlier, we obtain

$$G_R(\xi) \approx \min\left(R^4, \frac{R}{\xi_1^3}\right)$$

for all R. From this, the desired estimate $G_{aR}(\xi) = \Omega(a\cdot G_R(\xi))$ for any $a \ge 1$ follows. $\square$

The proof method shown for squares is quite general, but for more complicated shapes, technical problems arise in the proof of the Amplification lemma. It is seldom possible to calculate the Fourier transform explicitly. Even for such simple shapes as discs, one already has to deal with Bessel functions.

Illustrations. Let us try to illustrate pictorially the reasons for the averaging over ϑ and r in the above proof. The squared Fourier transform $g_{r,\vartheta}(\xi) = |\hat{I}_{Q(r,\vartheta)}(\xi)|^2$ for $r = 1$ and $\vartheta = 0$ is shown below:

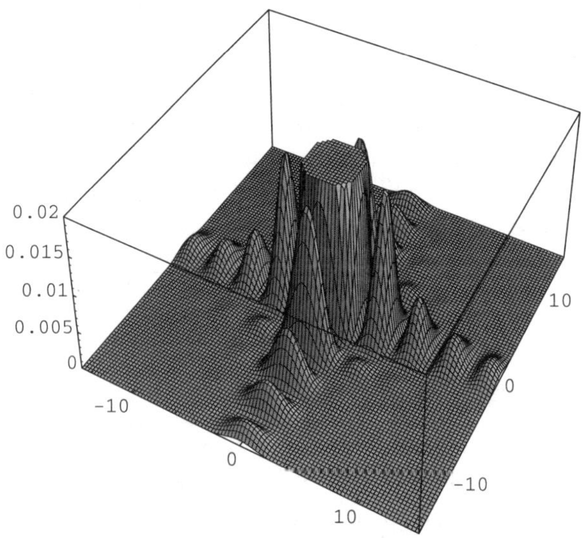

The central peak has height ≈ 0.4 and is trimmed. Choosing another ϑ simply means rotating the graph around the origin. The effect of making r 10 times smaller, say, is stretching the graph in the horizontal plane 10 times (so that the whole coordinate region shown in the picture, the $[-15, 15]^2$ square, would be in the area of the middle peak of the function $g_{0.1,0}$). The ratio of $g_{1,0}/g_{0.1,0}$ thus oscillates quite wildly and cannot be expected to be bounded away from 0 uniformly, as the following picture illutrates (the vertical range is trimmed to $[0, 10]$).

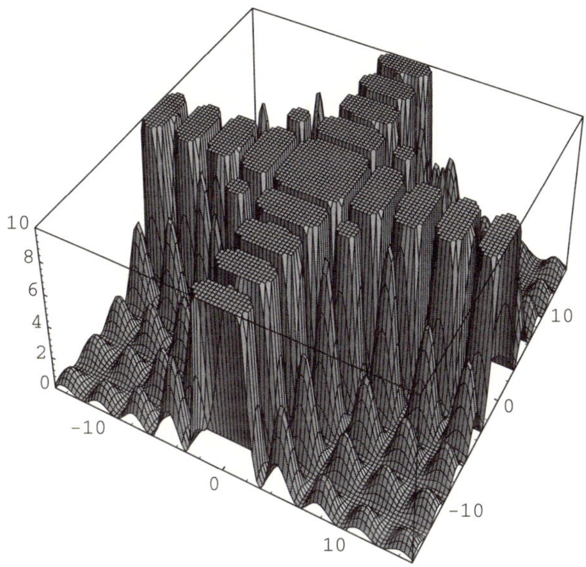

As was mentioned in the proof above, this has to be the case, because the average squared discrepancy for translates of a single square can be made quite small.

Also, if we attempted to work only with an average over r, trying to avoid the averaging over ϑ, then we cannot succeed in getting discrepancy of the order $n^{1/4}$, again because the discrepancy for axis-parallel squares is of much smaller order of magnitude, as we know. Finally, if we average $g_{r,\vartheta}(\xi)^2$ over ϑ only (and not over r), we get a function rotationally symmetric around the origin. It can be shown that the "amplification" effect still fails at some points, i.e. it is not true that $\operatorname*{ave}_{\vartheta} g_{r,\vartheta}(\xi)$ grows linearly with r for all ξ (Exercise 5). This time, the failure is an artifact of the proof method, since the discrepancy for arbitrarily rotated squares with side 1 (or any given positive constant) must be $\Omega(n^{1/4})$. This is because the intersection of any halfplane with the square $[0, 1]^2$ can be subdivided into $O(1)$ intersections of squares of side 1 with $[0, 1]^2$. The following diagram shows the graph of the ratio $\operatorname*{ave}_{\vartheta} g_{1,\vartheta}(\xi) / \operatorname*{ave}_{\vartheta} g_{0.1,\vartheta}(\xi)$ along the ξ_1-axis:

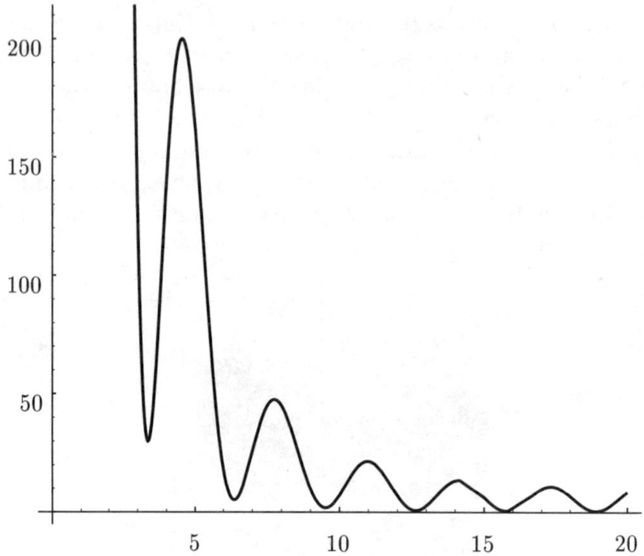

In this connection, it is also interesting to look at the "point component" of the Fourier transform, that is, the function $|\hat{D}(\xi)|^2$. The subsequent figures show this function for various 16-point sets in the unit square (note that the origin is in the front corner of the plots). For the 4×4 grid set, high peaks along the axes correspond to its bad discrepancy for certain axis-parallel squares (the peaks are trimmed; their height is about 6 in this picture):

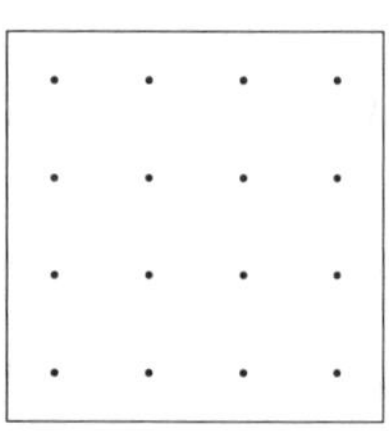

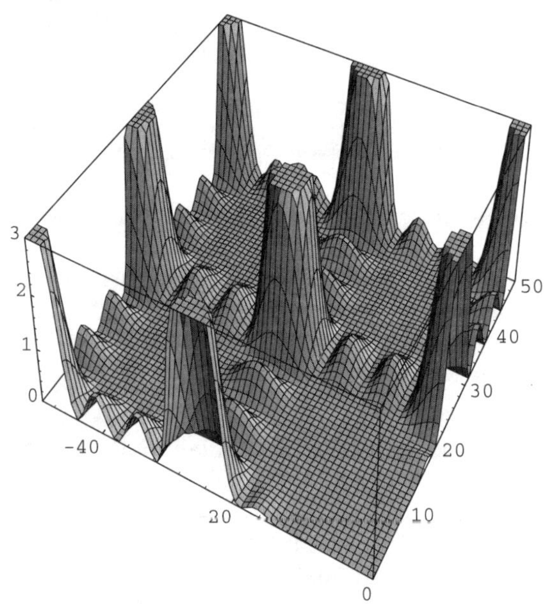

Another lattice set from Example 2.19, not surprisingly, has similar high peaks but off the axes:

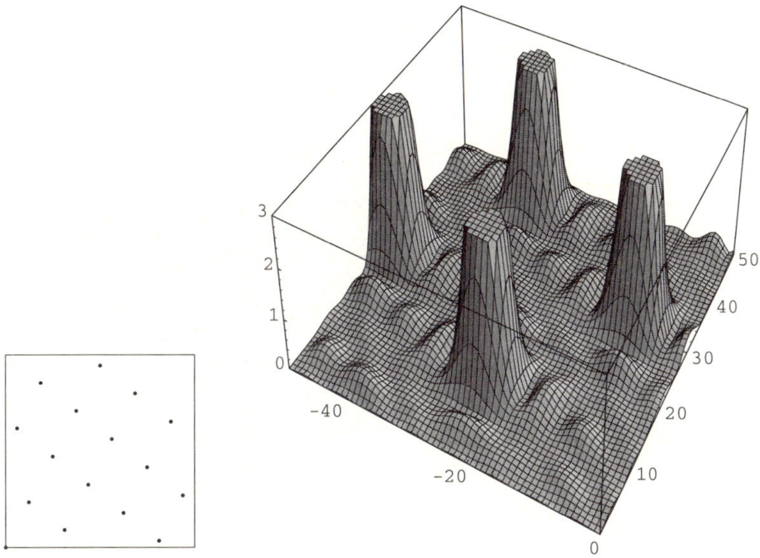

The Van der Corput set shows a more irregular structure but it also has lower regions along the axes:

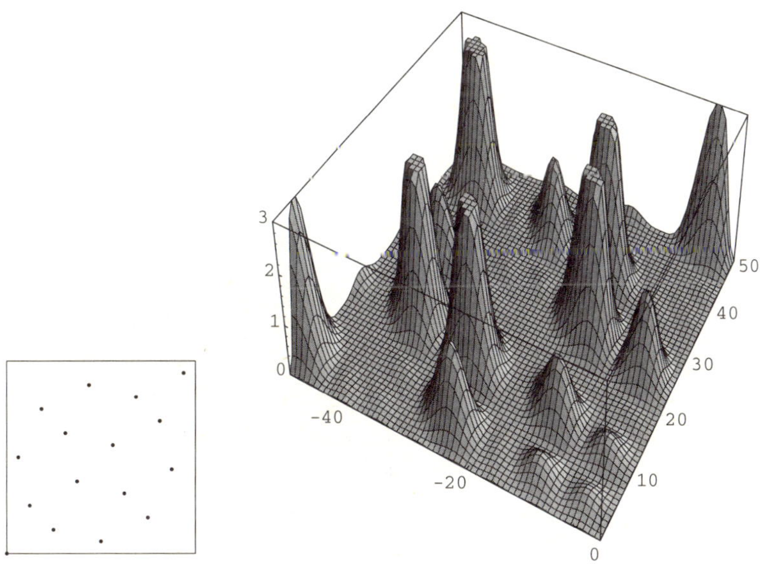

A randomly scrambled Van der Corput set (as in Section 2.4) looks still less regular and a little more leveled:

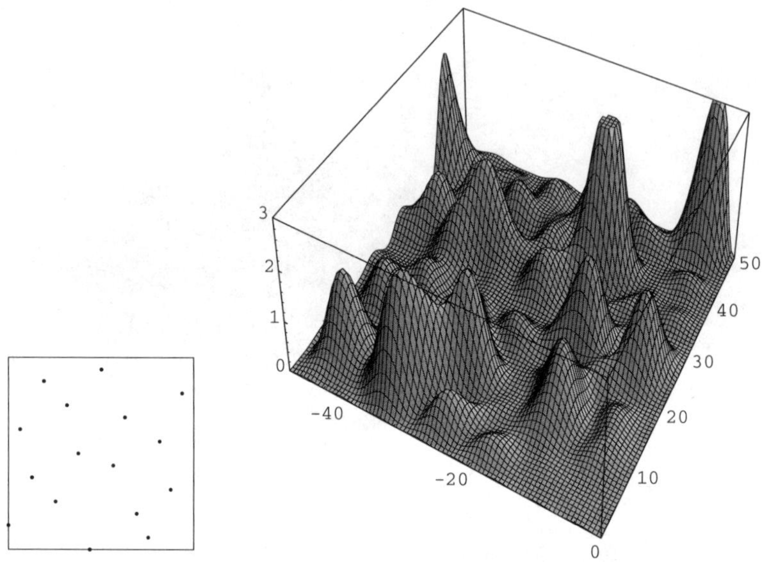

Finally a set constructed as in the proof of the $O(n^{1/4})$ bound for discs (Theorem 3.1), by choosing a random point in each square of a 4×4 grid, shows a generally more uniform behavior (since it is tailored for shapes with rotation allowed):

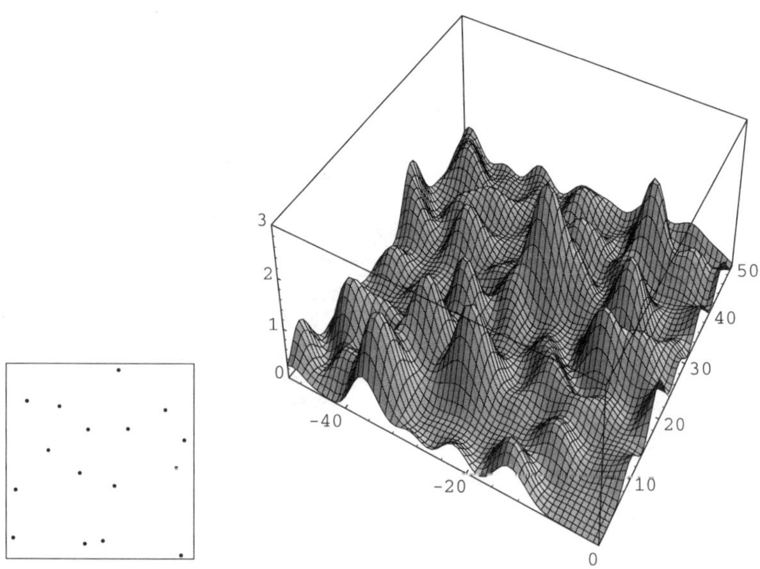

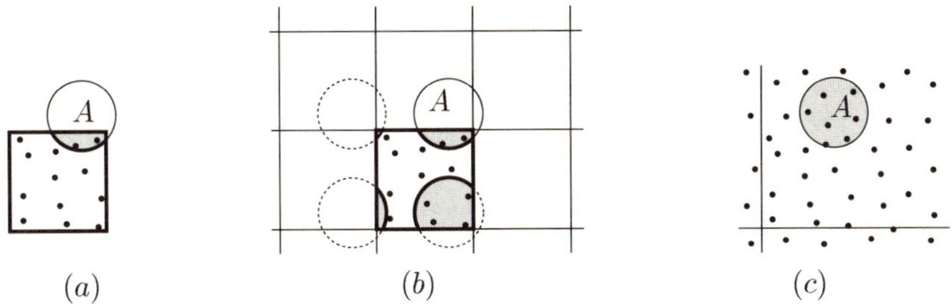

Fig. 7.1. Three kinds of discrepancy: for intersections with the unit square (a), toroidal (b), whole plane (c).

(I hope that all this may be a sufficient excuse for including several nice pictures.) Of course, for 16-point sets, the differences in discrepancies of various sets are not so drastic, but for larger sets the pictures of the Fourier transform have too erratic structure to be presented reasonably.

Bibliography and Remarks. The material of this section is based on Beck's results as presented in [BC87].

The Fourier transform method was introduced to discrepancy theory by Roth [Rot64], for proving a lower bound for the discrepancy of arithmetics progressions. Beck developed it to considerable extent for proving lower bounds in Lebesgue-measure discrepancy ([Bec87], [Bec88a]; most of the material is also in [BC87]). Independently, Montgomery [Mon89] obtained many similar results in the planar case (some are nicely presented in [Mon94]).

Let us give a brief overview of some known lower bounds for the Lebesgue-measure discrepancy (excluding the case of axis-parallel boxes). This is complicated, among others, by the fact that the lower bounds were proved in several similar but technically different settings. For introducing these settings, let us speak about discs in the plane for definiteness, with obvious possibilities of generalization to other shapes and higher dimensions.

Four Settings for Discrepancy Lower Bounds. The first setting is the one we have used throughout the text. Speaking about the discrepancy for discs, we actually mean the discrepancy for intersections of discs with the unit square, as in Fig. 7.1(a). Next, one can insist that the discs be completely contained in the unit square (so we restrict the considered family); lower bounds in this setting are stronger and generally more difficult. Let us refer to this setting by phrases like "the discrepancy for completely contained discs."

Another setting, intermediate between these two, is the so-called *toroidal discrepancy*. Here the point set and the discs reside in the *unit torus* $\mathbf{R}^d/\mathbf{Z}^d$. This can be visualized as in Fig. 7.1(b): we can imagine that the point set P still lies in the unit square but instead of intersecting the discs with the unit square, we take them "modulo 1." That is, instead of the set $A \cap [0,1]^2$ we consider the set $\{A\} = \{(\{x\},\{y\})\colon (x,y) \in A\}$.[3] Lower bounds for the toroidal discrepancy imply lower bounds in the first setting (for sets of bounded diameter, that is) but not for the completely contained case. The toroidal discrepancy is technically advantageous since the group of translations in the unit torus is compact, and as a consequence, one gets a discrete-valued Fourier transform .

Finally, Beck considered, in several of his works, what one might call a "whole-space setting" (Fig. 7.1(c)). Here the point set is infinite and spread out to the whole plane, and the discs have a constant-bounded diameter (in Beck's papers, the situation is also re-scaled by a factor of $n^{1/d}$, so that there is about one point per unit volume of space; here, for simplicity, we do not re-scale). Note that here n only enters the situation as the factor multiplying the volume in the definition of discrepancy. The toroidal discrepancy can be regarded as a special case of the whole-plane setting, where the plane is periodically tiled with copies of the unit square with the considered n-point set in it.

Discs and Balls. The crucial difference between the discrepancy for axis-parallel boxes (logarithmic behavior) and the discrepancy with rotation allowed (grows like a fractional power of n) was first shown by Schmidt, answering a question of Erdős [Erd64]. In [Sch69a], Schmidt proved that the toroidal discrepancy for arbitrarily rotated boxes in $\mathbf{R}^3$ is at least $\Omega(n^{1/6})$ (recall that the near-tight bound is $\Omega(n^{1/3})$). In [Sch69c], he established the bound $\Omega(n^{1/2-1/2d-\varepsilon})$, with an arbitrarily small fixed $\varepsilon > 0$, for the toroidal discrepancy for balls in $\mathbf{R}^d$. This is already tight up to the ε in the exponent. He also obtained a nontrivial bound, although with a much smaller exponent, for the discrepancy for balls completely contained in the unit cube. Here a near-tight bound of $\Omega(n^{1/2-1/2d-\varepsilon})$ is due to Beck [Bec87],[BC87].

Halfspaces and Spherical Caps. We recall that the tight $\Omega(n^{1/2-1/2d})$ lower bound for halfspaces was established by Alexander [Ale90], [Ale91]. A predecessor of this result was the bound of $\Omega(n^{1/4}\log^{-7/2} n)$ for *Roth's disc segment problem* proved by Beck. Here n points are placed in the disc of unit area (instead of the unit square), and the discrepancy for halfplanes is considered, or rather the discrepancy for the intersections of the disc with halfplanes—the "disc segments" (more details can be found in [BC87]). Note that this result implies the slightly suboptimal $\Omega(n^{1/4}\log^{-7/2} n)$ lower bound for the combinatorial discrepancy for halfplanes.

[3] Recall that $\{x\}$ denotes the fractional part of x. Toroidal discrepancy was already mentioned in the remarks to Section 1.2.

Schmidt [Sch69b], [Sch69c] considered the discrepancy for spherical caps on the sphere S^d, as well as the discrepancy for *spherical slices*, which are the intersections of two hemispheres in S^d. He proved the $\Omega(n^{1/2-1/2d-\varepsilon})$ bound in both cases. For caps, this was subsequently improved by Beck [Bec84] to the near-tight $\Omega(n^{1/2-1/2d})$. For spherical slices, the same lower bound was shown much later by Bümlinger [Blü91]. He defined a suitable measure on the slices and established the following surprising fact: for any point set $P \subset S^d$, the L_2-discrepancy of P for slices is at least a constant multiple of the L_2-discrepancy of P for caps (it would be interesting to find a simple proof). The lower bound for slices then follows from Beck's result for caps.

Copies of a Fixed Convex Set. Beck [Bec87],[BC87] showed that the family of all translated, rotated, and scaled-down copies of an arbitrary convex body C in $\mathbf{R}^d$ has discrepancy $\Omega(n^{1/2-1/2d}\sqrt{S})$ in the whole-space model, where S is the surface area of C. Here one has to assume that C is not too small or too flat, namely that it contains a ball of radius $n^{-1/d}$. Note that this strongly generalizes the results of this section. In the plane, similar results for the toroidal discrepancy were independently obtained by Montgomery [Mon89], [Mon94].

Discs of a Fixed Radius. In the paper [Bec88b], Beck studied the discrepancy for discs of a single (fixed) radius. He proved an $\Omega(n^{1/4})$ lower bound in the weakest setting (intersections with the unit square). For toroidal discrepancy, Montgomery ([Mon89], [Mon94]) earlier proved the existence of a disc of radius either $\frac{1}{4}$ or $\frac{1}{2}$ with $\Omega(n^{1/4})$ discrepancy. These mysterious two radii appear in the result since a certain linear combination of two Bessel functions turns out to be nonnegative, while the sign of a single Bessel function varies and hence the method doesn't work for a single radius.

Translated and Scaled Copies of a Fixed Convex Set. Let A be a compact convex set in $\mathbf{R}^d$, and let $\mathcal{T}_A$ be the family of all translated and scaled-down copies of A (no rotation allowed!). Beck [Bec88a] (also [BC87]) investigated the discrepancy for such families $\mathcal{T}_A$ in the plane, again in the whole-plane model. Here the discrepancy behavior strongly depends on the smoothness of the boundary of A.

As one extreme, one has a logarithmic upper bound for the square. Also for convex polygons with a fixed number of sides, the discrepancy is $O(\log n)$ (see the remarks to Section 4.5). From below, Beck shows an $\Omega\big(\sqrt{\log(n \operatorname{vol}(A))}\big)$ lower bound for any A with $\operatorname{vol}(A) \geq \frac{2}{n}$ (note that this includes the case of axis-parallel squares).

On the other hand, if the boundary curve of A is twice continuously differentiable and the ratio of its maximum and minimum curvatures is bounded by a constant, then the discrepancy is at least $\Omega\big(n^{1/4}\sqrt{\operatorname{vol}(A)/\log(n \operatorname{vol}(A))}\big)$, i.e. essentially the same as that for circular discs. In general, Beck's upper and lower bounds can be stated in terms of certain "approximability numbers" for A. Namely, let $\xi_n(A)$ denote the smallest integer $\ell \geq 3$ for which there exists a convex ℓ-gon A_ℓ inscribed into A such that the area of $A \setminus A_\ell$ is at most ℓ^2/n. Then we have $D(n, \mathcal{T}_A) = \Omega(\sqrt{\xi_n(A)} \cdot \log^{-1/4} n)$ from below and $D(n, \mathcal{T}_A) = O(\xi_n(A) \log^{4.5} n)$ from above [Bec88a]. The upper bound is obtained by the partial coloring method and a suitable approximation argument.

Károlyi [Kár95b] extended these investigations to translated and scaled-down copies of a d-dimensional convex body, establishing upper bounds in terms of suitable approximability by convex polytopes. The key new part of the proofs are certain geometric decomposition and approximation results for convex polytopes which become considerably more complicated than in the planar case. Drmota [Drm93] considered translated and scaled copies of a fixed smooth convex body. He removed the logarithmic factor in Beck's planar lower bound and extend it to higher dimensions: for a fixed sufficiently smooth convex body $A \subset \mathbf{R}^d$, he proved $D(n, \mathcal{T}_A) \geq c(A) n^{1/2-1/2d}$ for a suitable constant $c(A) > 0$.

L_1-Discrepancy and One-Sided Discrepancy. Most of these lower bounds are established for the L_2-discrepancy. As for the L_1-discrepancy, Beck [Bec89b] proves an $\Omega(\log^{1/2-\varepsilon} n)$ lower bound for discs in the plane (toroidal discrepancy) by modifying the method of Halász [Hal81]. The paper [Bec89b] actually aimed at showing that there is always a disc with large "positive discrepancy" (excess of area compared to the number of points) and a disc with large "negative discrepancy." Perhaps the correct order of magnitude for this question of Schmidt cannot be gained via the L_1-discrepancy.

We should also mention the paper Bourgain et al. [BLM89] discussed in Section 1.4 as another nice example of discrepancy-type lower bounds via harmonic analysis.

Exercises

1. Check that the application of Fubini's theorem in the proof of Theorem 7.1 is legitimate, i.e. that for each $\xi \in \mathbf{R}^2$, the function $F(x, y) = \left|I_{-A}(x-y)e^{-\mathrm{i}\langle x,\xi\rangle}\right|$ has a finite integral over $\mathbf{R}^2 \times \mathbf{R}^2$.
2. Let $f \in L_1(\mathbf{R}^2) \cap L_2(\mathbf{R}^2)$ be a real function for which the inversion formula for Fourier transform holds, i.e. such that

$$f(x) = \frac{1}{2\pi} \int_{\mathbf{R}^2} \hat{f}(\xi)\, e^{\mathrm{i}\langle x,\xi\rangle}\, \mathrm{d}\xi$$

for all $x \in \mathbf{R}^2$. Prove the Parseval–Plancherel equality $\|f\|_2 = \|\hat{f}\|_2$ under these assumptions. (Let us remark that he function for which we have applied Parseval–Plancherel in the proof is not in $L_2(\mathbf{R}^2)$! One has to use a suitable limit argument or something similar.)

3. By modifying the proof shown in this section, prove that for any n-point set $P \subset [0,1]^2$ there exists a rectangle R with one side of length $\frac{1}{\sqrt{n}}$ such that $|D(P,R)| = \Omega(n^{1/4})$. (See [Cha98b].)
4. Show that the lower bound in Theorem 7.1 is tight up to a factor of $\sqrt{\log n}$, i.e. prove that $D(n, \mathcal{Q}_R) = O(n^{1/4}\sqrt{R \log n})$ (where $\frac{1}{\sqrt{n}} \le R \le 1$).
5. The goal of this exercise is to show that the averaging over r used in the proof in this section is indeed essential for this proof method. With notation as in the proof, define a function
$$h_r(\xi) = \operatorname*{ave}_{\vartheta} g_{r,\vartheta}(\xi).$$
Show that for any positive real numbers $r_1 < r_2$ and any positive constant $C > 0$ there exists a $\xi \in \mathbf{R}^2$ such that the"amplification" fails at ξ by a factor of at least C, that is, $h_{r_2}(\xi)/h_{r_1}(\xi) \le \frac{1}{C}\frac{r_2}{r_1}$.
6.* (Diaphony) Recall the notion of diaphony of a point set and the class $\tilde{\mathcal{R}}_d$ of axis-parallel boxes modulo 1 introduced in the remarks to Section 1.2. Prove that the diaphony of any finite set $P \subset [0,1]^d$ is between two constant multiples of $D_2(P, \tilde{\mathcal{R}}_d)$, where the constants depend on the dimension. Use the Fourier transform on the compact group $[0,1)^d$ with componentwise addition modulo 1, i.e. the d-dimensional Fourier series of a d-variate periodic function.
This result and those in the next exercise are from [Lev95].
7. (More on diaphony)
(a)* From the remarks to Section 1.2, recall the notion of diaphony of a point set. Prove that the diaphony of P equals
$$\left(-n^2 + \sum_{p,q \in P} \prod_{k=1}^{d} \left(1 + 2\pi^2 B_2(\{p-q\})\right)\right)^{1/2},$$
where $B_2(x) = x^2 - x + \frac{1}{6}$ is the Bernoulli polynomial of degree 2. (This is analogous to Warnock's formula 2.14 for the L_2-discrepancy for corners.)
(b) Show that diaphony is translation-invariant; that is, the diaphony of P equals that of of $\{P + x\}$.
8.* (Toroidal L_2-discrepancy for boxes) Derive the following analogue of Warnock's formula 2.14 for the L_2-discrepancy for the class $\tilde{\mathcal{R}}_d$ of axis-parallel boxes "modulo 1" discussed in the remarks to Section 1.2:
$$D_2(P, \tilde{\mathcal{R}}_d)^2 = -\frac{n^2}{3^d} + \sum_{p,q \in P} \prod_{k=1}^{d} \left(\frac{1}{3} + B_2(\{p_k - q_k\})\right),$$
where $B_2(x) = x^2 - x + \frac{1}{6}$ is the Bernoulli polynomial of degree 2.
This formula was calculated by Lev (private communication).

7.2 Axis-Parallel Cubes

Here we prove the following result, thereby also providing an alternative proof of Roth's lower bound for axis-parallel boxes (Theorem 6.1):

7.6 Theorem. *For any n-point set P in the unit cube $[0,1]^d$, there exists an axis-parallel cube Q contained in $[0,1]^d$ with $|D(P,Q)| = \Omega(\log^{(d-1)/2} n)$.*

Let us remark that for $d = 2$, Theorem 6.3 gives a better (and tight) bound by a completely elementary argument, by a reduction from Schmidt's lower bound for rectangles. On the other hand, no similar reduction relating the discrepancy for cubes to that for axis-parallel boxes is known for higher dimensions. Theorem 7.6 can also be turned, with little more work, into an L_2-discrepancy lower bound.

As usual, we present the proof of Theorem 7.6 in the planar case, since the higher-dimensional generalization brings nothing really new.

A large part of the setup for the proof is taken over from the preceding section. We consider an arbitrary but fixed n-point set P in the unit square, and we let D be the corresponding signed measure, given by $D(A) = D(P,A) = n \cdot \mathrm{vol}_{\square}(A) - |P \cap A|$. The function $\hat{D}(\xi) = \int_{\mathbf{R}^2} e^{-\mathrm{i}\langle x,\xi\rangle}\, \mathrm{d}D(x)$ is the Fourier transform of D. For a set A, Δ_A denotes the discrepancy function for translates of A, i.e. $\Delta_A(x) = D(A + x)$. Let us write $Q(a)$ for the square $[-a,a]^2$.

In order to handle squares crossing the boundary of $[0,1]^2$, we will consider the discrepancy for squares that are not too big, namely with side at most $s = n^{2/5}$. The heart of the proof is

7.7 Lemma. *We have*

$$\operatorname*{ave}_{r\in(0,s)} \left\|\Delta_{Q(r)}\right\|_2^2 = \Omega(\log n).$$

Proof of Theorem 7.6 for $d = 2$ from Lemma 7.7. This is simple but there is a small subtlety in handling the squares on the boundary (recall that the theorem claims the existence of a square with large discrepancy lying *within* $[0,1]^2$). We distinguish two cases: either all axis-parallel squares of side at most s have discrepancy below $5ns^2$, or there is an axis-parallel square Q_0, not necessarily fully contained in $[0,1]^2$, with discrepancy at least $5ns^2$. In the former case, for $r \le s$, the set

$$\{x \in \mathbf{R}^2 \colon Q(r) + x \text{ intersects the boundary of } [0,1]^2\}$$

has measure $O(r)$ and so its contribution to $\|\Delta_{Q(r)}\|_2^2$ is at most $O(r) \cdot (5ns^2)^2 = O(n^2 s^5) = O(1)$. Hence there must be squares with $\Omega(\sqrt{\log n})$ discrepancy fully contained in $[0,1]^2$. In the latter case, illustrated in the following picture,

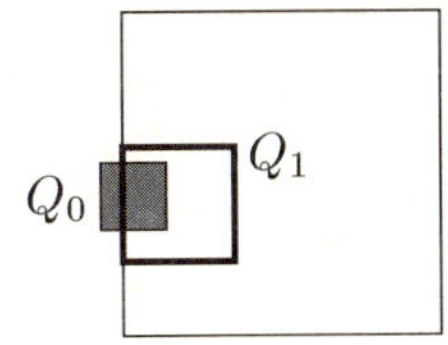

consider a square $Q_1 \subset [0,1]^2$ with side s and containing $Q_0 \cap [0,1]^2$. Since the discrepancy of Q_0 is larger than $n \cdot \mathrm{vol}_\square(Q_0) \le 4ns^2$, it must be caused by excess of points of P in Q_0, and in particular, $|Q_0 \cap P| \ge 5ns^2$. It follows that Q_1 has discrepancy at least $ns^2 = n^{1/5} = \Omega(\sqrt{\log n})$. □

Remark. Here we could get rid of the squares intersecting the boundary easily, since the discrepancy for small squares (of side about $s = n^{2/5}$) is still large enough. On the other hand, for squares with rotation allowed, say, we need squares with side $\Omega(1)$ to get the tight lower bound for discrepancy, and these large squares are harder to prevent from intersecting the boundary. Similar effects make it difficult to achieve tight lower bounds when the discrepancy grows as a power of n and we insist that the shape with large discrepancy be fully contained in the unit square.

Proof of Lemma 7.7. By the point component/shape component separation lemma 7.4 and by Fubini, we have

$$\operatorname*{ave}_{r\in(0,s)} \|\Delta_{Q(r)}\|_2^2 = \int_{\mathbf{R}^2} \left(\operatorname*{ave}_{r\in(0,s)} |\hat{I}_{Q(r)}(\xi)|^2 \right) \cdot |\hat{D}(\xi)|^2 \,\mathrm{d}\xi.$$

Write $G(\xi) = \operatorname*{ave}_{r\in(0,s)} |\hat{I}_{Q(r)}(\xi)|^2$. The general strategy is to exhibit a "magic" function $H(\xi)$ for which we know, on the one hand, that $\int_{\mathbf{R}^2} H(\xi) \cdot |\hat{D}(\xi)|^2 \,\mathrm{d}\xi$ is large, and for which we prove, on the other hand, that $G(\xi) = \Omega(H(\xi))$ uniformly for all ξ.

We define $H(\xi) = \sum_{j\in J} |\hat{h}_j(\xi)|^2$, where J is an index set with about $\log n$ elements and the h_j are certain suitable functions. More precisely, we let m be the smallest integer with $2^m \ge 40n$, and we let J denote the set of all ordered pairs $j = (j_1, j_2)$ of integers with $j_1 + j_2 = m$, $2^{-j_1} \ge s$ and $2^{-j_2} \ge s$. (For our planar case, we could obviously index the pairs in J by the j_1-component only, but for higher dimensions, we do need indexing by vectors.) The function h_j is given by

$$h_j(x) = \exp\left(-\tfrac{1}{2}(2^{2j_1}x_1^2 + 2^{2j_2}x_2^2)\right).$$

The level sets of h_j are ellipses; for example, the set $\{x \in \mathbf{R}^2\colon h_j(x) \ge e^{-1/2}\}$ is the ellipse with semiaxes 2^{-j_1} and 2^{-j_2}.

To see what is going on, let us imagine for a moment, just for the sake of illustration, that we used the characteristic function I_{R_j} of the rectangle $R_j = [-2^{-j_1}, 2^{-j_1}] \times [-2^{-j_2}, 2^{-j_2}]$ instead of h_j (on a very crude level, h_j and I_{R_j} are somewhat similar as density functions in the plane, as Fig. 7.2

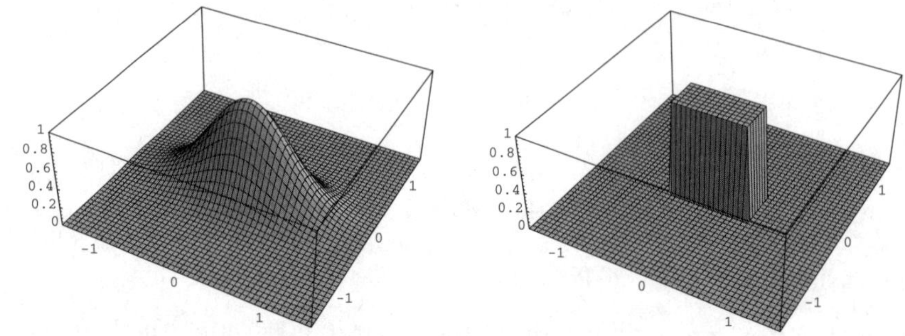

Fig. 7.2. The functions h_j and I_{R_j} for $j_1 = 1$ and $j_2 = 2$.

indicates[4]). The rectangle R_j has area $4 \cdot 2^{-m} \in [\frac{1}{10n}, \frac{1}{20n}]$. By a consideration similar to the proof of the Trivial discrepancy lemma 7.3, we see that a large fraction of the translated copies of R_j have discrepancy at least a small positive constant, and so $\|\Delta_{R_j}\|_2 = \Omega(1)$. By the point component/shape component separation, we also have

$$\|\Delta_{R_j}\|_2^2 = \int_{\mathbf{R}^2} |\hat{I}_{R_j}(\xi)|^2 \cdot |\hat{D}(\xi)|^2 \, \mathrm{d}\xi,$$

and so by putting $\tilde{H}(\xi) = \sum_{j \in J} |\hat{I}_{R_j}(\xi)|^2$, we obtain a function $\tilde{H}$ with

$$\int_{\mathbf{R}^2} \tilde{H}(\xi) \cdot |\hat{D}(\xi)|^2 \, \mathrm{d}\xi = \Omega(|J|) = \Omega(\log n).$$

This is all fine, but unfortunately, it is not true that $\tilde{H}(\xi) = O(G(\xi))$ for all ξ. On the other hand, the situation is not so bad, since the estimate of $\tilde{H}(\xi)$ by a multiple of $G(\xi)$ only fails for $\|\xi\|$ large. Here an old wisdom in harmonic analysis helps: the smoother a function f is the faster the Fourier transform $\hat{f}(\xi)$ converges to 0 as $\|\xi\| \to \infty$. This is the reason for replacing the highly non-smooth characteristic function I_{R_j} by the very smooth function h_j and defining $H(\xi) = \sum_{j \in J} |\hat{h}_j(\xi)|^2$.

In order to mimic the "trivial discrepancy" lower bound with the functions h_j, we need to generalize our notation a little. Namely, for a function $f\colon \mathbf{R}^2 \to \mathbf{R}$, we set

$$\Delta_f(x) = \int_{\mathbf{R}^2} f(y - x) \, \mathrm{d}D(y)$$

[4] And, the system of the rectangles R_j somewhat resembles the rectangles appearing in the proof of Roth's lower bound for corners (Theorem 6.1).

(note that for the characteristic function I_A of a set A, Δ_{I_A} in the new notation is the same as Δ_A in the previous notation). We can state two lemmas which together imply Lemma 7.7 in the manner indicated above.

7.8 Lemma (Trivial discrepancy for the h_j). *For all $j \in J$, we have* $\|\Delta_{h_j}\|_2 = \Omega(1)$.

7.9 Lemma (Pointwise bound). *For an absolute constant C, we have $H(\xi) \le C \cdot G(\xi)$ for all $\xi \in \mathbf{R}^2$.*

Before proving these lemmas, we recall two facts from calculus concerning the function e^{-x^2} (applied in our specific setting). These have already been addressed in Section 6.7 and so we omit further discussion here.

7.10 Fact.

(i) $$\hat{h}_j(\xi) = \frac{1}{2^{j_1}2^{j_2}} \exp\left(-\frac{1}{2}\left(\frac{\xi_1^2}{2^{2j_1}} + \frac{\xi_2^2}{2^{2j_2}}\right)\right).$$

(ii) $$\int_{\mathbf{R}^2} h_j(x)\,\mathrm{d}x = \frac{2\pi}{2^{j_1}2^{j_2}}.$$

Proof of Lemma 7.8. Let E_j denote the level set $\{h_j \ge e^{-1/2}\}$, i.e. the already mentioned ellipse with semiaxes 2^{-j_1} and 2^{-j_2}. We calculate

$$\begin{aligned} -\Delta_{h_j}(x) &= -\int_{\mathbf{R}^2} h_j(y-x)\,\mathrm{d}D(y) \\ &= \sum_{p\in P} h_j(p-x) - n\int_{[0,1]^2} h_j(y-x)\,\mathrm{d}y \\ &\ge |(P-x)\cap E_j| \cdot e^{-1/2} - n\int_{\mathbf{R}^2} h_j(z)\,\mathrm{d}z \\ &\ge \tfrac{1}{2}|(P-x)\cap E_j| - \tfrac{1}{4} \end{aligned}$$

(we have used Fact 7.10(ii) and $\frac{2\pi}{2^{j_1}2^{j_2}} = \frac{2\pi}{2^m} \le \frac{1}{4n}$). Hence $|\Delta_{h_j}(x)| \ge \frac{1}{4}|(P - x)\cap E_j|$ for all $x \in \mathbf{R}^2$, and integration gives

$$\begin{aligned} \|\Delta_{h_j}\|_2^2 &\ge \tfrac{1}{16}\int_{\mathbf{R}^2} |(P-x)\cap E_j|^2\,\mathrm{d}x \ge \tfrac{1}{16}\int_{\mathbf{R}^2} |(P-x)\cap E_j|\,\mathrm{d}x \\ &= \tfrac{1}{16} n \cdot \mathrm{vol}(E_j) = \tfrac{1}{16} n \cdot \pi \cdot 2^{-j_1}2^{-j_2} = \Omega(1). \end{aligned}$$

This proves Lemma 7.8. □

Sketch of Proof of Lemma 7.9. This is again a good training in estimates. We have, by Fact 7.10(i),

$$\begin{aligned} H(\xi) &= \sum_{j\in J} |\hat{h}_j(\xi)|^2 = \sum_{j\in J} \left(\frac{e^{-\xi_1^2/4^{j_1}}}{4^{j_1}}\right)\left(\frac{e^{-\xi_2^2/4^{j_2}}}{4^{j_2}}\right) \\ &\le \left(\sum_{2^{j_1}\ge 1/s} \frac{e^{-\xi_1^2/4^{j_1}}}{4^{j_1}}\right)\left(\sum_{2^{j_2}\ge 1/s} \frac{e^{-\xi_2^2/4^{j_2}}}{4^{j_2}}\right). \end{aligned}$$

A further calculation gives (see Exercise 1(a)) the estimate $H(\xi) = O(F(\xi))$ with

$$F(\xi) = \min\left(s^2, \frac{1}{\xi_1^2}\right) \min\left(s^2, \frac{1}{\xi_2^2}\right).$$

As the next step, we want to show that $F(\xi) = O(G(\xi))$. The Fourier transform of $I_{Q(r)}$ has already been calculated in the proof of the Amplification lemma 7.5, and also the estimates made there are similar to those needed here, so we leave the remaining calculation as Exercise 1(b). □

Bibliography and Remarks. This section again follows Beck and Chen [BC87]. The proof is also reproduced, with detailed calculations and with all constants made explicit, in Drmota and Tichy [DT97]. A slightly different proof for the planar case was given by Montgomery [Mon94].

Exercises

1. (a) Prove the estimate $\sum_{i \in \mathbf{N}:\, 2^i \geq a} e^{-4^{-i}t^2}/4^i \leq C \min(a^{-2}, t^{-2})$, $a \geq 1$, with some constant C.
   (b) Prove that

$$\operatorname*{ave}_{r \in (0,s)} [\sin^2(r\xi_1)\sin^2(r\xi_2)/\xi_1^2\xi_2^2] \geq c \min(s^2, \xi_1^{-2}) \min(s^2, \xi_2^{-2})$$

   for a suitable constant $c > 0$ and all ξ_1, ξ_2, and $s > 0$. Finish the proof of Lemma 7.9.
2. Generalize the proof of Theorem 7.6 to an arbitrary fixed dimension d.

7.3 An Excursion to Euclidean Ramsey Theory

This section is a detour from our main theme. But it shows a nice application of harmonic analysis similar to those in discrepancy lower bounds. The result is not so far away from discrepancy theory either, since it is a Ramsey-type theorem—see Section 1.4.

The result to be discussed belongs to the field of Euclidean Ramsey theory. Here is a simple illustrative problem in this area. Suppose that each point in the plane is colored either red or blue. Can we always find two points with unit distance having the same color? Oh yes, we can: consider the three vertices of an equilateral triangle. What if we have 3, 4, or more colors? For 3 colors, the answer is still positive (Exercise 1), but for 7 colors it is negative—there exists a 7-coloring of $\mathbf{R}^2$ such that no two points with unit distance have the same color. What happens for 4, 5, or 6 colors is still an open problem.

More generally, Euclidean Ramsey theory is concerned with problems of the following sort. Let K be a finite configuration of points in $\mathbf{R}^d$, and suppose that each point of $\mathbf{R}^d$ is colored by one of r colors. Does there necessarily exist a congruent copy of K (i.e. translated and rotated, no scaling) with all points having the same color? In particular, a finite set $K \subseteq \mathbf{R}^d$ is called *Ramsey* if for any number r of colors there exists a dimension $d' \geq d$ such that for any r-coloring of $\mathbf{R}^{d'}$ we can find a monochromatic congruent copy of K. Ramsey sets have been investigated intensively but so far only partial results are known. For a set to be Ramsey, it has to be *spherical*, meaning that all points of K lie on a common sphere. All affinely independent sets, vertex sets of regular n-gons, and Cartesian products of Ramsey sets are known to be Ramsey, but it is not known whether all spherical sets are Ramsey. The reader can try to discover some of the simpler results of the Euclidean Ramsey theory in the exercises, or look them up in [GRS90].

One may wonder what happens if we look for a monochromatic parallel translate of a given configuration K, or for a monochromatic similar copy of K (instead of a monochromatic congruent copy as above). Here the situation is much easier. For all configurations K with at least two points, there exists a coloring of $\mathbf{R}^d$ by two colors with no monochromatic parallel translate of K. On the other hand, a monochromatic similar copy exists for any coloring (for all d, r, and K), by so-called Gallai's theorem, which is a relatively easy generalization of Van der Waerden's theorem on arithmetic progressions.

After our brief overview, let us proceed to the main theme of this section. This is a "density" Ramsey-type result. Density results say that if one of the colors occupies a "big part" of space, then the desired configuration can be found in that color. In the discussed case, "big part" means a set of positive upper density, where the *upper density* $\delta(A)$ of a Lebesgue measurable set $A \subseteq \mathbf{R}^d$ is defined by

$$\delta(A) = \limsup_{R\to\infty} \frac{\mathrm{vol}(B(0,R)\cap A)}{\mathrm{vol}(B(0,R))},$$

with $B(0,R)$ denoting the ball of radius R centered at 0.

7.11 Theorem (Bourgain's density Ramsey theorem). *Let K be a set of d affinely independent points in $\mathbf{R}^d$ (for example, the vertices of a triangle in $\mathbf{R}^3$), and let $A \subseteq \mathbf{R}^d$ be a measurable set of positive upper density. Then there exists a number λ_0 such that for any $\lambda \geq \lambda_0$ the set A contains a congruent copy of the set λK.*

It is not known what other configurations K have this "density-Ramsey" property.

Although Bourgain emphasizes in his paper that the proof of this theorem only uses elementary harmonic analysis, I do not dare to present the proof of the general case. We only prove the planar case of this result, first established by Katznelson and Weiss by ergodic theory methods:

7.12 Theorem (Katznelson–Weiss theorem). *For every measurable set $A \subseteq \mathbf{R}^2$ of positive upper density there exists a number λ_0 such that for any $\lambda \geq \lambda_0$ the set A contains two points with distance exactly λ.*

This theorem will be proved from the following "bounded" version:

7.13 Proposition. *For any $\varepsilon > 0$ there exists a natural number $j_0 = j_0(\varepsilon)$ such that the following is true. Let $A \subseteq B(0,1)$ be a measurable set with $\mathrm{vol}(A) = \varepsilon$, and let $t_0 = 1 > t_1 > t_2 > \cdots$ be a decreasing sequence of real numbers such that $t_{j+1} \leq \frac{1}{2}t_j$ for all j. Then for some $j \leq j_0$, the set A contains two points with distance t_j.*

The proof of the Katznelson–Weiss theorem 7.12 from Proposition 7.13 proceeds by contradiction, and it is left as Exercise 7.

Fourier Transform of the Unit Circle. From Section 7.1, we will need the definition of the (two-dimensional) Fourier transform, the inversion formula (7.1), the Parseval–Plancherel identity $\|f\|_2 = \|\hat{f}\|_2$, and the convolution formula $\widehat{f * g} = \hat{f} \cdot \hat{g}$, where $(f * g)(y) = \frac{1}{2\pi}\int_{\mathbf{R}^2} f(x)g(y-x)\,\mathrm{d}x$.

Let S^1 denote the unit circle in $\mathbf{R}^2$ centered at 0, and let σ be the one-dimensional Lebesgue measure on S^1 (the whole S^1 has measure 2π). It will be convenient to consider σ as a measure in the whole $\mathbf{R}^2$ concentrated on S^1 (everything but the circle has σ-measure 0). The Fourier transform can also be defined for a measure in $\mathbf{R}^2$, instead of for a function. In particular, for the measure σ we have

$$\hat{\sigma}(\xi) = \frac{1}{2\pi}\int_{\mathbf{R}^2} e^{-\mathrm{i}\langle x,\xi\rangle}\,\mathrm{d}\sigma(x) = \frac{1}{2\pi}\int_0^{2\pi} e^{-\mathrm{i}(\xi_1\cos\vartheta + \xi_2\sin\vartheta)}\,\mathrm{d}\vartheta.$$

Since the Fourier transform commutes with rotations around the origin (right?), $\hat{\sigma}$ is rotationally symmetric, i.e. $\hat{\sigma}(\xi)$ only depends on $\|\xi\|$. And since for a real-valued function f (or for a measure, the argument is the same), $\hat{f}(-\xi)$ is the complex conjugate of $\hat{f}(\xi)$ (Exercise 8), $\hat{\sigma}$ is real-valued. Writing $x = \|\xi\|$, we get

$$\hat{\sigma}(\xi) = \hat{\sigma}(x,0) = \frac{1}{2\pi}\int_0^{2\pi} \mathrm{Re}\, e^{-\mathrm{i}x\cos\vartheta}\,\mathrm{d}\vartheta = \frac{1}{2\pi}\int_0^{2\pi} \cos(x\,\cos\vartheta)\,\mathrm{d}\vartheta.$$

Now this last integral happens to be known as the Bessel function $J_0(x)$ (shown in Fig. 7.3), and one can look up the following estimates (and much more precise ones) in almost any handbook of mathematical formulas:

$$|1 - \hat{\sigma}(\xi)| = O\left(\|\xi\|\right) \quad \text{as } \|\xi\| \to 0, \tag{7.4}$$

$$|\hat{\sigma}(\xi)| = O\left(\|\xi\|^{-1/2}\right) \quad \text{as } \|\xi\| \to \infty. \tag{7.5}$$

But one can also obtain these estimates without relying on the work of old masters on Bessel functions; it is not impossibly difficult (Exercises 9 and 10).

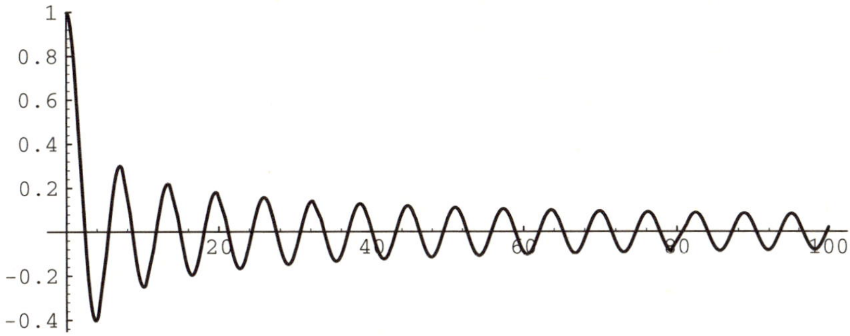

Fig. 7.3. Graph of the Bessel function $J_0(x)$.

Proof of Proposition 7.13. We have a set A of measure $\varepsilon > 0$ in the unit disc $B(0,1)$ and a sequence $1 = t_0 > t_1 > t_2 > \cdots$ with $t_{j+1} \leq \frac{1}{2}t_j$. The idea is to introduce a measure on the set of all pairs of points of distance t_j and show that for some j, the set of all pairs with distance t_j with both points in A has a positive measure.

Two points with distance t can be written as x and $x - tu$, where u is a point of the unit circle S^1 (the unnatural-looking sign "−" will pay off later, making the notation more convenient). A pair of points with distance t is thus represented by coordinates (x, u) with $x \in \mathbf{R}^2$ and $u \in S^1$, and this defines a natural product measure on the considered pairs. We use the usual planar Lebesgue measure for the x-component and the above-discussed Lebesgue measure σ for the u-component. The measure of the set of ordered pairs with distance t and with both points lying in A can be written as

$$M = \int_{\mathbf{R}^2}\int_{\mathbf{R}^2} I_A(x)I_A(x - tu)\,\mathrm{d}x\,\mathrm{d}\sigma(u), \tag{7.6}$$

with I_A standing for the characteristic function of A (recall that we regard σ as a measure on $\mathbf{R}^2$). We note that the inner integral is a convolution of two functions. Namely, if we put

$$F(y) = \int_{\mathbf{R}^2} I_A(x)I_{-A}(y - x)\,\mathrm{d}x,$$

then $F(y) = 2\pi \cdot (I_A * I_{-A})(y)$. This is where the two-dimensional Fourier transform enters the stage. By the convolution theorem, we have $\hat{F}(\xi) = 2\pi\hat{I}_A(\xi)\hat{I}_{-A}(\xi) = 2\pi|\hat{I}_A(\xi)|^2$, because $\hat{I}_{-A}(\xi) = \hat{I}_A(-\xi)$ is the complex conjugate of $\hat{I}_A(\xi)$. By the inversion formula,

$$F(y) = \int_{\mathbf{R}^2} |\hat{I}_A(\xi)|^2\, e^{\mathrm{i}\langle y,\xi\rangle}\,\mathrm{d}\xi.$$

We substitute this into (7.6) and change the order of integration, obtaining

$$\begin{aligned} M &= \int_{\mathbf{R}^2} F(tu)\,\mathrm{d}\sigma(u) \\ &= \int_{\mathbf{R}^2} |\hat{I}_A(\xi)|^2 \left(\int_{\mathbf{R}^2} \mathrm{e}^{\mathrm{i}\langle u,t\xi\rangle}\,\mathrm{d}\sigma(u)\right) \mathrm{d}\xi = \int_{\mathbf{R}^2} |\hat{I}_A(\xi)|^2 \hat{\sigma}(t\xi)\,\mathrm{d}\xi. \end{aligned}$$

We want to bound M away from 0 for a suitable value of $t \in \{t_1, t_2, \ldots, t_{j_0}\}$. More precisely, we show that $M = \Omega(\varepsilon^2)$ for some $t = t_j$.

Let us fix a parameter $\eta = c\varepsilon^2$, where c is a sufficiently small positive constant. Let t be some yet unspecified value for which we only assume $t \leq \eta$. We divide the plane into three regions R_1, R_2, R_3 (depending on t):

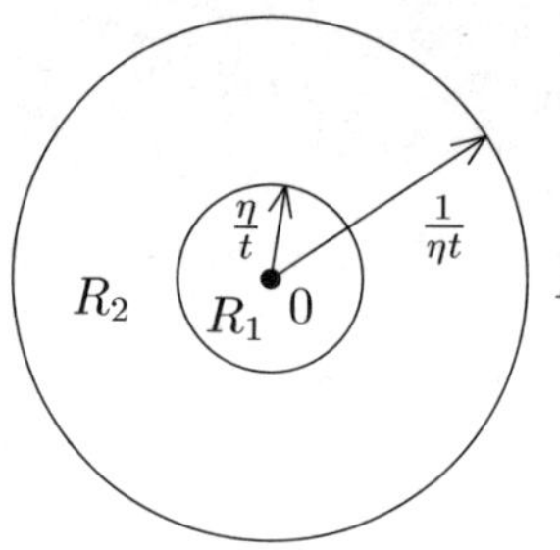

Let $M_i = \int_{R_i} |\hat{I}_A(\xi)|^2 \hat{\sigma}(t\xi)\,\mathrm{d}\xi$ be the integral of our function over R_i. The plan is to show that M_1 is positive and of the order $\Omega(\varepsilon^2)$, and that both M_2 and M_3 are much smaller in absolute value than M_1, whence $M = \Omega(\varepsilon^2)$.

For ξ lying in the inner region R_1, we have $\|t\xi\| \leq \eta \leq c$, and so by (7.4) we may assume $\hat{\sigma}(t\xi) \geq \frac{1}{2}$, say (recall that $\hat{\sigma}$ is real-valued). The disc $B(0,1)$ is contained in R_1, and thus

$$M_1 \geq \frac{1}{2} \int_{B(0,1)} |\hat{I}_A(\xi)|^2\,\mathrm{d}\xi. \tag{7.7}$$

We claim that

$$\left|\hat{I}_A(\xi) - \varepsilon\right| = O(\|\xi\|\varepsilon) \quad \text{as } \|\xi\| \to 0, \tag{7.8}$$

leaving verification to Exercise 9(c). Once one believes in (7.8) it is easy to see that the right-hand side of (7.7) is at least $c_1\varepsilon^2$ for an absolute positive constant c_1.

Next, we show that M_3, the integral over the outer region R_3, is negligible compared to ε^2, for all $t \leq \eta$. By (7.5), we have $|\hat{\sigma}(t\xi)| = O(\sqrt{\eta})$ uniformly for all ξ in R_3, and hence

$$\begin{aligned} |M_3| &= O(\sqrt{\eta}) \int_{R_3} |\hat{I}_A(\xi)|^2\,\mathrm{d}\xi \\ &\leq O(\sqrt{\eta}) \int_{\mathbf{R}^2} |\hat{I}_A(\xi)|^2\,\mathrm{d}\xi = O(\varepsilon\sqrt{\eta}) \end{aligned}$$

because $\int_{\mathbf{R}^2} |\hat{I}_A(\xi)|^2\,\mathrm{d}\xi = \int_{\mathbf{R}^2} |I_A(x)|^2\,\mathrm{d}x = \mathrm{vol}(A) = \varepsilon$ by Parseval–Plancherel. By the choice $\eta = c\varepsilon^2$ we get that $|M_3|$ is much smaller than M_1.

It remains to handle the middle region R_2. Here we cannot say much for one particular t. Certainly

$$|M_2| = O(1) \int_{R_2} |\hat{I}_A(\xi)|^2 \, \mathrm{d}\xi$$

since $|\hat{\sigma}|$ is uniformly bounded, but, in principle, the integral of $|\hat{I}_A|^2$ over R_2 could be nearly as large as the integral over the whole plane, which is ε. Here we must use the possibility of choosing t.

We recall that the annulus region $R_2 = R_2(t) = \{\xi\colon \frac{\eta}{t} \leq \|\xi\| \leq \frac{1}{\eta t}$ depends on t, and we note that if $t' < \eta^2 t$ then $R_2(t)$ and $R_2(t')$ are disjoint. Having sufficiently many such disjoint regions for various values of t, the integral of $|\hat{I}_A|^2$ over at least one of them must be small. More precisely, we fix integer parameters q and m (depending on ε) with $2^{-q} < \eta^2 = c^2\varepsilon^4$ and $m \geq \frac{C}{\varepsilon}$ for a large constant C, and we consider the m values $t_q, t_{2q}, \ldots, t_{mq}$. The parameter q was selected so that the respective regions $R_2(t_{jq})$ are mutually disjoint (recall the assumption $t_{j+1} \leq \frac{1}{2}t_j$), and hence there is a $j \leq mq$ such that $\int_{R_2(t_j)} |\hat{I}_A(\xi)|^2 \, \mathrm{d}\xi \leq \frac{\varepsilon}{m}$. This is much smaller than M_1 and so we get that M, the measure of pairs with distance t_j with both points in A, has the order $\Omega(\varepsilon^2)$ for this t_j, where $j \leq qm$. This concludes the proof. □

Bibliography and Remarks. The ergodic theory proof of Theorem 7.12, due to Katznelson and Weiss, was published in [FKW90] long after its discovery, together with several related results. The proof in this section is adapted from Bourgain [Bou86]. Yet another proof is due to Falconer and Marstrand [FM86].

An overview of the Euclidean Ramsey theory can be found in Graham et al. [GRS90]; more recent results are surveyed in Graham [Gra94]. The fact that all affinely independent sets are Ramsey was established by Frankl and Rödl [FR90], and regular n-gons being Ramsey is a particular case of results of Kříž [Kri91].

Exercises

1. Prove that if the plane is colored by 3 colors, one can always find two points with unit distance having the same color.
2.* Find a 7-coloring of the plane with no monochromatic unit-distance pair.
3.* Can you find an r-coloring of the plane such that no color contains two points with distance 2 together with their midpoint?
4. Find a coloring of the plane by 2 colors such that no color contains 3 points forming the vertex set of an equilateral triangle with unit side.
5. Prove that the vertex set of any regular simplex is Ramsey.

6.** Prove that if $K \subseteq \mathbf{R}^s$ and $L \subseteq \mathbf{R}^t$ are Ramsey (and finite) configurations, then the set

$$K \times L = \{(x_1, x_2, \ldots, x_s, y_1, y_2, \ldots, y_t) \in \mathbf{R}^{s+t} \colon (x_1, x_2, \ldots, x_s) \in K, \\ (y_1, y_2, \ldots, y_t) \in L\}$$

is Ramsey as well. Use Ramsey's theorem.

7. Prove the Katznelson–Weiss theorem 7.12 from Proposition 7.13.
8. Verify that if $f\colon \mathbf{R}^2 \to \mathbf{R}^2$ is a real-valued function then $\hat{f}(-\xi)$ is the complex conjugate of $\hat{f}(\xi)$. (And check that this fails for a complex-valued f in general.)
9. (a) Prove the estimate (7.4).
   (b) Prove $|e^{ix} - 1| = O(|x|)$ for all real x (in fact, $|e^{ix} - 1| \leq |x|$ is true).
   (c) Prove the estimate (7.8); part (b) may be useful.
10.* Prove the estimate $\hat{\sigma}(\xi) = \frac{1}{2\pi}\int_0^{2\pi} \cos(\|\xi\| \cos\vartheta)\, d\vartheta = O(\|\xi\|^{-1/2})$.

A. Tables of Selected Discrepancy Bounds

The tables on the next pages summarize some discrepancy bounds for various set systems and geometric families. Also citations of the original proofs are shown (where I could find them), as well as references to the relevant parts of this book. A reference in parentheses, like "(Th. 8.12)," means that the bound is a simple consequence of another result. For each result, only the first source (according to my knowledge) is shown, although several proofs may be known. Sources for earlier, weaker bounds are not given in the tables either.

An upper bound written as f actually means $O(f)$. A lower bound written as f means that the considered discrepancy is at least cf infinitely often, for some positive constant $c > 0$. But in many cases, one can get an $\Omega(f)$ lower bound as well.

The parameters d and p are considered fixed, and the constants of proportionality may depend on them. The number $\varepsilon > 0$ in the bounds is an arbitrarily small constant, and the constant of proportionality may again depend on it. Another fixed parameter is the size of the collection H in POL(H) on page 243.

Combinatorial discrepancy: asymptotic bounds		
Set system	Lower bound	Upper bound
Arbitrary m sets	$\sqrt{m}$ [OS78], Prop. 4.4	$\sqrt{m}$ [Spe85], Th. 4.9
Arbitrary m sets on n points, $m \geq n$	$\sqrt{n \log(2m/n)}$ Ex. 4.1.1	$\sqrt{n \log(2m/n)}$ [Spe85], Th. 4.2
Arbitrary m sets of size $\leq s$, $m \geq s$	$\sqrt{s \log(2m/s)}$ (Ex. 4.1.1)	$\sqrt{s \log(2m/s)}$ Ex. 4.6.4
Set system with max. degree t	$\sqrt{t}$ (Prop. 4.4)	$2t - 1$ [BF81], Th. 4.3
Set system on n points with max. degree t	$\sqrt{t}$ (Prop. 4.4)	$\sqrt{t} \log n$ [Ban98]
Set system on n points with primal shatter function $O(m^d)$	$n^{1/2-1/2d}$ (Th. 6.4)	$n^{1/2-1/2d}$ [Mat95], Th. 5.3
Set system on n points with dual shatter function $O(m^d)$	$n^{1/2-1/2d}\sqrt{\log n}$ [Mat97],[ARS99] Ex. 5.1.6	$n^{1/2-1/2d}\sqrt{\log n}$ [MWW93], Th. 5.4
Arithmetic progressions on $\{1, 2, \ldots, n\}$	$n^{1/4}$ [Rot64], Ex. 4.1.5	$n^{1/4}$ [MS96], Ex. 5.5.4
Intervals in k permutations on n points	$\sqrt{k}$ (Ex. 4.5.5)	$\sqrt{k} \cdot \log n$ [Sri97], Ex. 5.5.3

Geometric discrepancy: asymptotic bounds		
Discrepancy type	Lower bound	Upper bound
	Axis-parallel boxes ($\mathcal{R}_d$)	
$D(n, \mathcal{R}_2)$	$\log n$ [Sch72], Th. 6.2 [(a)]	$\log n$ Prop. 2.2
$D(n, \mathcal{R}_d)$, $d \geq 3$	$\log^{(d-1)/2} n \left(\frac{\log\log n}{\log\log\log n}\right)^{\frac{1}{2d-2}}$ [Bak]	$\log^{d-1} n$ [Hal60], Th. 2.4
$D_2(n, \mathcal{R}_d)$	$\log^{(d-1)/2} n$ [Rot54], Th. 6.1 [(b)]	$\log^{(d-1)/2} n$ [Dav56], [Rot80], [Fro80], Th. 2.5
$D_p(n, \mathcal{R}_d)$, $p > 1$	$\log^{(d-1)/2} n$ [Sch77a], Ex. 6.2.3	$\log^{(d-1)/2} n$ [Che81]
$D_1(n, \mathcal{R}_2)$	$\sqrt{\log n}$ [Hal81], Ex. 6.2.1	$\sqrt{\log n}$ from D_2 bound
$\mathrm{disc}(n, \mathcal{R}_2)$	$\log n$ [Bec81a] (Prop. 1.8+Th. 6.2)	$\log^{2.5} n$ [Sri97], Ex. 5.5.2
$\mathrm{disc}(n, \mathcal{R}_d)$, $d \geq 3$	$\log^{(d-1)/2} n \left(\frac{\log\log n}{\log\log\log n}\right)^{\frac{1}{2d-2}}$ (Prop. 1.8+[Bak])	$\log^{d+1/2} n\sqrt{\log\log n}$ Ex. 4.5.1
	Convex polytopes in $\mathbf{R}^d$ with given facet normals (POL(H); see p. 126)	
$D(n, \mathrm{POL}(H))$	same as for $D(n, \mathcal{R}_d)$	$\log^{d-1} n\,(\log\log n)^{1+\varepsilon}$ [Skr98]
$\mathrm{disc}(n, \mathrm{POL}(H))$	same as for $D(n, \mathcal{R}_d)$	$\log^{d+1/2} n\sqrt{\log\log n}$ [Mat], Ex. 4.5.3
	Halfspaces ($\mathcal{H}_d$)	
$D(n, \mathcal{H}_d)$	$n^{1/2-1/2d}$ [Ale90],[Ale91], Th. 6.9	$n^{1/2-1/2d}$ via $\mathrm{disc}(n, \mathcal{H}_d)$
$D_1(n, \mathcal{H}_2)$	??	$\log^2 n$ [BC93b], Th. 3.5
$\mathrm{disc}(n, \mathcal{H}_d)$	$n^{1/2-1/2d}$ [Ale90], Th. 6.4	$n^{1/2-1/2d}$ [Mat95], Th. 5.3

[(a)] The lower bound also holds for axis-parallel squares (Th. 6.3).

[(b)] The lower bound also holds for axis-parallel cubes (see Th. 7.6). Only the 2-dimensional case is explicitly treated in [Rot54].

Geometric discrepancy: asymptotic bounds (cont'd)		
Discrepancy type	Lower bound	Upper bound
	Balls ($\mathcal{B}_d$), arbitrary radius	
$D(n,\mathcal{B}_d)$	$n^{1/2-1/2d}$ via $D(n,\mathcal{H}_d)$ (c)	$n^{1/2-1/2d}\sqrt{\log n}$ [Bec87], Th. 3.1
$D_p(n,\mathcal{B}_d)$, $p \geq 2$	$n^{1/2-1/2d}$ [Bec87]	$n^{1/2-1/2d}$ [BC90], Ex. 5.4.3
$D_1(n,\mathcal{B}_2)$	$\log^{1/2-\varepsilon} n$ [Bec89b]	$n^{1/4}$
$\mathrm{disc}(n,\mathcal{B}_d)$	$n^{1/2-1/2d}$ from $D(n,\mathcal{B}_d)$	$n^{1/2-1/2d}\sqrt{\log n}$ [MWW93], Th. 5.4
	Circular discs, fixed radius ($\mathcal{BF}_2$)	
$D(n,\mathcal{BF}_2)$	$n^{1/4}$ [Bec88b]	$n^{1/4}$ [Mat95], Th. 5.3
	Scaled-down copies of a compact convex $A \subset \mathbf{R}^d$, with rotation ($\mathcal{S}_A$)	
$D(n,\mathcal{S}_A)$	$n^{1/2-1/2d}\sqrt{S}$ [Bec87] (d)	$n^{1/2-1/2d}\sqrt{S\log n}$ [Bec87] (d)
	Scaled-down copies of a compact convex $A \subset \mathbf{R}^d$, no rotation ($\mathcal{T}_A$)	
$D(n,\mathcal{T}_A)$, $d=2$	$\sqrt{\xi_n(A)}\cdot\log^{-1/4} n$ [Bec88a] (e)	$\xi_n(A)\log^{4+\varepsilon} n$ [Bec88a] (e)
$D(n,\mathcal{T}_A)$, A smooth	$c(A)n^{1/2-1/2d}$ [Drm93]	$C(A)n^{1/2-1/2d}\sqrt{\log n}$ [Drm93]
	Convex sets ($\mathcal{K}_d$)	
$D(n,\mathcal{K}_2)$	$n^{1/3}$ [Sch75], Ex. 3.1.6	$n^{1/3}\log^4 n$ [Bec88c]
$D(n,\mathcal{K}_d)$, $d \geq 3$	$n^{1-2/(d+1)}$ [Sch75], Ex. 3.1.6	$n^{1-2/(d+1)}\log^c n$ [Stu77] (f)

(c) The bound holds in the whole-space model as well. For balls completely contained in $[0,1]^d$, the lower bound is $n^{1/2-1/2d-\varepsilon}$ [Bec87].

(d) Here S is the surface area of A, and for the lower bound, it is assumed that A contains a ball of radius $n^{-1/d}$. Both bounds are in the whole-space model.

(e) Here $\xi_n(A)$ is the "approximability number" of A: the smallest $\ell \geq 3$ such that there is a convex ℓ-gon inscribed to A such that the area of the difference is $\leq \ell^2/n$. Károlyi [Kár95b] has upper bounds for the analogous problem in $\mathbf{R}^d$ which are too complicated to state here.

(f) Here $c = \frac{3}{2}$ for $d = 3$ and $c = 2/(d+1)$ for $d \geq 4$.

Bibliography

[AB92] M. Anthony and N. Biggs. *Computational Learning Theory.* Cambridge University Press, Cambridge, 1992. (refs: p. 150)

[ABC97] J. R. Alexander, J. Beck, and W. L. Chen. Geometric discrepancy theory and uniform distribution. In *Handbook of Discrete and Computational Geometry (J. E. Goodman and J. O'Rourke eds.)*, chapter 10, pages 185–207. CRC Press LLC, Boca Raton, FL, 1997. (refs: p. 8)

[Ada75] R. A. Adams. *Sobolev spaces.* Academic Press, London, 1975. (refs: p. 35)

[AE45] T. van Aardenne-Ehrenfest. Proof of the impossibility of a just distribution of an infinite sequence of points. *Nederl. Akad. Wet., Proc.*, 48:266–271, 1945. Also in *Indag. Math.* 7, 71-76 (1945). (refs: p. 6)

[AE49] T. van Aardenne-Ehrenfest. On the impossibility of a just distribution. *Nederl. Akad. Wet., Proc.*, 52:734–739, 1949. Also in *Indag. Math.* 11, 264-269 (1949). (refs: p. 6)

[AGHP92] N. Alon, O. Goldreich, J. Håstad, and R. Peralta. Simple construction of almost k-wise independent random variables. *Random Structures and Algorithms*, 3:289–304, 1992. (refs: p. 34)

[AKP+87] N. Alon, D. Kleitman, C. Pomerance, M. Saks, and P. Seymour. The smallest n-uniform hypergraph with positive discrepancy. *Combinatorica*, 7:151–160, 1987. (refs: p. 120)

[Ale90] R. Alexander. Geometric methods in the theory of uniform distribution. *Combinatorica*, 10(2):115–136, 1990. (refs: p. 191, 192, 201, 226, 243)

[Ale91] R. Alexander. Principles of a new method in the study of irregularities of distribution. *Invent. Math.*, 103:279–296, 1991. (refs: p. 201, 210, 226, 243)

[AM95] N. Alon and Y. Mansour. ε-discrepancy sets and their application for interpolation of sparse polynomials. *Inf. Process. Lett.*, 54(6):337–342, 1995. (refs: p. 35)

[AMM85] I. Aharoni, B. Maurey, and B. S. Mityagin. Uniform embeddings of metric spaces and of Banach spaces into Hilbert spaces. *Israel J. Math.*, 52:251–265, 1985. (refs: p. 209, 210)

[ARS99] N. Alon, L. Rónyai, and L. Szabó. Norm-graphs: variations and applications. *J. Combin. Theory Ser. B*, 1999. In press. (refs: p. 143, 242)

[AS79] I. A. Antonov and V. M. Saleev. An economic method of computing LP_r sequences (in Russian). *Zh. Vychisl. Mat. i Mat. Fiz.*, 19:243–245, 1979. (refs: p. 59)

[AS92] N. Alon and J. Spencer. *The Probabilistic Method.* J. Wiley and Sons, New York, NY, 1992. (refs: p. 8, 33, 34, 87, 102, 103, 104, 116, 133, 282)

[Ass83] P. Assouad. Density and dimension (in French). *Ann. Inst. Fourier (Grenoble)*, 33:233–282, 1983. (refs: p. 151)

[Bak] R. C. Baker. On irregularities of distribution, II. *J. Lond. Math. Soc.* To appear. (refs: p. 178, 243)

[Ban98] W. Banaszczyk. Balancing vectors and Gaussian measures of n-dimensional convex bodies. *Random Structures and Algorithms*, 12(4):351–360, 1998. (refs: p. 115, 168, 242)

[BC87] J. Beck and W. L. Chen. *Irregularities of Distribution.* Cambridge University Press, Cambridge, 1987. (refs: p. 7, 8, 50, 51, 70, 89, 90, 178, 179, 182, 225, 226, 227, 234)

[BC89] J. Beck and W. Chen. Irregularities of point distribution relative to convex polygons. In *Irregularities of partitions (G. Halász, V. T. Sós eds.)*, pages 1–22. Springer-Verlag, Berlin etc., 1989. (refs: p. 126)

[BC90] J. Beck and W. Chen. Note on irregularities of distribution II. *Proc. Lond. Math. Soc., III. Ser*, 61:251–272, 1990. (refs: p. 90, 244)

[BC93a] J. Beck and W. Chen. Irregularities of point distribution relative to convex polygons II. *Mathematika*, 40:127–136, 1993. (refs: p. 99)

[BC93b] J. Beck and W. Chen. Irregularities of point distribution relative to half-planes I. *Mathematika*, 40:102–126, 1993. (refs: p. 99, 243)

[BCM] H. Brönnimann, B. Chazelle, and J. Matoušek. Product range spaces, sensitive sampling, and derandomization. *SIAM J. Comput.* In press. Preliminary version in *Proc. 34th Annu. IEEE Sympos. Found. Comput. Sci.* 1993, 400–409. (refs: p. 152)

[BDCK66] J. Bretagnolle, D. Dacunha-Castelle, and J. L. Krivine. Lois stables et espaces L^p. *Ann. Inst. H.Poincaré, Sect. B*, 2:231–259, 1966. (refs: p. 210)

[Bec] J. Beck. Randomness in lattice point problems. In *Fifth Czech-Slovak Cominatorial Symposium (J. Kratochvíl and J. Nešetřil ed.), Prague 1998*. North-Holland, Amsterdam. To appear. (refs: p. 77)

[Bec81a] J. Beck. Balanced two-colorings of finite sets in the square. I. *Combinatorica*, 1:327–335, 1981. (refs: p. 22, 125, 243)

[Bec81b] J. Beck. Roth's estimate on the discrepancy of integer sequences is nearly sharp. *Combinatorica*, 1(4):319–325, 1981. (refs: p. 108, 125)

[Bec84] J. Beck. Sums of distances between points on a sphere: An application of the theory of irregularities of distribution to distance geometry. *Mathematika*, 31:33–41, 1984. (refs: p. 33, 227)

[Bec87] J. Beck. Irregularities of distribution I. *Acta Math.*, 159:1–49, 1987. (refs: p. 89, 90, 225, 226, 227, 244)

[Bec88a] J. Beck. Irregularities of distribution II. *Proc. London Math. Soc. (3)*, 56:1–50, 1988. (refs: p. 125, 126, 225, 227, 228, 244)

[Bec88b] J. Beck. On irregularities of point sets in the unit square. In *Combinatorics. Proc. 7th Hungarian colloquium held from July 5 to July 10, 1987 in Eger, Hungary (A. Hajnal, L. Lovász, V. T. Sós eds.), Colloq. Math. Soc. Janos Bolyai, 52*, pages 63–74. North-Holland, Amsterdam, 1988. (refs: p. 213, 227, 244)

[Bec88c] J. Beck. On the discrepancy of convex plane sets. *Monatsh. Math.*, 105:91–106, 1988. (refs: p. 92, 244)

[Bec89a] J. Beck. Balanced two-colorings of finite sets in the cube. *Discrete Mathematics*, 73:13–25, 1989. (refs: p. 126)

[Bec89b] J. Beck. On a problem of W. M. Schmidt concerning one-sided irregularities of point distributions. *Math. Ann.*, 285:29–55, 1989. (refs: p. 228, 244)

[Bec89c] J. Beck. A two-dimensional van Aardenne-Ehrenfest theorem in irregularities of distribution. *Compositio Math.*, 72:269–339, 1989. (refs: p. 178)

[Bec91a] J. Beck. Flat polynomials on the unit circle – note on a problem of Littlewood. *Bull. Lond. Math. Soc., III. Ser*, 23:269–277, 1991. (refs: p. 32)

[Bec91b] J. Beck. Quasi-random 2-colorings of point sets. *Random Structures and Algorithms*, 2:289–302, 1991. (refs: p. 143)

[Bec94] J. Beck. Probabilistic diophantine approximation, I Kronecker sequences. *Ann. Math.*, 140:451–502, 1994. (refs: p. 77)

[Beh22] H. Behnke. Über die Vertailung von Irrationalitäten mod 1. *Abh. Math. Semin. Univ. Hamburg*, 1:252–267, 1922. (refs: p. 42, 76)

[Beh24] H. Behnke. Zur Theorie der diophantischen Approximationen I. *Abh. Math. Semin. Univ. Hamburg*, 3:261–318, 1924. (refs: p. 42, 76)

[BEHW89] A. Blumer, A. Ehrenfeucht, D. Haussler, and M. Warmuth. Learnability and the Vapnik-Chervonenkis dimension. *J. ACM*, 36:929–965, 1989. (refs: p. 150, 151)

[BF81] J. Beck and T. Fiala. "Integer making" theorems. *Discr. Appl. Math*, 3:1–8, 1981. (refs: p. 104, 242)

[BF88] P. Bratley and B. L. Fox. Implementing Sobol's quasirandom sequence generator. *ACM Trans. Math. Software*, 14:88–100, 1988. (refs: p. 59)

[BF92] L. Babai and P. Frankl. *Linear algebra methods in combinatorics (Preliminary version 2).* Department of Computer Science, The University of Chicago, 1992. (refs: p. 150)

[BFR89] R.L Burden, J. D. Faires, and A.C. Reynolds. *Numerical Analysis.* PWS-KENT Pub. Co., Florence, Kentucky, 1989. (refs: p. 286)

[BG81] I. Bárány and V. S. Grinberg. On some combinatorial questions in finite-dimensional spaces. *Linear Algebra Appl.*, 41:1–9, 1981. (refs: p. 116)

[BH97] D. Bednarchak and M. Helm. A note on the Beck-Fiala theorem. *Combinatorica*, 17:147–149, 1997. (refs: p. 104)

[BL88] J. Bourgain and J. Lindenstrauss. Distribution of points on the sphere and approximation by zonotopes. *Isr. J. Math.*, 64:25–31, 1988. (refs: p. 33)

[BL93] J. Bourgain and J. Lindenstrauss. Approximating the ball by a Minkowski sum of segments with equal length. *Discr. & Comput. Geom.*, 9:131–144, 1993. (refs: p. 33)

[BLM89] J. Bourgain, J. Lindenstrauss, and V. Milman. Approximation of zonoids by zonotopes. *Acta Math.*, 162:73–141, 1989. (refs: p. 33, 228)

[Blü91] M. Blümlinger. Slice discrepancy and irregularities of distribution on spheres. *Mathematika*, 38:105–116, 1991. (refs: p. 227)

[BM83] U. Betke and P. McMullen. Estimating the sizes of convex bodies by projections. *J. London Math. Soc.*, 27:525–538, 1983. (refs: p. 33)

[Boh90] G. Bohus. On the discrepancy of 3 permutations. *Random Struct. Algo.*, 1:215–220, 1990. (refs: p. 126, 168)

[Bou86] J. Bourgain. A Szemerédi type theorem for sets of positive density in R^k. *Israel J. Math.*, 54:307–316, 1986. (refs: p. 239)

[BPR96] S. Basu, R. Pollack, and M.-F. Roy. On the number of cells defined by a family of polynomials on a variety. *Mathematika*, 43:120–126, 1996. (refs: p. 143)

[BR91] B. Berger and J. Rompel. Simulating $(\log n)^c$-wise independence in NC. *Journal of the ACM*, 38(4):1028–1046, 1991. (refs: p. 104)

[BS95] J. Beck and V. Sós. Discrepancy theory. In *Handbook of Combinatorics*, pages 1405–1446. North-Holland, Amsterdam, 1995. (refs: p. 7, 8, 32, 76, 108, 114, 116, 119)

[Cau50] A. Cauchy. Mémoire sur la rectification des courbes et la quadrature des surfaces courbes. *Mem. Acad. Sci. Paris*, 22:3–15, 1850. (refs: p. 192)

[CF90] B. Chazelle and J. Friedman. A deterministic view of random sampling and its use in geometry. *Combinatorica*, 10(3):229–249, 1990. (refs: p. 150)

[Cha92] B. Chazelle. A note on Haussler's packing lemma. Unpublished manuscript, Princeton, 1992. (refs: p. 159)

[Cha93] B. Chazelle. Cutting hyperplanes for divide-and-conquer. *Discrete Comput. Geom.*, 9(2):145–158, 1993. (refs: p. 151)

[Cha94] B. Chazelle. A spectral approach to lower bounds. In *Proc. 35th Ann. IEEE Symp. on Foundations of Computer Science*, pages 674–682, 1994. (refs: p. 34)

[Cha98a] B. Chazelle. Discrepancy bounds for geometric set systems with square incidence matrices. In *Discrete and Computational Geometry: Ten Years Later (B. Chazelle, E. Goodman, and R. Pollack eds.)*. American Mathematical Society, Providence, 1998. In press. (refs: p. 115)

[Cha98b] B. Chazelle. *The Discrepancy Method.* Book in preparation, Princeton University, 1998. (refs: p. 8, 34, 50, 92, 229)

[Che81] W. L. Chen. On irregularities of distribution. *Mathematika*, 27:153–170, 1981. (refs: p. 50, 243)

[Che83] W. L. Chen. On irregularities of distribution II. *Quart. J. Math. Oxford Ser. (2)*, 34:257–279, 1983. (refs: p. 50, 59, 70)

[Chu97] F. Chung. *Spectral Graph Theory (Regional Conference Series in Mathematics 92)*. Amer. Math. Soc., Providence, 1997. (refs: p. 33)

[CMS95] B. Chazelle, J. Matoušek, and M. Sharir. An elementary approach to lower bounds in geometric discrepancy. *Discr. Comput. Geom.*, 13:363–381, 1995. (refs: p. 191, 192, 197, 201, 287)

[Cor35a] J. G. van der Corput. Verteilungsfunktionen I. *Akad. Wetensch. Amsterdam, Proc.*, 38:813–821, 1935. (refs: p. 6, 42)

[Cor35b] J. G. van der Corput. Verteilungsfunktionen II. *Akad. Wetensch. Amsterdam, Proc.*, 38:1058–1066, 1935. (refs: p. 6, 42)

[CS88] J. H. Conway and N. J. A. Sloane. *Sphere packings, lattices and groups*. Springer-Verlag, New York etc., 1988. (refs: p. 78)

[CW89] B. Chazelle and E. Welzl. Quasi-optimal range searching in spaces of finite VC-dimension. *Discrete Comput. Geom.*, 4:467–489, 1989. (refs: p. 159, 164)

[Dav56] H. Davenport. Note on irregularities of distribution. *Mathematika*, 3:131–135, 1956. (refs: p. 50, 243)

[DL97] M. M. Deza and M. Laurent. *Geometry of cuts and metrics (Algorithms and Combinatorics 15)*. Springer-Verlag, Berlin etc., 1997. (refs: p. 210)

[Doe98a] B. Doerr. Discrepancy of direct products of hypergraphs. Manuscript, Tech. Univ. Kiel, Germany, 1998. (refs: p. 104)

[Doe98b] B. Doerr. Linear and hereditary discrepancy. Manuscript, Tech. Univ. Kiel, Germany, 1998. (refs: p. 114)

[Drm93] M. Drmota. Irregularities of distribution and convex sets. *Grazer Math. Ber.*, 318:9–16, 1993. (refs: p. 228, 244)

[DSW94] G.-L. Ding, P. Seymour, and P. Winkler. Bounding the vertex cover number of a hypergraph. *Combinatorica*, 14:23–34, 1994. (refs: p. 150)

[DT97] M. Drmota and R. F. Tichy. *Sequences, discrepancies and applications (Lecture Notes in Mathematics 1651)*. Springer-Verlag, Berlin etc., 1997. (refs: p. 7, 8, 76, 92, 178, 179, 234)

[Dud78] R. M. Dudley. Central limit theorems for empirical measures. *Ann. Probab.*, 6:899–929, 1978. (refs: p. 144, 150)

[Dud84] R. M. Dudley. *A course on empirical measures.* Lecture Notes in Math. 1097. Springer-Verlag, Berlin etc., 1984. (refs: p. 150)

[Dud85] R. M. Dudley. The structure of some Vapnik-Chervonenkis classes. In *Proc. of Berkeley Conference in honor of Jerzy Neyman and Jack Kiefer, Volume II, (LeCam and Olshen eds.)*, pages 495–507, 1985. (refs: p. 150)

[EGL+92] G. Even, O. Goldreich, M. Luby, N. Nisan, and B. Veliković. Approximations of general independent distributions. In *Proc. 24th ACM Symp. on Theory of Computing*, pages 10–16, 1992. (refs: p. 34)

[EH77] P. Erdős and A. Hajnal. On spanned subgraphs of graphs. In *Contributions to graph theory and its applications (Internat. Colloq., Oberhof 1977)*, pages 80–96. Tech. Hochschule Ilmenau, 1977. (refs: p. 154)

[EH89] P. Erdős and A. Hajnal. Ramsey-type theorems. *Discrete Applied Mathematics*, 25:37–52, 1989. (refs: p. 154)

[Erd64] P. Erdős. Problems and results on Diophantine approximation. *Compositio Math.*, 16:52–66, 1964. (refs: p. 14, 179, 226)

[ES74] P. Erdős and J. Spencer. *Probabilistic Methods in Combinatorics.* Academic Press, 1974. (refs: p. 108)

[Fau81] H. Faure. Discrepancy of sequences associated with a number system (in dimension one) (in French). *J. Bull. Soc. Math. Fr.*, 109:143–182, 1981. (refs: p. 43)

[Fau82] H. Faure. Discrepancy of sequences associated with a number system (in dimension s) (in French). *Acta Arith.*, 41(4):337–351, 1982. (refs: p. 59)

[Fau92] H. Faure. Good permutations for extreme discrepancy. *J. Number Theory*, 42:47–56, 1992. (refs: p. 43)

[Fel43] W. Feller. Generalization of a probability limit theorem of Cramér. *Trans. Amer. Math. Soc.*, 54:361–372, 1943. (refs: p. 104)

[FH96] K. Frank and S. Heinrich. Computing discrepancies of Smolyak quadrature rules. *J. Complexity*, 12:287–314, 1996. (refs: p. 72)

[FKW90] H. Furstenberg, Y. Katznelson, and B. Weiss. Ergodic theory and configurations in sets of positive density. In *Mathematics of Ramsey theory (J. Nešetřil, V. Rödl eds.)*, pages 184–198. Springer, Berlin, 1990. (refs: p. 239)

[FM86] K. J. Falconer and J. M. Marstrand. Plane sets with positive density at infinity contain all large distances. *Bull. London Math. Soc.*, 18:471–474, 1986. (refs: p. 239)

[FP83] P. Frankl and J. Pach. On the number of sets in a null t-design. *Eur. J. Comb.*, 4:21–23, 1983. (refs: p. 150)

[FR90] P. Frankl and V. Rödl. A partition property of simplices in Euclidean space. *J. Amer. Math. Soc.*, 3:1–7, 1990. (refs: p. 239)

[Fra83] P. Frankl. On the trace of finite sets. *J. Comb. Theory, Ser. A*, 34:41–45, 1983. (refs: p. 150)

[Fro80] K. K. Frolov. An upper estimate for the discrepancy in the L_p-metric, $2 \leq p \leq \infty$ (in Russian). *Dokl. Akad. Nauk SSSR*, 252:805–807, 1980. English translation in *Soviet. Math. Dokl.* 21(1980) 840–842. (refs: p. 30, 50, 78, 79, 243)

[FW94] K.-T. Fang and Y. Wang. *Number-theoretic methods in statistics.* Chapman & Hall, London etc., 1994. (refs: p. 78)

[GH62] A. Ghouila-Houri. Caractérisation des matrices totalement unimodulaires. *C. R. Acad. Sci. Paris*, 254:1192–1194, 1962. (refs: p. 119)

[Gia97] A. Giannopoulos. On some vector balancing problems. *Studia Math.*, 122(3):225–234, 1997. (refs: p. 104, 116, 133)

[GKT97] P. J. Grabner, B. Klinger, and R. F. Tichy. Discrepancies of point sequences on the sphere and numerical integration. In *Multivariate approximation. Recent trends and results (W. Haussmann et al. eds.), Math. Research Vol. 101*, pages 95–112. Akademie Verlag, Berlin, 1997. (refs: p. 16)

[Glu89] E. D. Gluskin. Extremal properties of orthogonal parallelepipeds and their applications to the geometry of Banach spaces. *Math. USSR Sbornik*, 64(1):85–96, 1989. (refs: p. 104, 133)

[Gra94] R. L. Graham. Recent trends in Euclidean Ramsey theory. *Discrete Math.*, 136:119–127, 1994. (refs: p. 239)

[GRS90] R. L. Graham, B. L. Rothschild, and J. Spencer. *Ramsey Theory.* J. Wiley & Sons, New York, 1990. (refs: p. 32, 235, 239, 288)

[Hal60] J.H. Halton. On the efficiency of certain quasi-random sequences of points in evaluating multi-dimensional integrals. *Numer. Math.*, 2:84–90, 1960. (refs: p. 43, 243)

[Hal81] G. Halász. On Roth's method in the theory of irregularities of point distributions. In *Recent progress in analytic number theory, Vol. 2 (Durham, 1979)*, pages 79–94. Academic Press, London-New York, 1981. (refs: p. 178, 228, 243)

[Ham60] J. M. Hammersley. Monte Carlo methods for solving multivariable problems. *Ann. New York Acad. Sci.*, 86:844–874, 1960. (refs: p. 43)

[Hau92] D. Haussler. Decision theoretic generalizations of the PAC model for neural net and other learning applications. *Information and Computation*, 100(1):78–150, 1992. (refs: p. 150)

[Hau95] D. Haussler. Sphere packing numbers for subsets of the Boolean n-cube with bounded Vapnik-Chervonenkis dimension. *Journal of Combinatorial Theory Ser. A*, 69:217–232, 1995. (refs: p. 159)

[Hei96] S. Heinrich. Efficient algorithms for computing the L_2 discrepancy. *Math. Comput.*, 65:1621–1633, 1996. (refs: p. 72)

[Hic96] F. J. Hickernell. The mean square discrepancy of randomized nets. *ACM Transactions on Modeling and Computer Simulation*, 6(4):274–296, 1996. (refs: p. 14, 70)

[Hic98] F. Hickernell. A generalized discrepancy and quadrature error bound. *Math. Comput.*, 67(221):299–322, 1998. (refs: p. 14, 29, 30, 71)

[HK96] J. Hoogland and R. Kleiss. Discrepancy-based error estimates for quasi-Monte Carlo. I: General formalism. *Comput. Phys. Comm.*, 98:111–127, 1996. (refs: p. 29)

[HL22a] G. H. Hardy and J. E. Littlewood. Some problems of diophantine approximation: the lattice points of a right-angled triangle. Part I. *Proc. London Math. Soc. (2)*, 20:15–36, 1922. (refs: p. 14, 42)

[HL22b] G. H. Hardy and J. E. Littlewood. Some problems of diophantine approximation: the lattice points of a right-angled triangle. Part II. *Abh. Math. Sem. Hamburg*, 1:212–249, 1922. (refs: p. 14)

[HL95] D. Haussler and P. M. Long. A generalization of Sauer's lemma. *J. Comb. Theory, Ser. A*, 71:219–240, 1995. (refs: p. 151)

[Hla61] E. Hlawka. Funktionen von beschränkter Variation in der Theorie der Gleichverteilung. *Ann. Mat. Pura Appl.*, 54:325–333, 1961. (refs: p. 28)

[Hla75] E. Hlawka. Zur Theorie der Gleichverteilung I,II. *Anz. Österr. Akad. Wiss., Math-naturw., Kl. 1975, No. 2*, pages 13–14, 1975. Ibid., Kl. 1975, No. 3, 23–24. (refs: p. 28)

[Hla84] E. Hlawka. *The theory of uniform distribution.* A B Academic Publ., Berkhamsted, Herts., 1984. (refs: p. 6, 7, 80)

[HW87] D. Haussler and E. Welzl. Epsilon-nets and simplex range queries. *Discrete Comput. Geom.*, 2:127–151, 1987. (refs: p. 22, 150, 151)

[JHK97] F. James, J. Hoogland, and R. Kleiss. Multidimensional sampling for simulation and integration: measures, discrepancies, and quasi-random numbers. *Comp. Phys. Comm.*, 99:180–220, 1997. (refs: p. 26, 29, 32, 77)

[JT93] E. J. Janse van Rensburg and G. M. Torrie. Estimation of multidimensional integrals: Is Monte Carlo the best method? *J. Phys. A: Math. Gen.*, 26:943–953, 1993. (refs: p. 27)

[Kár95a] Gy. Károlyi. Geometric discrepancy theorems in higher dimensions. *Studia Sci. Math. Hungarica*, 30:59–94, 1995. (refs: p. 126)

[Kár95b] Gy. Károlyi. Irregularities of point distributions relative to homothetic convex bodies I. *Monatsh. Math.*, 120:247–279, 1995. (refs: p. 228, 244)

[Kes66] H. Kesten. On a conjecture of Erdős and Szüsz related to uniform distribution mod 1. *Acta Arith.*, 12:193–212, 1966. (refs: p. 76)

[Khi23] A. Khintchine. Ein Satz über Kettenbrüche mit arithmetischen Anwendungen. *Math. Z.*, 18:289–306, 1923. (refs: p. 77)

[Khi24] A. Khintchine. Einige Sätze über Kettenbrüche mit Anwendungen auf die Theorie der Diophantischen Approximationen. *Math. Ann.*, 92:115–125, 1924. (refs: p. 77)

[Kle66] D. Kleitman. On a combinatorial problem of Erdős. *J. Combinatorial Theory*, 1:209–214, 1966. (refs: p. 134)

[KM97a] M. Karpinski and A. Macintyre. Approximating the volume of general Pfaffian bodies. In *Structures in logic and computer science (J. Mycielski et al. eds.) Lect. Notes Comput. Sci. 1261*, pages 162–171. Springer-Verlag, Berlin etc., 1997. (refs: p. 143)

[KM97b] M. Karpinski and A. Macintyre. Polynomial bounds for VC dimension of sigmoidal and general Pfaffian neural networks. *J. Syst. Comput. Sci.*, 54(1):169–176, 1997. (refs: p. 143)

[KMT75] J. Komlós, P. Major, and G. Tusnády. Weak convergence and embedding. In *Limit Theorems Probab. Theor, Keszthely 1974*, pages 149–165. J. Bolyai Math. Soc., Budapest, 1975. (refs: p. 125)

[KN74] L. Kuipers and H. Niederreiter. *Uniform Distribution of Sequences.* J. Wiley & Sons, New York, 1974. (refs: p. 7, 76, 92)

[Kni98] P. Knieper. *Discrepancy of arithmetic progressions.* Shaker Verlag, Aachen, 1998. Doctoral dissertation, Humboldt U. Berlin. (refs: p. 108)

[Kok43] J. F. Koksma. A general theorem from the theory of the uniform distribution modulo 1 (in Dutch). *Mathematica B (Zutphen)*, 1:7–11, 1942/43. (refs: p. 28)

[Kor59] N. M. Korobov. The approximate computation of multiple integrals (in Russian). *Dokl. Akad. Nauk SSSR*, 124:1207–1210, 1959. (refs: p. 77, 78)

[KPW92] J. Komlós, J. Pach, and G. Woeginger. Almost tight bounds for ϵ-nets. *Discrete Comput. Geom.*, 7:163–173, 1992. (refs: p. 151)

[Kri91] I. Kriz. Permutation groups in Euclidean Ramsey theory. *Proc. Amer. Math. Soc.*, 112:899–907, 1991. (refs: p. 239)

[KRS96] J. Kollár, L. Rónyai, and T. Szabó. Norm-graphs and bipartite Turán numbers. *Combinatorica*, 16(3):399–406, 1996. (refs: p. 143)

[KT97] B. Klinger and R. F. Tichy. Polynomial discrepancy of sequences. *J. Comput. Appl. Math*, 84:107–117, 1997. (refs: p. 28)

[Lac90] M. Laczkovich. Equidecomposability and discrepancy; a solution of Tarski's circle-squaring problem. *J. Reine Angew. Math.*, 404:77–117, 1990. (refs: p. 32)

[Lac95] M. Laczkovich. Discrepancy estimates for sets with small boundary. *Studia Sci. Math. Hungarica*, 30:105–109, 1995. (refs: p. 92)

[Lag85] J. C. Lagarias. The computational complexity of simultaneous diophantine approximation problems. *SIAM J. Computing*, 14:196–209, 1985. (refs: p. 80)

[Lam85] J. P. Lambert. Quasi-Monte Carlo, low discrepancy sequences, and ergodic transformations. *J. Comput. Appl. Math*, 12/13:419–423, 1985. (refs: p. 8)

[Ler04] M. Lerch. Question 1547 (in French). *L'Intermediaire des Mathematiciens*, 11:144–145, 1904. (refs: p. 12)

[Lev95] V. F. Lev. On two versions of L^2-discrepancy and geometrical interpretation of diaphony. *Acta Math. Hung.*, 69(4):281–300, 1995. (refs: p. 14, 15, 229)

[Lev96] V. F. Lev. Translations of nets and relationship between supreme and L^κ-discrepancies. *Acta Math. Hung.*, 70(1-2):1–12, 1996. (refs: p. 14)

[LLSZ97] N. Linial, M. Luby, M. Saks, and D. Zuckerman. Efficient construction of a small hitting set for combinatorial rectangles in high dimension. *Combinatorica*, 17:215–234, 1997. (refs: p. 34)

[Lov86] L. Lovász. *An algorithmic theory of numbers, graphs and convexity.* SIAM regional series in applied mathematics, SIAM, Philadelphia, 1986. (refs: p. 80, 81)

[LPS86] A. Lubotzky, R. Phillips, and P. Sarnak. Hecke operators and distributing points on the sphere. *Commun. Pure Appl. Math*, 39:149–186, 1986. (refs: p. 90)

[LPS87] A. Lubotzky, R. Phillips, and P. Sarnak. Hecke operators and distributing points on S^2. *Commun. Pure Appl. Math*, 40:401–420, 1987. (refs: p. 90)

[LSV86] L. Lovász, J. Spencer, and K. Vesztergombi. Discrepancy of set-systems and matrices. *European J. Combin.*, 7:151–160, 1986. (refs: p. 22, 114, 119)

[LV89] L. Lovász and K. Vesztergombi. Extremal problems for discrepancy. In *Irregularities of partitions (G. Halász, V. T. Sós eds.)*, pages 107–113. Springer, Berlin, 1989. (refs: p. 114, 153)

[Mat] J. Matoušek. On the discrepancy for boxes and polytopes. *Monatsh. Math.* In press. (refs: p. 126, 127, 243, 280)

[Mat95] J. Matoušek. Tight upper bounds for the discrepancy of halfspaces. *Discr. & Comput. Geom.*, 13:593–601, 1995. (refs: p. 133, 143, 168, 242, 243, 244)

[Mat96a] J. Matoušek. Derandomization in computational geometry. *J. Algorithms*, 20:545–580, 1996. (refs: p. 34, 151)

[Mat96b] J. Matoušek. Improved upper bounds for approximation by zonotopes. *Acta Math.*, 177:55–73, 1996. (refs: p. 33, 134)

[Mat97] J. Matoušek. On discrepancy bounds via dual shatter function. *Mathematika*, 44:42–49, 1997. (refs: p. 99, 143, 242)

[Mat98a] J. Matoušek. An L_p version of the Beck-Fiala conjecture. *European J. Combinatorics*, 19:175–182, 1998. (refs: p. 134)

[Mat98b] J. Matoušek. On the L_2-discrepancy for anchored boxes. *J. of Complexity*, 14:527–556, 1998. (refs: p. 14, 70, 71)

[MC94] W. Morokoff and R. Caflisch. Quasi-random sequences and their discrepancies. *SIAM J. Sci. Comput.*, 15:1251–1279, 1994. (refs: p. 32, 71)

[MC95] W. Morokoff and R. Caflisch. Quasi-Monte Carlo integration. *J. Comp. Phys*, 122:218–230, 1995. (refs: p. 26, 28)

[Mil64] J. W. Milnor. On the Betti numbers of real algebraic varieties. *Proc. Amer. Math. Soc.*, 15:275–280, 1964. (refs: p. 143)

[MMN95] G. Mullen, A. Mahalanabis, and H. Niederreiter. Tables of (t, m, s)-net and (t, s)-sequence parameters. In *Monte Carlo and quasi-Monte Carlo methods in scientific computing. Proceedings of a conference at the University of Nevada, Las Vegas, Nevada, USA, June 23-25, 1994, H. Niederreiter et al. eds.* Springer-Verlag, Berlin etc., 1995. (refs: p. 59)

[Mon89] H. L. Montgomery. On irregularities of distribution. In *Congress of Number Theory (Zarautz, 1984)*, pages 11–27. Univ. del País Vasco, Bilbao, 1989. (refs: p. 225, 227)

[Mon94] H. L. Montgomery. *Ten lectures on the interface between analytic number theory and harmonic analysis (CBMS Regional conference series in mathematics No. 84).* Amer. Math. Soc., Providence, 1994. (refs: p. 8, 225, 227, 234)

[MR95] R. Motwani and P. Raghavan. *Randomized Algorithms.* Cambridge University Press, Cambridge, 1995. (refs: p. 34)

[MS96] J. Matoušek and J. Spencer. Discrepancy in arithmetic progressions. *J. Amer. Math. Soc.*, 9:195–204, 1996. (refs: p. 108, 134, 169, 242)

[MWW93] J. Matoušek, E. Welzl, and L. Wernisch. Discrepancy and ε-approximations for bounded VC-dimension. *Combinatorica*, 13:455–466, 1993. (refs: p. 143, 168, 242, 244)

[Neš95] J. Nešetřil. Ramsey theory. In *Handbook of Combinatorics*, pages 1331–1403. North-Holland, Amsterdam, 1995. (refs: p. 32)

[Nie87] H. Niederreiter. Point sets and sequences with small discrepancy. *Monatsh. Math.*, 104:273–337, 1987. (refs: p. 59)

[Nie92] H. Niederreiter. *Random number generation and quasi-Monte Carlo methods.* CBMS-NSF Regional Conference Series in Applied Mathematics. 63. SIAM, Philadelphia, PA, 1992. (refs: p. 26, 43, 51, 59, 61, 78)

[Nie98] H. Niederreiter. Nets, (t, s)-sequences, and algebraic curves over finite fields with many rational points. *Documenta Math. J. DMV*, Extra volume ICM 1998, vol. III:377–386, 1998. (refs: p. 59)

[NT96] S. Ninomiya and S. Tezuka. Towards real-time pricing of complex financial derivatives. *Applied Mathematical Finance*, 3:1–20, 1996. (refs: p. 27)

[NX96] H. Niederreiter and C.P. Xing. Low-discrepancy sequences and global function fields with many rational places. *Finite Fields and Their Appl.*, 2:241–273, 1996. (refs: p. 59)

[Ole51] O. A. Oleinik. Estimates of the Betti numbers of real algebraic hypersurfaces (in Russian). *Mat. Sbornik (N. S.)*, 28(70):635–640, 1951. (refs: p. 143)

[OP49] O. A. Oleinik and I. B. Petrovskiĭ. On the topology of of real algebraic surfaces (in Russian). *Izv. Akad. Nauk SSSR*, 13:389–402, 1949. (refs: p. 143)

[OS78] J. Olson and J. Spencer. Balancing families of sets. *J. Comb. Theory, Ser. A*, 25:29–37, 1978. (refs: p. 108, 109, 115, 242)

[Ost22] A. Ostrowski. Bemerkungen zur Theorie der Diophantischen Approximationen I. *Abh. Math. Semin. Univ. Hamburg*, 1:77–98, 1922. Part II ibid., pp. 250–251. (refs: p. 42)

[Owe97] A. B. Owen. Monte-Carlo variance of scrambled net quadrature. *SIAM J. Numer. Analysis*, 34(5):1884–1910, 1997. (refs: p. 70)

[PA95] J. Pach and P. K. Agarwal. *Combinatorial Geometry.* John Wiley & Sons, New York, NY, 1995. (refs: p. 8)

[Pas93] S. H. Paskov. Average case complexity of multivariate integration for smooth functions. *J. Complexity*, 9:291–312, 1993. (refs: p. 28, 32)

[Pol90] D. Pollard. *Empirical processes: Theory and applications.* NSF-CBMS Regional Conference Series in Probability and Statistics vol. 2. Institute of Math. Stat. and Am. Stat. Assoc., Hayward, CA, Alexandria, VA, 1990. (refs: p. 150)

[PR93] R. Pollack and M.-F. Roy. On the number of cells defined by a set of polynomials. *C. R. Acad. Sci. Paris*, 316:573–577, 1993. (refs: p. 143)

[PT95] S. H. Paskov and J. R. Traub. Faster valuation of financial derivatives. *The Journal of Portfolio Management*, pages 113–120, 1995. (refs: p. 27)

[Ran69] B. Randol. On the Fourier transform of the indicator function of a planar set. *Trans. Amer. Math. Soc.*, 139:271–278, 1969. (refs: p. 99)

[Ric51] R. D. Richtmyer. The evaluation of definite integrals, and quasi-Monte Carlo method based on the properties of algebraic numbers. Report LA-1342, Los Alamos Scientific Laboratory, Los Alamos, NM, 1951. (refs: p. 77)

[Rog94] A. D. Rogers. A functional from geometry with applications to discrepancy estimates and the Radon transform. *Trans. Amer. Math. Soc.*, 341:275–313, 1994. (refs: p. 202)

[Rot54] K. F. Roth. On irregularities of distribution. *Mathematika*, 1:73–79, 1954. (refs: p. 6, 14, 175, 243)

[Rot64] K. F. Roth. Remark concerning integer sequences. *Acta Arithmetica*, 9:257–260, 1964. (refs: p. 32, 108, 225, 242)

[Rot79] K. F. Roth. Irregularities of distribution III. *Acta Arith.*, 35:373–384, 1979. (refs: p. 50)

[Rot80] K. F. Roth. On irregularities of distribution IV. *Acta Arith.*, 37:67–75, 1980. (refs: p. 50, 243)

[RSW93] A. Razborov, E. Szemerédi, and A. Wigderson. Constructing small sets that are uniform in arithmetic progressions. *Comb. Probab. Comput.*, 2(4):513–518, 1993. (refs: p. 34)

[Rud74] W. Rudin. *Real and Complex Analysis.* McGraw-Hill, New York, 1974. (refs: p. 203, 215)

[Rud91] W. Rudin. *Functional Analysis (2nd edition).* McGraw-Hill, New York, 1991. (refs: p. 215)

[Ruz93] I. Ruzsa. The discrepancy of rectangles and squares. In *Österreichisch-Ungarisch-Slowakisches Kolloquium über Zahlentheorie (Maria Trost, 1992)*, pages 135–140. Karl-Franzens-Univ. Graz, Graz, 1993. (refs: p. 182)

[SA95] M. Sharir and P. K. Agarwal. *Davenport-Schinzel Sequences and Their Geometric Applications.* Cambridge University Press, Cambridge, 1995. (refs: p. 92)

[San76] L. A. Santaló. *Integral Geometry and Geometric Probability.* Addison-Wesley, Reading, MA, 1976. (refs: p. 33, 192)

[Sau72] N. Sauer. On the density of families of sets. *Journal of Combinatorial Theory Ser. A*, 13:145–147, 1972. (refs: p. 150)

[Sch37] I. J. Schoenberg. On certain metric spaces arising from Euclidean spaces by change of metric and their embedding in Hilbert space. *Ann. Math.*, 38:787–793, 1937. (refs: p. 210)

[Sch38] I. J. Schoenberg. Metric spaces and positive definite functions. *Trans. Amer. Math. Soc.*, 44:522–53, 1938. (refs: p. 210)

[Sch64] W. M. Schmidt. Metrical theorems on fractional parts of sequences. *Trans. Amer. Math. Soc.*, 110:493–518, 1964. (refs: p. 77)

[Sch69a] W. M. Schmidt. On irregularities of distribution II. *Trans. Amer. Math. Soc.*, 136:347–360, 1969. (refs: p. 226)

[Sch69b] W. M. Schmidt. On irregularities of distribution III. *Pacific J. Math.*, 29:225–234, 1969. (refs: p. 227)

[Sch69c] W. M. Schmidt. On irregularities of distribution IV. *Invent. Math.*, 7:55–82, 1969. (refs: p. 16, 226, 227)

[Sch72] W. M. Schmidt. On irregularities of distribution VII. *Acta Arith.*, 21:45–50, 1972. (refs: p. 178, 243)

[Sch75] W. M. Schmidt. On irregularities of distribution IX. *Acta Arith.*, 27:385–396, 1975. (refs: p. 92, 244)

[Sch77a] M. W. Schmidt. Irregularities of distribution X. In *Number theory and algebra*, pages 311–329. Academic Press, New York, 1977. (refs: p. 178, 243)

[Sch77b] W. M. Schmidt. *Lectures on irregularities of distribution.* Tata Institute of Fundamental Research, Bombay, 1977. (refs: p. 179)

[Sch95] A. Schrijver. Polyhedral combinatorics. In *Handbook of Combinatorics*, pages 1649–1704. North-Holland, Amsterdam, 1995. (refs: p. 119)

[She72] S. Shelah. A combinatorial problem, stability and order for models and theories in infinitary languages. *Pacific J. Math.*, 41:247–261, 1972. (refs: p. 150)

[Sie89] C. L. Siegel. *Lectures on the geometry of numbers. Notes by B. Friedman. Rewritten by Komaravolu Chandrasekharan with the assistance of Rudolf Suter.* Springer-Verlag, Berlin etc., 1989. (refs: p. 78, 81)

[SJ94] I.S. Sloan and S. Joe. *Lattice methods for multiple integration.* Clarendon Press, Oxford, 1994. (refs: p. 78)

[Skr90] M. M. Skriganov. Lattices in algebraic number fields and uniform distributions modulo 1. *Leningrad Math. J.*, 1:535–558, 1990. Russian version in *Algebra i Analiz* 1(2):207-228, 1989. (refs: p. 78, 79)

[Skr94] M. M. Skriganov. Constructions of uniform distributions in terms of geometry of numbers. *Algebra i Analiz*, 6:200–23, 1994. Also in *St. Petersburg Math. J.* 6:635–664,1995. (refs: p. 78, 79, 81)

[Skr98] M. M. Skriganov. Ergodic theory on $SL(n)$, diophantine approximations and anomalies in the lattice point problem. *Invent. Math.*, 132:1–72, 1998. (refs: p. 14, 79, 126, 243)

[SM94] J. Spanier and E. H. Maize. Quasi-random methods for estimating integrals using relatively small samples. *SIAM Rev.*, 36:18–44, 1994. (refs: p. 26, 78)

[Sob67] I. M. Sobol. Distribution of points in a cube and approximate evaluation of integrals (in Russian). *Zh. Vychisl. Mat. i Mat. Fiz.*, 7:784–802, 1967. (refs: p. 59)

[Sós58] V. T. Sós. On the distribution mod 1 of the sequence $\{n\alpha\}$. *Ann. Univ. Sci. Budapest*, 1:127–234, 1958. (refs: p. 9)

[Sós76] V. T. Sós. On the discrepancy of the sequence $\{n\alpha\}$. In *Top. Number Theory, Debrecen 1974, Colloq. Math. Soc. Janos Bolyai 13*, pages 359–367. 1976. (refs: p. 179)

[Sós83a] V. T. Sós. Irregularities of partitions. In *Surveys in Combinatorics 82, 9th British Combinatorial Conference (E. K. Loyd ed.)*, pages 201–245. Cambridge University Press, Cambridge, 1983. (refs: p. 7, 8, 32)

[Sós83b] V. T. Sós. On strong irregularities of the distribution of $\{n\alpha\}$ sequences. In *To the memory of Paul Turán (Studies in Pure Mathematics; L. Alpar et al. eds.)*, pages 685–700. Birkhäuser, Basel and Akad. Kiadó, Budapest, 1983. (refs: p. 76)

[Spe85] J. Spencer. Six standard deviations suffice. *Trans. Amer. Math. Soc.*, 289:679–706, 1985. (refs: p. 104, 115, 133, 134, 242)

[Spe87] J. Spencer. *Ten lectures on the probabilistic method.* CBMS-NSF. SIAM, Philadelphia, PA, 1987. (refs: p. 8, 104, 110, 116, 168)

[Sri97] A. Srinivasan. Improving the discrepancy bound for sparse matrices: better approximations for sparse lattice approximation problems. In *Proc. 8th ACM-SIAM Symposium on Discrete Algorithms*, pages 692–701, 1997. (refs: p. 104, 126, 134, 168, 242, 243)

[Sti94] J. Stillwell. *Elements of Algebra: Geometry, Numbers, Equations.* Undergraduate Texts in Mathematics. Springer-Verlag, Berlin etc., 1994. (refs: p. 76, 81)

[Sto73] K. B. Stolarsky. Sums of distances between points on a sphere II. *Proc. Amer. Math. Soc.*, 41:575–582, 1973. (refs: p. 32, 191)

[Stu77] W. Stute. Convergence rates for the isotrope discrepancy. *Ann. Prob.*, 5:707–723, 1977. (refs: p. 92, 244)

[SW93] R. Schneider and J. A. Wieacker. Integral geometry. In *Handbook of Convex Geometry, vol. B*, pages 1349–1390. North-Holland, Amsterdam, 1993. (refs: p. 192)

[SW97] I. H. Sloan and H. Woźniakowski. An intractability result for multiple integration. *Math. Comput.*, 66:1119–1124, 1997. (refs: p. 27)

[SW98] I. H. Sloan and H. Woźniakowski. When are quasi-Monte Carlo algorithms efficient for high dimensional integrals? *J. Complexity*, 14:1–33, 1998. (refs: p. 28, 29, 30)

[SY70] J. Sacks and D. Ylvisaker. Statistical designs and integral approximation. In *Proc. 12th Biennal Seminar of the Canad. Math. Congress (R. Pyke ed.), Montreal*, pages 115–136. Canad. Math. Soc., 1970. (refs: p. 32)

[Tez95] S. Tezuka. *Uniform random numbers. Theory and practice.* Kluwer Academic Publishers, Dordrecht, 1995. (refs: p. 26, 59, 71)

[Tho65] R. Thom. On the homology of real algebraic varieties (in French). In *Differential and Combinatorial Topology (S.S. Cairns ed.).* Princeton Univ. Press, 1965. (refs: p. 143)

[TW80] R. Tijdeman and G. Wagner. A sequence has almost nowhere small discrepancy. *Monatsh. Math.*, 90:315–329, 1980. (refs: p. 179)

[TW98] J. F. Traub and A. G. Werschulz. *Complexity and Information.* Cambridge University Press, Cambridge, 1998. (refs: p. 26)

[TWW88] J. F. Traub, G. W. Wasilkowski, and H. Woźniakowski. *Information-based Complexity.* Academic Press, New York, NY, 1988. (refs: p. 26, 32)

[Vap82] V. N. Vapnik. *Estimation of dependences based on empirical data.* Springer-Verlag, Berlin, 1982. (refs: p. 150)

[VC71] V. N. Vapnik and A. Ya. Chervonenkis. On the uniform convergence of relative frequencies of events to their probabilities. *Theory Probab. Appl.*, 16:264–280, 1971. (refs: p. 22, 150, 151)

[Wag93] G. Wagner. On a new method for constructing good point sets on spheres. *Discr. & Comput. Geom.*, 9:111–129, 1993. (refs: p. 33)

[Wah90] G. Wahba. *Spline models for observational data.* CBMS-NSF Regional Conference Series in Applied Mathematics 59. SIAM, Philadelphia, PA, 1990. (refs: p. 29, 32, 35)

[War72] T. T. Warnock. Computational investigations of low-discrepancy point sets. In *Applications of number theory to numerical analysis (S. K. Zaremba ed.)*, pages 319–343. Academic Press, New York, 1972. (refs: p. 71)

[Wel88] E. Welzl. Partition trees for triangle counting and other range searching problems. In *Proc. 4th Annu. ACM Sympos. Comput. Geom.*, pages 23–33, 1988. (refs: p. 150, 159, 163)

[Wer92] L. Wernisch. Manuscript, FU Berlin, 1992. (refs: p. 44)

[Wey16] H. Weyl. Über die Gleichverteilung von Zahlen mod Eins. *Math. Ann*, 77:313–352, 1916. (refs: p. 7, 15, 73)

[WF94] M. Watanabe and P. Frankl. Some best possible bounds concerning the traces of finite sets. *Graphs Comb.*, 10:283–292, 1994. (refs: p. 150)

[Wil96] A. J. Wilkie. A general theorem of the complement and some new o-minimal structures. Manuscript, Oxford University. Also see *Documenta Math. J. DMV, Extra volume ICM 1998*, vol. I, 633–636, 1996. (refs: p. 143)

[Woź91] H. Woźniakowski. Average case complexity of multivariate integration. *Bull. Amer. Math. Soc.*, 24:185–194, 1991. (refs: p. 30, 32)

[WW95] G. Wasilkowski and H. Woźniakowski. Explicit cost bounds of algorithms for multivariate tensor product problems. *J. Complexity*, 11:1–56, 1995. (refs: p. 30)

[WW97] G. Wasilkowski and H. Woźniakowski. The exponent of discrepancy is at most 1.4778.... *Math. Comput*, 66:1125–1132, 1997. (refs: p. 30)

[Zar68] S. K. Zaremba. Some applications of multidimensional integration by parts. *Ann. Pol. Math*, 21:85–96, 1968. (refs: p. 28)

[Zar70] S. K. Zaremba. Isotropic discrepancy and numerical integration (in French). *Ann. di Mat. Pura et Appl. (4)*, 87:125–136, 1970. (refs: p. 92)

[Zin76] P. Zinterhof. Über einige Abschätzungen bei der Approximation von Funktionen mit Gleichverteilungsmethoden. *Sitzungsber. Österr. Akad. Wiss. Math.-Naturwiss. Kl. II*, 185:121–132, 1976. (refs: p. 15)

Index

Hints

1.1.1(b). First note that the x-coordinate of the kth point of P lies in the interval $[\frac{k}{n} - \frac{M}{n}, \frac{k}{n} + \frac{M}{n}]$, where $M = D(P, \mathcal{R}_2)$.

1.1.1(c). If $f(n)$ is bounded, then use (b) and a compactness argument. Otherwise let $n_i = \min\{n\colon f(n) \geq 2^i\}$; construct initial segments of length n_i with discrepancy $\leq 2f(n_i)$ using (b) and concatenate them.

1.2.1. By induction on d.

1.2.2(b). No, take $\mathcal{A} = \{[0,1] \times [0,1/2]\}$, $n = 1, 2$.

1.3.2. For $D(.)$, you may use a rectangular grid of points. To lower-bound disc(.), consider the vertex set of a convex n-gon.

1.3.3. The complements of finite sets.

1.4.2(a). We have $\langle f, \eta_x \rangle = \int_0^1 f'(y) \cdot \frac{\partial \min(1-x,1-y)}{\partial y} \, \mathrm{d}y = -\int_x^1 f'(x) \, \mathrm{d}x = f(x)$.

2.1.2. Define $P_n' \subset [0,1]^{d+1}$ by appending the $(d+1)$st coordinate equal to $\frac{i}{n}$ to the ith point of P_n, and use the result of Example 2.3.

2.1.3. Show that any point distinct from 0 has vertical distance at least $\Omega(n^{-1/2})$ from the line $x = y$, and take R as a long narrow rectangle along that line.

2.3.2. In order that $x(h)$ lie in a b-ary canonical box of size $b^{-m_1} \times \cdots \times b^{-m_d}$, $m_1 + \cdots + m_d = \rho$, a system of linear equations must be satisfied whose matrix consists of $\mathcal{C}[\leq m_1, \ldots, \leq m_{d-1}]$ plus m_d more rows, which are the last m_d rows of the $m \times m$ identity matrix. By the assumption on $\mathcal{C}|_j$ for $j = m_1 + \cdots + m_{d-1}$, the upper left $j \times j$ submatrix is nonsingular, and so the whole matrix of the linear system is nonsingular.

2.3.3(a). Let $x_i \in [0,1]^d$ be the ith point of a b-ary net, $i = 1, 2, \ldots, b^2$. Define the jth b^2-tuple as $\left(\lfloor b \cdot (x_1)_j \rfloor, \lfloor b \cdot (x_2)_j \rfloor, \ldots, \lfloor b \cdot (x_{b^2})_j \rfloor\right)$.

2.3.3(b). Each coordinate of x_i, the ith point of the b-ary net, will have only two b-ary digits after the b-ary point. The first digit of $(x_i)_j$ is the ith entry of the jth b^2-tuple; this guarantees that all b-ary canonical boxes with two sides b^{-1} and others 1 are fine. The second digit of $(x_i)_j$ may be chosen to be the number of indices $\ell < i$ for which $(x_\ell)_j$ has the same first digit as $(x_i)_j$; this takes care of the canonical boxes with one side b^{-2} and others 1.

2.3.4. The dimension reduction is clear (use orthogonal projection). To reduce the size parameter m, use an affine map mapping a b-ary canonical box of volume $b^{m'-m}$ onto the unit cube.

2.3.5. Induction on m.

2.3.6(a). The total number of $k \times m$ matrices is b^{km}. Counting the full-rank ones: if i rows have already been fixed to linearly independent vectors, then their linear span has b^i vectors, and these must be avoided by the $(i+1)$st row, hence the number of full-rank matrices is $\prod_{i=1}^{k}(b^m - b^{i-1})$.

2.3.6(b). There are $M = O(\rho^{d-1})$ partitions of the integer ρ into d nonnegative summands. The matrix for each partition is random, and by (a), the probability that one such matrix does not have a full rank is $1 - \prod_{i=1}^{\rho}(1 - b^{-m+i-1}) \leq \sum_{i=1}^{\rho} b^{-m+i-1} \leq b^{-m+\rho}$. Hence if $M/b^{m-\rho} \leq \frac{1}{2}$ then at least half of the choices of $\mathcal{C}$ have $\rho(\mathcal{C}) \geq \rho$.

2.3.7(a). One obtains a system of equations for the coefficients b_j of the inverse. The first equation is $a_0 b_0 = 1$, showing the necessity of $a_0 \neq 0$, and the sufficiency follows by induction (having determined $b_0, b_1, \ldots, b_{j_0}$, the term b_{j_0+1} can always be expressed from the first equation it occurs in).

2.3.8(b). Use all the b linear polynomials $z - a$, $a \in GF(b)$, in the construction above Theorem 2.10. Apply Theorem 2.9.

2.4.1(a). Multiplying a matrix $C^{(k)}$ by a lower triangular matrix from the left is equivalent to a sequence of row operations on $C^{(k)}$, where a row may be multiplied by a nonzero number or a multiple of some row may be added to a row *below it*. These operations do not decrease the rank of any matrix $\mathcal{C}[\leq m_1, \leq m_2, \ldots, \leq m_k]$.

2.4.1(b). Yes. Here each permutation $\pi_{a_1,a_2,\ldots,a_{j-1}}(a_j)$ in the description of a scrambling (for the kth coordinate, say) is a linear function (over $GF(b)$) of $a_1, \ldots, a_j$ of the form $\sum_{i=1}^{j} \ell_{ji}^{(k)} a_i$, where $\ell_{ji}^{(k)}$ denotes an entry of the lower-triangular matrix $L^{(k)}$.

2.4.4. Use Lemma 2.14. The result is $n(2^{-d} - 3^{-d})$.

2.4.5(a). $n^2 3^{-d}$.

2.4.6. The formula is $n^2(4/3)^d - 2n\sum_{p\in P}\prod_{k=1}^{d}((3-p_k^2)/2) + \sum_{p,q\in P}\prod_{k=1}^{d}(2 - \max(p_k, q_k))$.

2.4.8(b). For any given $a \neq a' \in GF(b)$ and $b \neq b' \in GF(b)$, the system of equations $ah + g = b$, $a'h + g = b'$ has a unique solution (h, g), $h, g \in GF(b)$, $h \neq 0$. From this, with notation as in the proof of Lemma 2.15(ii), one can conclude that for $a_{t+1} \neq a'_{t+1}$ and for h and g_{t+1} random, the pair (b_{t+1}, b'_{t+1}) is uniformly distributed on the set of all pairs (q, q') with $q \neq q'$, $q, q' \in GF(b)$, and hence $\mathbf{E}\left[\,|b_{t+1} - b'_{t+1}|\,\right] = \frac{b+1}{3}$. Moreover, for any $j > t+1$, any a_j, a'_j, and any fixed value of h, we find that the expectation of $b_j - b'_j$ for a random choice of g_j is 0.

2.4.9. An easy proof is via generating functions. Let

$$F(x_1, x_2, \ldots, x_s) = \sum_{t_1, t_2, \ldots, t_s} f(t_1, \ldots, t_s) x_1^{t_1} x_2^{t_2} \cdots x_s^{t_s}$$

be the multivariate generating function for f, and similarly for $G(x_1, \ldots, x_s)$. We get $G(x) = F(x)/\prod_{i=1}^{s}(1 - x_i)$.

2.4.10. Use $\binom{t+s-1}{s-1}/\binom{t-t_0+s-1}{s-1} = \frac{(t+s-1)(t+s-2)\cdots(t-t_0+s)}{t(t-1)\cdots(t-t_0+1)} \leq \binom{t_0+s-1}{s-1}$ and $\sum_{t=t_0}^{\infty} \binom{t-t_0+s-1}{s-1} b^{-t} = b^{-t_0}(1 - 1/b)^{-s}$.

2.4.11(a). Generalize the problem as follows. Given point sets P and Q and weight functions $v: P \to \mathbf{R}$, $w: Q \to \mathbf{R}$, calculate $f_d(P, v, Q, w) = \sum_{p \in P} \sum_{q \in Q} v(p) w(q) \prod_{k=1}^{d} \min(p_k, q_k)$. Let c be the median of the x_d-coordinates of the points of P, and let $P_{\leq} = \{p \in P \colon p_d \leq c\}$, $Q_{\leq} = \{q \in Q \colon q_d \leq c\}$, and similarly for $P_>$ and $Q_>$. Then

$$\begin{aligned} f_d(P, v, Q, w) &= f_d(P_{\leq}, v_{\leq}, Q_{\leq}, w_{\leq}) + f_d(P_>, v_>, Q_>, w_>) \\ &\quad + f_{d-1}(\bar{P}_{\leq}, \tilde{v}_{\leq}, \bar{Q}_>, v_>) + f_{d-1}(\bar{P}_>, v_>, \bar{Q}_{\leq}, \tilde{w}_{\leq}), \end{aligned}$$

where $v_{\leq}$ denotes the restriction of v to $P_{\leq}$ (and similarly for $w_{\leq}$, $v_>$, $w_>$), a bar above a set $X \subseteq \mathbf{R}^d$ denotes its projection onto the first $d-1$ coordinates, and $\tilde{v}_{\leq}(\bar{p}) = p_d v(p)$ for $p \in P_{\leq}$, $\tilde{w}_{\leq}(\bar{q}) = q_d w(q)$ for $q \in Q_{\leq}$. Prove by induction that using this formula, $f_d(P, v, Q, w)$ can be evaluated in time $O((m+n)\log^d(m+n))$, where $m = |P|$, $n = |Q|$.

2.5.2(c). If $(a_0, a_1, \ldots)$ is periodic, then $\alpha = \alpha_i$ for some i; derive a quadratic equation for α.

2.5.3. Show that the discrepancy of any rectangle $[0, x) \times [k/q_j, (k+1)/q_j)$ is $O(1)$, and that a general corner can be sliced into $O(\log n)$ of these plus a small remainder.

2.5.4. Let $\Lambda^*_{(i)}$ stand for the Λ^* from (i), and similarly for $\Lambda^*_{(ii)}$. Clearly, $\Lambda^*_{(ii)} \subseteq \Lambda^*_{(i)}$. As for $\supseteq$, a $y \in \Lambda^*_{(i)}$ can be written as $\sum_{k=1}^{d} \alpha_k b_k^*$, where $\alpha_k \in \mathbf{R}$ and b_k^* is the kth row of $(B^{-1})^T$. And $\langle y, b_k \rangle = \alpha_k$ must be integral by (i).

2.5.6. If $x \in \Lambda$ and we put $\delta = |x_1 \cdots x_d|^{1/d}$ then for the D with $\frac{\delta}{x_1}, \ldots, \frac{\delta}{x_d}$ on the diagonal, the point $Dx^T = (\delta, \ldots, \delta) \in D\Lambda$ has length $\delta\sqrt{d}$.

2.5.7. If $z \in \Lambda^*$ satisfies $\|z\| < \frac{1}{r}$ then $|\langle v_i, z \rangle|$ is an integer bounded above by $\|v_i\| \cdot \|z\| < 1$, and hence it is 0. Then $z = 0$ by the linear independence of the v_i.

2.5.8. By the theorem, $B(0, r)$ encloses $\Omega(r^d)$ lattice points, with the constant of proportionality only depending on d. If there are no d linearly independent vectors in $B(0, r)$, $\Lambda \cap B(0, r)$ is contained in a hyperplane, and since any two points of Λ are at least ε apart, a volume argument with balls of radius $\frac{\varepsilon}{2}$ around the points gives $|\Lambda \cap B(0, r)| = O((r/\varepsilon)^{d-1})$. For r sufficiently large, we get a contradiction.

2.5.9. Prove (ii) in Exercise 6 for Λ^* using Exercise 7, establishing the assumption via Exercise 8.

2.5.11(a). Clearly $\alpha_j \in \mathbf{R}$. The automorphism of $\mathbf{Q}(\omega)$ given by $\omega \mapsto \omega^j$ sends α_1 to α_j, and hence all the α_j are conjugate. Thus, the degree of each α_j over $\mathbf{Q}$ is at least d. On the other hand, the degree of $Q(\omega)$ over $\mathbf{Q}(\alpha_1)$ is 2, because we have $g(\omega) = 0$ for $g(x) = x^2 - \alpha_1 x + 1$ and clearly $\omega \notin \mathbf{Q}(\alpha_1) \subset \mathbf{R}$. Since the degree of $\mathbf{Q}(\omega)$ over $\mathbf{Q}$ is $p - 1 = 2d$, the degree of each α_j is d. Hence $\alpha_1, \ldots, \alpha_d$ are the roots of a monic irreducible polynomial of degree d with rational coefficients (the minimal polynomial), and this polynomial must be $q(x)$. Finally, since ω is an algebraic integer, meaning that its (monic) minimal polynomial has integer coefficients, the α_j are algebraic integers too, because algebraic integers form a ring. The solution in (b) applies as well, but it requires a little more of the Galois theory.

2.5.11(b). To see that the α_j are real, note that $r^{md} = r^{(p-1)/2} \equiv -1 \pmod{p}$. The automorphism given by $\omega \mapsto \omega^{r^{j-1}}$ sends α_1 to α_j, and each of the automorphisms of $\mathbf{Q}(\omega)$ (given by $\omega \mapsto \omega^{r^\ell}$ for some ℓ) maps α_1 to some α_j. Since there are no other automorphisms a subfield of $\mathbf{Q}(\omega)$ may have, the minimal polynomial has degree d and it equals $q(x)$.

3.1.1. Count the intersections between the circle and the sides of the small squares, using the fact that any line is only intersected twice by the circle. Or observe that all intersected squares lie in an annulus of width $a \cdot 2\sqrt{2}$, where a is the side of a grid square, and use a volume argument.

3.1.2. Decompose n into summands of the form 4^k, take a small-discrepancy set for each of them, perturb a little to have them disjoint, and observe that $|D(P_1 \dot{\cup} P_2, A)| \leq |D(P_1, A)| + |D(P_2, A)|$.

3.1.5(a). For discs B of not too large radius, use discs with center coordinates and radii being integer multiples of n^{-2}, say. For huge discs B, use a suitable discrete collection of discs of a fixed large radius.

3.1.5(b). The proof that all discs in $\mathcal{F}_1$ as in (a) have, with a positive probability, the right discrepancy is almost identical to the proof in the text. It remains to prove that if B_{in} and B_{out} have discrepancy at most Δ then B has discrepancy at most $\Delta + 1$.

3.1.6(a). Choose a collection of as many disjoint caps (parts of the disc C cut off by suitable chords), each of area $\frac{1}{2n}$, as possible; calculation shows that about $n^{1/3}$ can be chosen. Let C_1 be C minus the caps containing at least one point of P, and let C_2 be C minus the caps containing no points of P. Check that $(|P \cap C_1| - n\,\mathrm{vol}(C_1)) - (|P \cap C_2| - n\,\mathrm{vol}(C_2)) \geq M$, where $M = \Omega(n^{1/3})$ is the number of caps.

3.2.1(a). It suffices to show that the function $g(a, \beta)$ is $\Omega(m)$ for all $(a, \beta) \in [0, \frac{c}{m}] \times I$, where $c > 0$ is a positive constant and I is an interval of length $\Omega(1)$.

3.2.1(c). Same as (b).

3.2.1(d). As in the proof of Proposition 3.4, but use $\int |g|^p \leq \left(\int g^2\right)^{p/2}$.

4.1.1(a). An elementary proof: assuming n even, the probability is at least $2^{-n}\sum_{j=n/2+\Delta}^{n/2+2\Delta}\binom{n}{j} \geq 2^{-n}\Delta\binom{n}{n/2}\left(\frac{n-2\Delta}{n+2\Delta}\right)^{2\Delta}$. Using the estimate $\frac{1-x}{1+x} \geq e^{-3x}$ (valid for x small enough) and $\binom{n}{n/2} \geq 2^n/2\sqrt{n}$ (easily verified from tail estimates for the binomial distribution), this is $\geq (\Delta/2\sqrt{n})\exp(-12\Delta^2/n)$. For $\Delta \leq \sqrt{n}$, use monotonicity and this bound for $\Delta = \sqrt{n}$.

4.1.1(b). Let $u = |\chi^{-1}(-1)|$. Then $\chi(R)$ is the sum of u independent random variables with values $-1, 0$ and $n-u$ variables with values $0, 1$. Thus $S' = 2\chi(R) + 2u - n$ behaves as a sum of n independent random ± 1 variables. If e.g., $u \leq n/2$ then $\Pr[\chi(R) \geq \Delta] = \Pr[S' \geq 2\Delta - 2u - n] \geq \Pr[S' \geq 2\Delta]$.

4.1.1(c). Calculate that $\Pr[\mathrm{disc}(\mathcal{R},\chi) \leq c_1\sqrt{n\log(2m/n)}] < 2^{-n}$ for any fixed χ; then there is a choice of $\mathcal{R}$ for which no χ works.

4.1.2. If X is the ground set of $\mathcal{S}$ and Y the ground set of $\mathcal{T}$, color $(x, y) \in X \times Y$ by $\chi(x)\xi(y)$, where χ is a coloring witnessing $\mathrm{disc}(\mathcal{S})$ and ξ is a coloring witnessing $\mathrm{disc}(\mathcal{T})$.

4.1.2(b). $\{\{0,1\},\{0,2\},\{0,3\},\{0,4\},\{1,2,\ldots,6\}\}$.

4.2.3(b). By (a), we have $A^TA = nI + (n-1)J$.

4.2.4. By the proof of Proposition 4.4, it suffices to find a vector x with $\|x\| = 1$ and $1 - x_1^2 + (2x_1 + x_2 + \cdots + x_n)^2$ small. A suitable choice is $x_1 = \sqrt{1 - 4/(n+3)}$ and $x_2 = x_3 = \cdots = x_n = -\frac{2}{n-1}x_1$.

4.2.5. Show that A^TA, where A is the incidence matrix, is a circulant matrix. Its eigenvectors are of the form $(1, \omega, \omega^2, \ldots, \omega^{n-1})$, where ω is an nth root of 1. The corresponding eigenvalues are $\sum_{d=1}^{6k}\left|\sum_{j=0}^{k-1}\omega^{jd}\right|^2$. It is enough to show that there is a d_0 (for each root ω) such that the inner sum is $\Omega(k^2)$. By the pigeonhole principle, there is a $d_0 \leq 6k$ with $-\frac{\pi}{3k} \leq \arg(\omega^{d_0}) \leq \frac{\pi}{3k}$, so the real part of each ω^{jd_0} is at least $\frac{1}{2}$.

4.2.6(b). There are 2^n index sets I and only at most $(n+1)^m$ possible values of the row sum.

4.3.2. Let A and B be disjoint n-point sets. Put $\mathcal{S} = \{S \subseteq A \cup B\colon |S \cap A| = |S \cap B|\}$.

4.3.3. Use weights $1 - \frac{2}{n+1}$ for all points.

4.3.5. The proof of Theorem 4.6 goes through unchanged.

4.3.6. A case analysis shows that the linear discrepancy is $\frac{4}{3}$. For hereditary discrepancy, delete the point 4.

4.3.7(a). Binet–Cauchy implies the existence of an $n \times n$ submatrix B of A with $|\det(B)|^{1/n} \geq \Delta\sqrt{\frac{m}{n}}/\binom{m}{n}^{1/2n}$. A calculation using the estimate $\binom{m}{n} \leq (em/n)^n$ (see the proof of Lemma 4.13) gives $|\det(B)|^{1/n} \geq \Delta/\sqrt{e}$. Let $\mathcal{S}_0$ be the set system with incidence matrix B and apply Theorem 4.7.

4.3.7(b). Since $\det(A^TA)$ is the product of the eigenvalues of A^TA, we have $\Delta_{eig} \leq \Delta = \left(\frac{n}{m}\det(A^TA)^{1/n}\right)^{1/2}$. And from the proof in (a) we know that $\Delta = O(\Delta_{det})$.

4.3.7(c). Spencer's upper bound 4.2. A direct proof is possible using Hadamard's bound for the determinant of a $k \times k$ matrix: $\det(B) \le \prod_{i=1}^{k} \|b_i\|$, where b_i is the ith row of B.

4.3.8. We have $A^T A = kI + (k-1)J$, where $k = \frac{n+2}{4}$, and $B^T B = A^T A - v^T v$, where v is a row vector with $2k-1$ ones and $2k$ zeros. Hence $x^T B^T B x = (k-1)\left(\sum_{i=1}^{n} x_i\right)^2 + k\|x\|^2 - \left(\sum_{i \in K} x_i\right)^2$, where $|K| = 2k-1$. Find an x such that this is $O(1)$.

4.3.10. Let $f(k) = |\det(H_k)|$ and $g(k) = |\det((H_k + J)/2)|$. We have $f(k) = 2^{k-1} f(k-1)^2$ and $g(k) = g(k-1) f(k-1)$. We get $\det(H_k) = n^{n/2}/2^{n-1}$, $n = 2^k$.

4.4.1. To get a coloring for $\mathcal{S}_1|_Y$, contract all edges not belonging to Y. Color the edges of the resulting tree by levels: even level red, odd level blue.

4.4.2. Show that the set system dual to Hoffmann's example is totally unimodular (has hereditary discrepancy at most 1).

4.4.3. The bound would be the same for $\mathcal{S}$ and for $\mathcal{S}^*$; use Exercise 2.

4.4.4(a). The image of $\mathbf{Z}^n$ is a sublattice of $\mathbf{Z}^n$. A volume argument shows that a large ball must contain about the same number of points of $\mathbf{Z}^n$ and of $A\mathbf{Z}^n$ and hence the lattices coincide. Alternatively, the total unimodularity of A easily implies that A^{-1} is integral.

4.4.4(b). We need to show that each integral point $b \in A\mathbf{R}^n$ is the image of an integer point. Let $\bar{A}$ be a regular $k \times k$ submatrix of A with $k = \operatorname{rank}(A)$; we may assume that $\bar{A}$ is contained in the first k rows and in the first k columns of A. Let $\bar{b}$ consist of the first k components of b; then $\bar{x} = \bar{A}^{-1}\bar{b}$ is integral by (a). Append $n-k$ zero components to $\bar{x}$. The resulting vector is mapped to b by A.

4.4.4(c). The solution set is bounded, and so if it is nonempty then it has some vertices. A vertex is determined by some n of the inequalities holding with equality. Use (b).

4.4.4(d). The desired zero-discrepancy coloring is equivalent to a 0/1 solution to the system $Ax = b$ with $b = A\frac{1}{2} \in \mathbf{Z}^m$, where A is the incidence matrix and $\frac{1}{2}$ stands for the vector of $\frac{1}{2}$'s. Apply (c) with $u = \mathbf{0}$, $v = \mathbf{1}$, and $w = z = b$.

4.4.4(e). Add a new point to each set of odd size; this preserves the total unimodularity. Use (d). Or, (c) can be used directly with $w = \lfloor \frac{1}{2} A \rfloor$ and $z = \lceil \frac{1}{2} A \rceil$.

4.5.1. To produce $\mathcal{F}$ in dimension d, start with the canonical intervals in the x_1-direction; for each of them, take the canonical intervals in the x_2-direction; for each of these, consider the canonical intervals in the x_3-direction, etc. The logarithm of the product of the sizes of the canonical sets larger than t created in this manner is $O(\frac{n}{t} \log^{d-1} n \log t)$. The parameter t is set to $K \log^{d-1} n \log\log n$, and the size of sets in $\mathcal{M}$ is $O(t \log^{d-1} n)$. See [Mat] for details.

4.5.2(a). Let $T_0 = \gamma_1 \cap \gamma_2 \cap \gamma_3$, where γ_i is the halfplane given by $\gamma_i = \{(x,y) \in \mathbf{R}^2\colon a_i x + b_i y \le c_i\}$. Define the embedding $f\colon \mathbf{R}^2 \to \mathbf{R}^3$ by $f(x,y)_i = a_i x + b_i y$, $i = 1,2,3$, and set $\rho = f(\mathbf{R}^2)$.

4.5.2(b). Similar to (a), if $h \in H$ is given by $\langle a, x\rangle \le b$, the coordinate of the embedding $\mathbf{R}^d \to \mathbf{R}^{|H|}$ corresponding to h is given by $\langle a, x\rangle$.

4.5.3(a). Treat each family $\mathrm{POL}(H_i)$ independently in the same way as the axis-parallel rectangles, defining auxiliary set systems $\mathcal{F}_i$ and $\mathcal{M}_i$. To obtain a partial coloring, combine all the $\mathcal{F}_i$'s into $\mathcal{F}$ and all the $\mathcal{M}_i$'s into $\mathcal{M}$.

4.5.3(b). Let T_0 be the triangle in the definition of $\mathcal{T}$, let a, b, c be its sides (segments), and let h_a be the line extending the segment a (and similarly for h_b, h_c). Suppose that a is vertical and that b lies above c, say. Then T_0 can be expressed as the difference of two semiinfinite trapezoids $T_1 \setminus T_2$, where T_1 is the set of points lying vertically below the segment b and T_2 are the points vertically below c. Similarly, any $T \in \mathcal{T}$ can be written using a set from $\mathrm{POL}(\{h_a, h_b\})$ and a set from $\mathrm{POL}(\{h_a, h_c\})$.

4.5.3(c). Given a convex polygon $A \in \mathrm{POL}(H)$, let B be the set of all points lying vertically below A. Write A as the difference $(A \cup B) \setminus B$. Each of the sets B and $A \cup B$ is a disjoint union of at most k semiinfinite trapezoids. These trapezoids belong to $\mathrm{POL}(H_1) \cup \mathrm{POL}(H_2) \cup \cdots \cup \mathrm{POL}(H_k)$, where each H_i consists of one vertical line and one line of H.

4.5.3(d). Similar to (c), define B as the set of points lying vertically below a given convex polytope $A \in \mathrm{POL}(H)$. Decompose both the sets B and $A \cup B$ into semiinfinite vertical prisms, each bounded from above by a facet of A. Decompose the vertical projection of each of these prisms using semiinfinite trapezoids in the xy-plane, as in (c), and lift vertically to get a decomposition for each vertical prism. Each set in the resulting expression for A belongs to a family $\mathrm{POL}(G)$ for a set G of 3 planes. It remains to check that the total number of such triples G needed in the decompositions for all $A \in \mathrm{POL}(H)$ is bounded by a function of k. A similar approach works in any fixed dimension d, but some basic knowledge about convex polytopes may be needed for the proof.

4.5.4. Take the system $\mathcal{F}$ of canonical intervals as was done in the proof using the Partial coloring lemma, but with threshold $t = 1$ (all canonical intervals on both levels). Apply Beck–Fiala on $\mathcal{F}$ ($\max \deg_{\mathcal{F}}(x) = O(\log^2 n)$). Each rectangle is a disjoint union of $O(\log^2 n)$ sets of $\mathcal{F}$.

4.5.5(a). Define a graph on $\{1, \ldots, n\}$ with edge set

$$\{\{\pi_i(2j-1), \pi_i(2j)\}\colon 1 \le j \le \frac{n}{2}, i = 1, 2\}.$$

Show that it is bipartite; color one class by $+1$ and one by -1.

4.5.5(b). Let u be a suitable parameter (depending on k), and let $\mathcal{F}$ consist of canonical intervals of length 2^u along each π_i, i.e. sets of the form $\{\pi_i(j2^u + 1), \pi_i(j2^u + 2), \ldots, \pi_i((j+1)2^u)\}$. Each set of $\mathcal{P}$ can be written as a disjoint union of sets of $\mathcal{F}$ plus $O(2^u)$ remaining points. The Partial coloring lemma with $\mathcal{M} = \emptyset$ gives a partial coloring with zero discrepancy for all sets of $\mathcal{F}$. Iterate. The resulting bound is $O(k \log n)$ by this method.

4.5.5(c). The permutations can be chosen so that the resulting set system contains some arbitrary k prescribed sets. Use some set system of k sets with discrepancy about $\sqrt{k}$ (such as the set system obtained from the Hadamard matrix).

4.5.6. Consider the system of all canonical intervals along each permutation. The maximum degree is $O(\log n)$, so this system has discrepancy $O(\log n)$, and each interval is a disjoint union of $O(\log n)$ canonical intervals.

4.5.7(a). Let t be a suitable threshold parameter (the appropriate value turns out to be $C\sqrt{n \log n}$). Define $\mathcal{F}$ as the system of all canonical intervals of length t in the arithmetic progressions of $\mathcal{A}_n$, i.e. sets of the form $\{ktd + r, (kt+1)d+r, (kt+2)d+r, \ldots, (k+1)td+r\} \subseteq \{1, 2, \ldots, n\}$, where $d = 1, 2, \ldots$, $r = 1, 2, \ldots, d$, and $k = 0, 1, \ldots$. Each set in $\mathcal{A}_n$ is a disjoint union of sets in $\mathcal{F}$ plus a remainder of $< 2t$ points. Let the remainders be $\mathcal{M}$, and use the Partial coloring lemma.

4.5.7(b). $\mathcal{A}_n$ restricted to an m-point subset of $\{1, 2, \ldots, n\}$ no longer behaves as $\mathcal{A}_m$, the arithmetic progressions on m points.

4.5.7(c). Proceed as in (a), but take intervals consisting of t points from X. Observe that the total number of such intervals for all canonical progressions with one given difference d is at most $\frac{m}{t}$.

4.6.2. If there are two values with distinct probabilities p_1, p_2, show that the entropy decreases by replacing both probabilities by $(p_1 + p_2)/2$.

4.6.3. The problem is with the iteration of the partial coloring, since one cannot guarantee that the size of the sets decreases as points get colored. As we know, if $n = s^2$, say, then we can take n sets of size s on $2s$ points with discrepancy $\Omega(\sqrt{s \log s})$.

4.6.4(a). Exercise 4.1.1.

4.6.4(b). For $n < m$, add dummy points, and for $n > m$, use Theorem 4.9. If $m = O(s)$, Spencer's upper bound 4.2 will do, and for $m > s^{1+\varepsilon}$, the Random coloring lemma 4.1 is sufficient.

4.6.4(c). For $\Delta_S = C\sqrt{s \log(2m/s)}$ and set size s, the total entropy required for all sets is $\leq m(s/m)^A$ with a constant A as large as desired (if C is sufficiently large). Hence the number of colorings giving the same discrepancy, up to Δ_S, to all the sets, is at least $2^{m(1-(s/m)^A)}$ (recall that $n = m$). By Kleitman's inequality, a partial coloring with at most $2\alpha m$ uncolored points exists provided that $H(\frac{1}{2} - \alpha) \leq 1 - (s/m)^A$. Since we have $1 - H(\frac{1}{2} - \alpha) = \frac{2}{\ln 2}\alpha^2 + O(\alpha^4)$ (Taylor series), we conclude that $\alpha = O((s/m)^{A/2})$. Also see [AS92] for similar calculations. (Alternatively, one could use the elementary tail estimates derived in Exercise 4.1.1 for estimating the sum of binomial coefficients in Kleitman's inequality.)

5.1.4(a). Each cell in the Venn diagram of m sets of $\mathcal{T}$ is a disjoint union of some cells in the Venn diagram of the $\leq tm$ sets of $\mathcal{S}$ used in the definition of the considered m sets.

5.1.4(b). For any $Y, S_1, \ldots, S_t \subseteq X$, we have

$$\Phi(S_1, \ldots, S_t) \cap Y = \Phi(S_1 \cap Y, \ldots, S_t \cap Y)$$

(verify by induction on the structure of Φ). From N sets on an m-point $Y \subseteq X$, at most N^t sets can be defined using Φ.

5.1.5(b). If A is the incidence matrix of the projective plane, we get $A^T A = qI + J$, where I is the identity matrix and J is the matrix of all 1's, hence $x^T A^T A x = q\langle x, x\rangle + \left(\sum x_i\right)^2 \geq q\|x\|^2$.

5.1.5(c). Fix one point, x, and color half of the $q+1$ lines passing through x by $+1$ and the other half by -1 (x is colored arbitrarily). All lines not containing x have discrepancy 0.

5.1.5(d). The dual of a projective d-space is also a projective d-space, so it suffices to consider the primal shatter function, say. Let $A \subseteq X$ be an m-point subset of the projective d-space. For a hyperplane h, fix a maximal subset $B_h \subseteq A$ of affinely independent points in $h \cap A$. Then $|B_h| \leq d$, and $h \cap A$ equals the intersection of the affine hull of B_h with A. Hence the primal shatter function is $O(m^d)$.

5.2.1(b). $\lfloor \log_2(2^{d_1} + 2^{d_2}) \rfloor \leq 1 + \max(d_1, d_2)$ by (a).

5.2.1(c). We have $\pi^*_{\mathcal{S}}(m) \leq \pi^*_{\mathcal{S}_1}(m)\pi^*_{\mathcal{S}_2}(m)$; a little more precise bound is $\pi^*_{\mathcal{S}}(m) \leq \max\{\pi^*_{\mathcal{S}_1}(m_1)\pi^*_{\mathcal{S}_2}(m_2)\colon m_1 + m_2 = m\}$.

5.2.2(a). Any set in convex position is shattered.

5.2.2(b). Place the d-point set A to be shattered on a short arc of a circle. Make C nearly a circular disc, and for each $B \subseteq A$, construct a piece of the boundary of C that cuts off exactly the points of B from A when C is rotated appropriately.

5.2.2(c). Argue that the dual shatter function is $O(m^2)$.

5.2.3. By induction on d, prove that there exists a d-point set $A \subset \mathbf{R}$ and an $\varepsilon_d > 0$ such that for any $B \subseteq A$, $S_a \cap A = B$ holds for all a from an interval of length ε_d. In the inductive step, add a number to A much larger than $\max A$.

5.2.4(a). Let A be the incidence matrix of $\mathcal{S}$. A shattered subset in $\mathcal{S}^*$ of size 2^d means a $2^d \times 2^{2^d}$ submatrix M in A whose columns are all possible 0/1 vectors of length 2^d. One can select a $2^d \times d$ submatrix from M whose rows are all possible 0/1 vectors of length d, which yields a d-element shattered subset for $\mathcal{S}$.

5.2.4(b). Consider the set system dual to $(X, 2^X)$.

5.2.5(b). Take all subsets of $\{1, 2, \ldots, n\}$ consisting of at most d intervals of consecutive numbers.

5.2.6. For (i), lower-bound the probability q that a random set of size s misses a fixed k-point set. This q can be made at least $n^{-\delta}$ for any prescribed $\delta > 0$ by choosing c small enough. The probability that (i) fails is $\leq n^k(1-q)^m$. For (ii), consider a fixed 3-point set and independent random sets $R_1, \ldots, R_4$ of size s; estimate the probability q_1 that R_1 contains all the 3 points and R_2, R_3, R_4 each contain at least 2 of the 3 points. One gets $q_1 = O((s/n)^9)$. The probability that (ii) fails is $\leq n^3 m^4 q_1$.

5.2.8(a). For each fixed s-element multiset N that is not a $\frac{1}{r}$-net, fix a "witness set," i.e. a set $S_N \in \mathcal{S}$ with $\mu(S_N) \geq \frac{1}{r}$ and $S_N \cap N = \emptyset$. Show that the probability of (random) M hitting S_N in $\geq \frac{s}{r}$ elements is $\Omega(1)$. This is a calculation with a binomial distribution, and Chernoff type estimates can be used.

5.2.8(c). The random N and M as in (a) can be obtained by first selecting a random N_0 by $2s$ draws and then randomly dividing it into N and M. For every fixed N_0, there are $O(s^d)$ distinct intersections of the form $R = S \cap N_0$ with $S \in \mathcal{S}$. For each such intersection R, use (b). This yields that for any fixed N_0, the probability of "$\exists S \in \mathcal{S}$: $S \cap N = \emptyset$ and $|S \cap M| \geq \frac{s}{r}$" conditioned on $N \cup M = N_0$ is $o(1)$. Hence the r.h.s. in (a) is $o(1)$.

5.2.9(a). In terms of the incidence matrix: we are given b a-element 0/1 vectors. Append $d-a$ elements to each of these b vectors in such a way that b distinct d-element vectors are obtained. These all occur as rows in the $2^d \times d$ incidence matrix of G.

5.2.9(b). Call two vertices from A, the class with n vertices, equivalent if the have the same neighborhood in the other class B. If any equivalence class has at least m vertices, we find a homogeneous subgraph on $m+m$ vertices. Otherwise, there are more than $\Phi_{d-1}(2m-1)$ equivalence classes. Each of them defines a subset of B, and by the Shatter function lemma 5.9, B contains a d-element set shattered by these subsets. By (a), this gives an induced copy of H, a contradiction.

5.2.10(b). To each function f in a given family $\mathcal{F}$, assign the polynomial $p_f = \prod_{i=1}^n \prod_{1\leq j\leq k,\, j\neq f(i)} (x_i - j)$. Check that they are linearly independent in the vector space of real functions with domain $\mathcal{F}$, and consider their linear span. Show that this space is generated by the monomials in $x_1, \ldots, x_n$ in which at most d of the x_i occur in power $k-1$ and the others in powers at most $k-2$. To reduce a monomial in which the x_i with $i \in A$ have powers $k-1$, $|A| = d+1$, use a suitable multiple of $\prod_{i\in A} \prod_{1\leq j\leq k,\, j\neq f_A(i)} (x_i - j)$, where $f_A\colon A \to \{1, 2, \ldots, k\}$ witnesses that A is not k-shattered by $\mathcal{F}$.

5.2.10(c). Induction on n for d fixed: Choose some $x \in X$. Let $\mathcal{F}_1$ consist of the functions of $\mathcal{F}$ restricted to $X \setminus \{x\}$, and for each 2-element subset $\{i, j\} \subseteq \{1, 2, \ldots, k\}$, let $\mathcal{F}_{i,j}$ consist of the functions on $X \setminus \{x\}$ such that their extension to X with value i at x lies in $\mathcal{F}$ and also the extension with value j at x is in $\mathcal{F}$. No $\mathcal{F}_{i,j}$ has a 2-shattered subset of size d.

5.2.10(d). Consider all functions attaining at most d values $\neq 1$.

5.3.1(a). Fix an intersection v at distance $\frac{r}{2}$ from x and a line ℓ passing through v. Go from v along ℓ in a suitable direction, and count $\frac{r}{2}$ intersections $v_1, \ldots, v_{r/2}$. From each v_i, follow the line intersecting ℓ at v_i and mark $\frac{r}{2}$ intersections along that line. Quadratically many intersections are marked, each at most twice.

5.4.1. Take about $\sqrt{n}$ halfplanes in general position, and place a point into each 2-dimensional cell of the arrangement of their bounding lines. Each edge of any matching crosses some boundary, so some boundary crosses $\Omega(\sqrt{n})$ edges.

5.4.2. Construct a matching with the appropriate crossing number, delete one endpoint of each edge, and iterate.

5.4.3(a). We have $\mathbf{E}[S_iS_j] = \frac{1}{4}$ for $i \neq j$, and $\mathbf{E}\left[S_i^2\right] = \frac{1}{2}$. Thus, $\mathbf{E}\left[S^2\right] = \frac{1}{4}n(n+1)$.

5.4.3(b). Use Theorem 5.17 and a random coloring as in the proof of Theorem 5.4. Compute the expected value of the L_2-discrepancy using (a).

5.4.3(c). The expectation of S^p can be calculated as in (a) for p an even integer. A perhaps simpler approach is to use the Chernoff tail estimate as in Section 4.1 for S.

5.5.2. Let $\mathcal{F}_i$ be the sets of size 2^i; we have $|\mathcal{F}_i| = O((n/2^i)\log n)$, and the decomposition of a rectangle uses at most about $\log n$ sets of each $\mathcal{F}_i$. Hence the resulting discrepancy of the partial coloring for rectangles is $O(\log n \sum_i \Delta_i)$. Let i_0 be such that $2^{i_0} \approx \log n$ (this is the critical size since we have about n sets in $\mathcal{F}_{i_0}$), and set $\Delta_i = C\sqrt{\log n}\,\varphi(i)$ with $\varphi(i)$ as in the proof of Theorem 5.3 and C a sufficiently large constant. This leads to a partial coloring with $O(\log^{3/2} n)$ discrepancy.

5.5.3. Let $\mathcal{S}_i$ be the canonical intervals of size 2^i; we have $|\mathcal{S}_i| \leq kn/2^i$. For i_0 with $2^{i_0} \approx k$ we have about n sets in $\mathcal{S}_{i_0}$; this is the critical size. Set the discrepancy bound Δ_i for $\mathcal{S}_i$ to $C\sqrt{k}\,\varphi(i)$ with $\varphi(i)$ as in the proof of Theorem 5.3, and show the existence of a partial coloring with $O(\sqrt{k})$ discrepancy. Iteration presents no problem since $\mathcal{P}_k$ restricted to a subset is again induced by k permutations on that subset.

5.5.4(a). Define canonical arithmetic progressions as sets the form

$$\{k2^q d + r, (k2^q + 1)d + r, (k2^q + 2)d + r, \ldots, (k+1)2^q d + r\},$$

where $d = 1, 2, \ldots$, $r = 1, 2, \ldots, d$, $q = 0, 1, 2 \ldots$. Each set of $\mathcal{A}_n$ can be decomposed into canonical progressions using at most two canonical progressions of each size. Let $\mathcal{S}_j$ be the system of canonical progressions of size 2^j. Check that $|\mathcal{S}_j| = O(n^2/2^{2i})$. Hence, substituting $i = \log_2 n - j$, the number of sets of a given size is exactly as in the proof shown in the text (for $d = 2$), and the choice of Δ_i and the entropy calculation can be just copied.

5.5.4(b). Use canonical progressions similar to (a), but only "counting" the numbers present in X. Show that this time the number of canonical progressions of size 2^j is $O(nm/2^{2j})$, and the entropy calculation works out with the same Δ_j.

6.1.1. Let N and m be as in the planar proof, and for each d-tuple of indices $(i_1,\ldots,i_d)$ with $i_k \geq 0$ and $\sum i_k = m$, define one function $f_{i_1,\ldots,i_d}$ on the $2^{i_1} \times \cdots \times 2^{i_d}$ grid of small boxes, analogously to the planar case. The number of these functions is $M = \binom{m+d-1}{d-1}$, and the resulting bound is $\int D^2 \geq c2^{-4d}M$ with an absolute constant $c > 0$.

6.2.2(a). Consider the coordinate index k in which the condition is violated, and let s be the largest integer occurring among $r_k^{(1)},\ldots,r_k^{(\ell)}$ an odd number of times. The one-dimensional integral of f with all coordinates except for x_k fixed and with x_k running through a binary canonical interval of length 2^{-s} is always 0.

6.2.2(b). Example: $(6,6,3)$, $(6,1,8)$, $(12,1,2)$, $(7,6,2)$, $(12,0,3)$, $(7,0,8)$.

6.2.3(a). Expanding the $(2t)$-th power, we need to count the number of ordered $(2t)$-tuples of nonnegative integer vectors with component sum m that satisfy the condition in (a) of the previous exercise. It suffices to fix the first $d-1$ coordinates of each vector (the dth one can be calculated). In each coordinate, there are t pairs of equal entries, each entry between 0 and m. First fix the pairing (at most $(2t)^t$ choices) and then the entries in each pair (at most $(m+1)^t$ choices).

6.2.3(b). Let q be the conjugate exponent to p, i.e. with $\frac{1}{p} + \frac{1}{q} = 1$. We may assume that $q = 2t$ is an even integer. Hölder gives $\left(\int |D|^p\right)^{1/p} \geq \int FD/\left(\int |F|^q\right)^{1/q}$. Use (a) to estimate the denominator.

6.3.1. If R_0 cannot be completed to a square contained in $[0,1]^2$ then its shorter side is at least $\frac{1}{3}$. Subtract 1 or 2 squares from R_0, obtaining a good starting rectangle.

6.3.2(a). Let $P = [0,1)^2 \cap \{ia + \frac{j}{na}\colon i,j \in \mathbf{Z}\}$.

6.3.2(b). Let ℓ be a line passing through one vertex of T and through the midpoint of the opposite side. Another line ℓ' is parallel to ℓ and passes through another vertex of T. Consider the family of evenly spaced parallel lines containing ℓ and ℓ' as neighboring lines, and place points of P equidistantly along lines of this family with a suitable spacing.

6.4.1(a). By motion invariance, s may be taken as the segment from $(0,0)$ to $(0,a)$.

6.4.1(b). Each line intersecting K, up to a set of measure 0, intersects exactly 2 sides, so the sets for sides form a "double cover."

6.4.1(c). Use (b) and a limit argument.

6.4.2(b). 8.

6.4.3(c). See e.g. the book [BFR89].

6.6.2(a). Define $\tilde{D}$ exactly as in the proof in the text. A formula analogous to (6.10) is obtained, with the integration according to the Lebesgue measure replaced by integration according to μ. Only the "mixed" term has to be estimated, and here one has $\int_{\mathbf{R}^2} |J(p,x)|\,\mathrm{d}\mu(x) \le O(w)\mu(B(p,4w)) + \int_{\|p-x\|>4w} O\big(\frac{w^4}{\|p-x\|^3}\big)\,\mathrm{d}\mu(x) = O\big(w^{1+\alpha} + \sum_{k=0}^{\infty} \frac{w^4}{(2^{k+2}w)^3}\,\mu(B(p,2^{k+3}w))\big) = O(w^{1+\alpha})$. The total contribution of the mixed term is $O(n^2 w^{1+\alpha})$, and setting $w = cn^{-1/\alpha}$ will do.

6.6.2(b). Note that a ball of radius mr can be covered by $O(m^2 r)$ balls of radius r. Thus, the sum $\sum_{k=0}^{\infty} \frac{w^4}{(2^{k+2}w)^3}\,\mu(B(p,2^{k+3}w))$ is still dominated by the first term, and the estimate of the mixed term remains the same.

6.6.3. In this case, $J(p,q) = -4\|p-q\|\Delta_1^2 g_2(0)$, where $g_2(x) = \sqrt{1+xh^2}$ with $h = w/\|p-q\|$. By (6.3), we get $\Delta_1^2 g_2(0) = g_2''(\xi)$ for some $\xi \in (0,2)$, and since $g_2''(x) = -h^4/4(1+xh^2)^{3/2}$ is clearly ≤ 0 for all x and all h (this is the advantage of this approach—the derivatives are very simple), $J(p,q) \ge 0$ follows. For the proof in dimension d, let $t = \lceil d/2 \rceil + 1$, embed $\mathbf{R}^d$ into $\mathbf{R}^{d+t}$, and set $\tilde{D}(A) = \sum_{b\in\{0,1\}^t} (-1)^{\sum_{j=1}^t b_j} D(A - \sum_{j=1}^t b_j \mathbf{w}_j)$, where $\mathbf{w}_j$ has w at position $d+j$ and 0's elsewhere. This leads to $J(p,q) = (-1)^{t+1}2^t\|p-q\|\Delta_1^t g_2(0)$. The derivative $g_2^{(t)}(x)$ is $(-1)^{t+1}C_t h^{2t}(1+xh^2)^{-t+1/2}$ with a positive constant C_t, hence $J(p,q) \ge 0$. See [CMS95] for details (in that paper, the proof is presented with a discrete set Q replacing the continuous Lebesgue measure, but conceptually it is the same).

6.6.4. Three times the measure of the planes intersecting a unit segment.

6.6.5. Take $q = p + \mathbf{w}$.

6.7.1. We get $g_{ij} = \max(i,j)$. The Gram matrix has rank $d+1$, and this is not possible for vectors in $\mathbf{R}^d$.

6.7.2. Let $D = \{x_1, x_2, \ldots\}$ be a countable dense set in X. Put $X_n = \{x_1, x_2, \ldots, x_n\}$, and by induction on n, construct isometries $f_n\colon X_n \to \ell_2$ such that f_{n+1} restricted to X_n coincides with f_n, using the fact stated in the exercise. Then $f = \bigcup_{n=1}^{\infty} f_n$ is an isometric embedding $D \to \ell_2$. Extend f to the whole X by sending the limit of each Cauchy sequence to the limit of the image of the sequence.

6.7.3. Use the inequality $\|y\|^2 \ge 0$ with $y = \sum_{i=1}^n x_i v_i$.

6.7.5(a). Follow the proof of Lemma 6.15 but use the function $\varphi(x) = (1-e^{-x^2})/x^{1+\alpha}$.

7.1.2. First prove that if g,h are such that $g,h,gh \in L_1(\mathbf{R}^2)$, then $\int_{\mathbf{R}^2} g\hat{h} = \int_{\mathbf{R}^2} \hat{g}h$. Apply this with $g = f$, $h = \overline{\hat{f}}$ (the complex conjugate of the Fourier transform).

7.1.4. Proceed as in the proof of Theorem 3.1 and take advantage of the fact that a small square intersects fewer squares of the grid.

7.1.5. Choose $\xi = (\xi_1, 0)$ with ξ_1 large, and such that $r_2\xi_1$ is an integer multiple of π and $(r_1\xi_1 \bmod \pi) \in [\pi/4, 3\pi/4]$, say. By similar estimates as in the proof of the Amplification lemma, show that $h_{r_1}(\xi) \approx r_1/\xi_1^3$ (use that

$\sin^2(r_1\xi_1\cos\vartheta) \approx 1$ for $0 \le \vartheta \le 1/r_1\xi_1)$ but $h_{r_2}(\xi)$ is much smaller than r_2/ξ_1^3 (here $\sin^2(r_2\xi_1\cos\vartheta)$ is close to 0 for $0 \le \vartheta \le C/r_2\xi_1$, for C sufficiently large, and the integral over $C/r_2\xi_1 \le \vartheta \le \pi/4$ is always small).

7.1.6. Evaluate the Fourier series of $f(x) = D(P, B_{x,a})$ (with a fixed). Use the Parseval equality and then integrate over $a \in [0,1]^d$. This leads to the expression

$$D_2(P,\tilde{R}_d)^2 = 3^{-d} \sum_{m\in\mathbf{Z}^d\setminus\{0\}} \frac{|\hat{P}(m)|^2}{\prod_{k=1}^d \max(|m_k|^2,1)} \cdot \left(\frac{3}{2\pi^2}\right)^{\mathrm{nnz}(m)},$$

where $\mathrm{nnz}(m)$ is the number of nonzero components of m.

7.1.7(a). Use the Fourier expansion $B_2(x) = \sum_{m\in\mathbf{Z}\setminus\{0\}} \frac{1}{2\pi^2m^2} e^{2\pi i m x}$.

7.1.8. Use the expansion of $B_2(x)$ in the hint to Exercise 7(a) and the hint to Exercise 6.

7.2.1(a). For $t^2 \le a^2$, simply use $e^{-4^{-i}t^2} \le 1$. Otherwise, the largest term is the one with $2^i \approx t$, giving about t^{-2}, and for i getting larger or smaller the terms decrease fast enough.

7.2.1(b). For $\xi_i \le s^{-1}$ use $\sin(r\xi_i) \approx r\xi_i$. In the case $\xi_1, \xi_2 > s^{-1}$, one can argue (for instance) that for at least $\frac{3}{4}$ of the r in $(0,s)$, we have $\sin^2(r\xi_1) \ge \frac{1}{10}$, say, and similarly for at least $\frac{3}{4}$ of the r we have $\sin^2(r\xi_2) \ge \frac{1}{10}$.

7.3.1. Use a suitable 7-point configuration (arising by gluing 4 equilateral triangles).

7.3.3. Use a pattern of concentric annuli whose width decreases with radius; see [GRS90].

7.3.4. Use strips of width $\sqrt{3}/2$.

7.3.5. Consider the regular simplex of a suitable larger dimension.

7.3.6. See [GRS90].

7.3.7. The negation of the theorem means that there exists a set A with $\delta(A) > 2\varepsilon$ and a sequence $\lambda_1 < \lambda_2 < \cdots$ of numbers tending to infinity such that no two points of A have distance λ_j for any j. We may assume $\lambda_{j+1} \ge 2\lambda_j$. For $j_0(\varepsilon)$ as in the proposition, let $R > \lambda_{j_0}$ be such that $\mathrm{vol}(B(0,R) \cap A)/\mathrm{vol}(B(0,R)) \ge \varepsilon$. Apply the proposition with $(\frac{1}{R}A) \cap B(0,1)$ in the role of A and with $t_j = \lambda_j/R$, obtaining a contradiction.

7.3.9(a). For instance: Set $f(r) = \hat{\sigma}(r,0)$. We have $f(0) = 1$, and f is differentiable at 0 (even analytic); by the symmetry of $\hat{\sigma}$, the derivative at $r = 0$ must be 0. Therefore even $|f(r) - f(0)| = o(r)$ as $r \to 0$.

7.3.9(c). We have $\hat{I}_A(0) = \varepsilon$, and $\left|\hat{I}_A(\xi) - \hat{I}_A(0)\right| = \frac{1}{2\pi}\left|\int_A e^{i\langle x,\xi\rangle} - 1\,\mathrm{d}x\right| \le \frac{1}{2\pi}\int_A |\langle x,\xi\rangle|\,\mathrm{d}x$ by (b), and since $A \subseteq B(0,1)$ the last integral is bounded by $\|\xi\|\,\mathrm{vol}(A)$. Part (a) can also be done in much the same way.

Printing (computer to plate): Mercedes Druck, Berlin
Binding: Buchbinderei Lüderitz & Bauer, Berlin